Palms *to* Pines

GEOLOGICAL AND HISTORICAL EXCURSIONS THROUGH THE PALM SPRINGS REGION

GEOLOGIC TIME SCALE

EON	ERA	PERIOD (SYSTEM)		EPOCH (SERIES)	Age estimates of boundaries (in mega-annum [Ma])
PHANEROZOIC	Cenozoic	Quaternary (Q)		Holocene	0.01
				Pleistocene	2.6
		Tertiary (T)	Neogene	Pliocene	5.3
				Miocene	23.0
			Paleogene	Oligocene	33.9
				Eocene	56.0
				Paleocene	66.0
	Mesozoic	Cretaceous (K)		Late	100
				Early	145
		Jurassic (J)		Late	164
				Middle	174
				Early	201
		Triassic (TR)		Late	237
				Middle	247
				Early	252
	Paleozoic	Permian		Lopingian / Guadalupian / Cisuralian	299
		Carboniferous	Pennsylvanian	Late / Middle / Early	323
			Mississippian	Late / Middle / Early	359
		Devonian		Late / Middle / Early	419
		Silurian		Pridoli / Ludlow / Wenlock / Llandovery	444
		Ordovician		Late / Middle / Early	485
		Cambrian		Late / Middle / Early	541
Precambrian					~4600

Age estimates and picks of boundaries are rounded to the nearest whole number (1 Ma) for pre-Late Cretaceous, and rounded to one decimal place (100 ka) for the Late Cretaceous to Pleistocene interval.

Ma = mega annum or one million years ago

ka = kilo annum or one thousand years ago

Reference: Walker, J.D., Geissman, J.W., Bowring, S.A., and Babcock, L.E., compilers, 2012, Geologic Time Scale v. 4.0: Geological Society of America, doi: 10.1130/2012.CTS004R3C.

Palms *to* Pines

GEOLOGICAL AND HISTORICAL EXCURSIONS THROUGH THE PALM SPRINGS REGION, RIVERSIDE COUNTY, CALIFORNIA

SDAG • 2008/2009 ~ SCGS VOLUME 39 • 2012

FOR FIELD TRIPS HELD

October 17–19, 2008

November 6–8, 2009

October 12–14, 2012

FIELD TRIP GUIDE EDITORS

Scott Snyder, 2008

Bryan Miller-Hicks, 2009

Jeffrey A. Miller, 2012

SAN DIEGO ASSOCIATION OF GEOLOGISTS

and

SOUTH COAST GEOLOGICAL SOCIETY

Available from Sunbelt Publications

www.sunbeltpublications.com

Palms to Pines: Geological and Historical Excursions through the Palm Springs Region, Riverside County, California

First edition 2013, seventh printing 2025.

Edited by Scott Snyder, Bryan Miller-Hicks, and Jeffrey Miller
Cover and Book Design/Composition by Michael Schrauzer
Project management by David M. Bloom
Printed in the United States of America

San Diego Geological Society
3130 North Evergreen Street
San Diego, California 92110
www.sandiegogeologists.org

South Coast Geological Society
P.O. Box 10244
Santa Ana, California 92711-0244
www.southcoastgeo.org

Available from:
Sunbelt Publications
664 Marsat Ct., Suite A
Chula Vista, CA 91911
(619) 258-4911, fax: (619) 258-4916
www.sunbeltpublications.com info@sunbeltpub.com

28 27 26 25 10 9 8 7

Library of Congress Cataloging-in-Publication Data

Palms to pines : geological and historical excursions through the Palm Springs Region, Riverside County, California / edited by Scott Snyder, Bryan Miller-Hicks, and Jeffrey Miller. -- First edition.
pages cm
ISBN 978-0-916251-24-6 (alk. paper)
1. Geology--California--Palm Springs Region--Guidebooks. 2. Palm Springs (Calif.)--Guidebooks. 3. Geology--Fieldwork. I. Snyder, Scott W., editor of compilation. II. Miller-Hicks, Bryan, editor of compilation. III. Miller, Jeffrey E., editor of compilation.
QE90.P28P35 2012
557.94'97--dc23

2013044155

The San Diego Association of Geologists and the South Coast Geological Society gratefully acknowledge the generous contributions of our sponsors to the publication of this book.

Front Cover: Images used under Creative Commons license.
Inside front cover: Group photographs of field trip attendees by Monte Murbach (2008), Scott Snyder (2009), and Jo'l M. Chapman (2012).
Inside back cover: Route Maps created by David M. Bloom.
Back Cover: Miles Kenney, Dave Bloom, and George Morgan on Coachella Fan, 2008. Photograph by Diane Murbach.

Photographs on pages i, 5, 11, 12, 13, 14, 19, 20 (upper), 25, 86, 146, 150, 171, 212, and 228 courtesy of Scott Snyder.
Photographs on pages vii, 6, 18, 20 (lower), 27, and 161 courtesy of David M. Bloom.
Photographs on pages 15, 16, and 23 courtesy of Bryan Miller-Hicks.
Photograph on page 22 courtesy of Diana Lindsay.
Photograph on page 94 courtesy of Bill Elliott.
Photograph on page 173 courtesy of Diane Murbach.

TABLE OF CONTENTS

PART II: PAPERS

Dedication of *Palms to Pines Guidebook* to HARRY M. QUINN

I came across a gem. Not a gemstone, but a gem of a southern California native. Harry M. Quinn. Born February 24, 1939, Harry has lived in the Pinyon Pines area on and off since he was 2 years old. He's a geologist, paleontologist, archeologist, and historian. Harry lived with his grandparents, who would head into the mountains above Coachella Valley to camp for several weeks. Harry would spend time at Hurkey Creek Campground (Trip 1, Stop 2) and Pinyon Flats Campground.

As a child, Harry became intimately familiar with the area by hiking and wandering, playing in streams, ponds, lakes, and springs ... he knew where the springs were. He met and befriended all sorts of folks. At a young age he began writing down stories told to him by the local Indians. Harry eventually went to college at University of Southern California and became a geologist, the first of several "ologists" he would become in his career. He worked for many years in oil and gas, and in mining in Alaska and Nevada. A couple of local mines are named for him and his grandfather. The Shadow Mountains/Victorville Dibblee Geologic Map was dedicated to him. He has done many paleontological studies for Environmental Impact Reports in southern California.

Harry served in the army, teaching survival school. He should know. He is a survivor of three helicopter and two airplane crashes. Harry knows every plant and its beneficial uses at different times during the year. While it may not look like a buffet as you drive the Pines to Palms Highway, Harry can show you all the local edible offerings, the ones the native people used to survive and thrive.

Harry is an authority on local Native American archaeology, history, and culture. He has been a friend of the local tribes for many years and is a bridge between them and us transplants. He shares his knowledge with anyone who cares to listen, to pass it along and ensure its survival. He does it because he is probably one of the few people who can; I doubt there is someone who knows more.

Many local historical, nature, and tribal societies acknowledge Harry's community involvement, including Coachella Valley Archeological Society, La Quinta Cove Neighborhood Association, Agua Caliente Cultural Museum, Coachella Valley Historical Society, College of the Desert, San Jacinto Santa Rosa National Monument, and others.

I can think of no better person to whom this guidebook should be dedicated. Thank you, Harry, for your years of learning and for educating us.

—Scott Snyder

2008 Field Trip Leader and 2009 President of the San Diego Association of Geologists

MEMORIALS

DR. FREEMAN "BEACH" LEIGHTON

Beach was born December 19, 1924 and passed away October 3, 2012 at the age of 87. Beach grew up in Illinois where his father was Chief of the Illinois Geological Survey. He obtained his Bachelor of Science in 1946 from the University of Virginia with a degree in general engineering. At Caltech, Beach earned his Master of Science in 1949 with a thesis in Glaciology, and his PhD in 1952 with a thesis on vermiculite deposits. Beach taught geology at Whittier College from 1951 until 1972, reaching Professor Emeritus status.

Beach founded Leighton and Associates in 1961. His consulting firm quickly became a leader in landslide investigations and prevention for hillside development. Beach's experience using rope ladders to examine glaciers during his studies at Caltech led to his using rope ladders in the 1960s to descend into bucket auger holes to map the subsurface geology of hillside developments. "Down-hole logging" is now recognized as one of the best techniques to identify landslide slip surfaces and design mitigation. Before his retirement in 1988, his company grew to a staff of 385 employees in twelve California offices. He located one of his offices in Palm Desert, where he discovered the beauty of the Coachella Valley.

Beach was best known for removing the guesswork from slope instability and for restoration of large landslides such as the Laguna Beach Landslide. In 1994 the Association of Engineering Geologists selected Beach as an Honorary Member for his contributions to the practice and progress of engineering geology.

RUSSELL GORDON GASTIL

Gordon was born June 25, 1928 and passed September 29, 2012 at the age of 84. Gordon grew up in eastern San Diego County, in the communities of Encanto and Alpine. He enrolled at San Diego State Teachers College, which is now San Diego State University. The college did not offer a geology degree at that time so Gordon was directed to the University of California, Berkeley, where he earned his B.S. and Ph.D. in geology in 1953. Gordon's doctoral thesis was a study of Proterozoic crystalline basement rocks in south-central Arizona.

Baylor Brooks hired Gordon in 1959 to teach full time at San Diego State College, where he taught geology until 1993. Gordon received his first NSF grant at State in 1961 for Pb-alpha dating. In 1963, he and Ned Allison secured NSF funding to produce a reconnaissance geologic map of the state of Baja California, Mexico.

Gordon was awarded the prestigious Dibblee Medal on August 21, 2002 in recognition of his extraordinary accomplishments in field geology and geologic mapping. Gordon's insatiable curiosity and geologic insight served as a model for many hundreds of San Diego State University students across five decades who learned the art of geologic mapping under his direction. He ultimately supervised 58 Master's theses at SDSU along with many dozens of undergraduate research projects known as "senior reports."

Gordon has published and co-authored many works. His most renowned is his Baja mapping project in Geological Society of America, Memoir 140, Reconnaissance Geologic Map of the State of Baja California, 1971.

OFFICERS

2012 SOUTH COAST GEOLOGICAL SOCIETY

President: Jeff Miller
Vice President: Paul Parmentier
Historian: Stephen E Jacobs
Secretary: Alexander "Sandy" Jack
Treasurer: Dan Pankratz
Publications: Chris Baker

2008 SAN DIEGO ASSOCIATION OF GEOLOGISTS

President: Sarah Gray
Vice President: Scott Snyder
Secretary: Bryan Miller-Hicks
Treasurer: Chuck Houser
Publications: Lowell Lindsay
Webmaster: Carolyn Glockhoff

2009 SAN DIEGO ASSOCIATION OF GEOLOGISTS

President: Scott Snyder
Vice President: Bryan Miller-Hicks
Secretary: Phil Farquharson
Treasurer: Chuck Houser
Publications: Lowell Lindsay
Webmaster: Carolyn Glockhoff

SAN DIEGO GEOLOGICAL SOCIETY, INC.

President: Dave Bloom
Vice President: Mike Hart
Secretary: Diane Murbach
Treasurer: Greg Cranham
Director: Lowell Lindsay
Director: Monte Murbach
Director (ex officio): Cari Gomes

San Diego Association of Geologists is a program of San Diego Geological Society, Inc., a California 501(c)(3) nonprofit corporation.

San Diego Association of Geologists SPONSORS

RUBY SPONSORS ($500)

Steve Zigan, *Student Sponsor*

William J. Elliott, Engineering Geologist, *Guidebook Sponsor*

MISSION GEOSCIENCE, INC.
ENVIRONMENTAL & GEOTECHNICAL CONSULTANTS

Jim Ashby/Mission Geoscience, Inc.
(949) 955-9086 • www.missiongeoscience.com

TETRA TECH

David Bloom/Tetra Tech
(619) 702-6059 • www.tetratech.com

H&P Mobile Geochemistry Inc.

Suzie Nawikas/H&P Mobile Geochemistry, Inc.
(800) 834-9888 • www.handpmg.com

EMERALD SPONSORS ($100)

Anderson Drilling
Dr. Pat Abbott, Professor Emeritus, San Diego State University
AMEC Earth & Environmental
Sally & Dennis Avery
Dr. Richard W. Berry
Sherry Bloom, Coldwell Banker Real Estate
Curtis Burdett
Todd Clark
George Copenhaver, PG, CEG
Joe Corones
Greg Cranham
Dr. Margaret Eggers, Eggers Environmental, Inc.
Phil Farquharson, CG-Squared Productions
Katherine Freese
Geocon, Inc.
Carolyn Glockhoff, Caro-Lion Enterprises
Dr. Sarah Gray, University of San Diego
Hargis + Associates
Jonas & Associates
Diana and Lowell Lindsay, Sunbelt Publications
Julie Moore, SPL Laboratories
Ninyo & Moore
Dr. Monte Marshall
Monte and Diane Murbach, Murbach Geotech
Mike Palmer
Southwest Geophysics, Inc.
Stone Brewing Company
Dr. Anne Sturz
Sue Tanges, Southland Geotechnical Consultants
TerraCosta Consulting Group, Inc.
David and Jan Steller
Carole L. Ziegler

The generosity of these sponsors allows our organizations to continue to keep membership dues low, reduce rates for students, produce monthly newsletters, purchase supplies, host the annual field trip, produce our geologic guidebooks, and provide scholarships to local geology students. To contact any of the above sponsors, please visit our websites at www.sandiegogeologists.org and www.southcoastgeo.org.

South Coast Geological Society SPONSORS

Steve Zigan, *Student Sponsor*

Larry Cann, *Supplies Sponsor*

Earth Consultants International
(714) 544-5321 • www.earthconsultants.com

Cardno ERI
(949) 457-8950 • www.cardnoeri.com

Albus-Keefe & Associates
(714) 630-1626 • www.albus-keefe.net

Conestoga-Rovers & Associates Inc.
(949) 648-5200 • www.craworld.com

Golder Associates
(714) 508-4400 • www.golder.com

NMG Geotechnical, Inc.
(949) 442-2442 • www.nmggeotechnical.com

E2 ManageTech, Inc.
(562) 740-1060 • www.e2managetech.com

Leighton Consulting, Inc
(949) 250-1421
www.leightongeo.com

Regenesis
(949) 366-8000 • www.regenesis.com

H&P Mobile Geochemistry, Inc.
(760) 804-9678 • www.handpmg.com

BC2 Environmental
(714) 744-2990 • www.bc2env.com

The Source Group, Inc.
(562) 597-1055 • www.thesourcegroup.net

Saudi Aramco
(713) 432-4675 • www.aramco.jobs

San Diego Association of Geologists PUBLICATIONS

1973 *Studies on the Geology and Geologic Hazards of the Greater San Diego Area, California*, Arnold Ross and Robert J. Dowlen, eds.

1974 *Recent Geological and Hydrological Studies, Eastern San Diego County and Adjacent Areas: a Guidebook*, Michael W. Hart and Robert J. Dowlen, eds.

1975 *Studies on the Geology of Camp Pendleton, and Western San Diego County, California*, Arnold Ross and Robert J. Dowlen, eds.

1977 *Geology of Southwestern San Diego County and Northwestern Baja California*, Gregory T. Farrand, ed.

1978 *Natural History of the Coronado Islands, Baja California, Mexico*, Herman T. Kuper, ed.

1979 *Earthquakes and other Perils, San Diego Region*, Patrick L. Abbott and William J. Elliott, eds. *(Geological Society of America Annual Meeting Field Trip Guidebook)*

1981 *Geologic Investigations of the Coastal Plain, San Diego County, California*, Patrick L. Abbott and Shannon O'Dunn, eds.

1982 *Geologic Studies in San Diego*, Patrick L. Abbott, ed.

1985 *On the manner of Deposition of Eocene Strata in Northern San Diego County*, Patrick L. Abbott, ed.

1988 *Landslides in Crystalline Basement Terrain*, James R. Evans, ed.

1989 *The Seismic Risk in the San Diego Region: Special Focus on the Rose Canyon Fault System, Proceedings & SDAG Field Trip*, Glenn Roquemore, Susan Tanges, and Marion Wright, eds.

1990 *Geotechnical Engineering Case Histories in San Diego County*, John H. Hoobs, ed.

1991 *Environmental Perils, San Diego Region,* Patrick L. Abbott and William J. Elliott, eds. (Geological Society of America Annual Meeting Field Trip Guidebook).

1992 *Natural History of the Coronado Islands, Baja California, Mexico*, Lynn Perry, ed.

1993 *Colorado Desert and Salton Trough Geology*, Joe Corones, ed.

1994 *Geology and Natural History of Camp Pendleton, U.S.M.C. Base, San Diego County, California*, Phillip S. Rosenberg, ed.

1995 *Paleontology and Geology of the Western Salton Trough Detachment, Anza-Borrego Desert State Park, California*, Paul Remeika and Anne Sturz, eds.

1996 *Geology and Natural Resources of Coastal San Diego County, California*, Tissa Munasinghe and Phillip Rosenberg, eds.

1997 *Santa Cruz Island: Geology Field Trip Guide*, James R. Boles and Werner Landry, eds.

1998 *Geology and Geothermal Resources of the Imperial and Mexicali Valleys*, Lowell E. Lindsay and William Hample, eds.

1999 *Water for Southern California: Water Resources Development at the Close of the Century*, Greg T. Cranham, ed.

2000 *Geology and Enology of the Temecula Valley, Riverside County, California*, Barbara B. Birnbaum and Kerry D. Cato, eds.

2001 *Coastal Processes and Engineering Geology of San Diego, California*, Robert C. Stroh, ed.

2003 *Geology of the Elsinore Fault Zone, San Diego Region*, Monte L. Murbach and Michael W. Hart, eds. (joint publication with South Coast Geological Society).

2004 *Mining History and Geology of Joshua Tree National Park*, Margaret R. Eggers, ed.

2005/06 *Geology and History of Southeastern San Diego County, California*, David M. Bloom, Phillip T. Farquharson, and Carole L. Ziegler, eds.

2006 *Mines and Geology of the Randsburg Area, An Historical Gem of the Mojave Desert*, by D.D. Trent.

2008/09 *Palms to Pines: Geological and Historical Excursions through the Palm Springs Region*, Jeff Miller, Bryan Miller-Hicks, and Scott Snyder, eds.

2010 *Geology and Natural History of Camp Pendleton, U.S.M.C. Base, San Diego County, California*, 2nd Edition, Phillip S. Rosenberg, ed.

2010 *Geology and Lore of the Northern Anza-Borrego Desert Region*, Monte L. Murbach and Charles C. Houser, eds. (joint publication with South Coast Geological Society).

2011 *Picacho and the Cargo Muchachos: Gold, Guns, and Geology of Eastern Imperial County, California*, Todd A. Wirths, ed.

2012 *Roadside Geology along Sunrise Highway*, by Michael J. Walawender

2012 *Waiting for Tsunami: Coastal Hazards of Northern San Diego County, California*, Cari Gomes and David M. Bloom, eds.

2013 *San Luis Rey on Display, Geoscience in Northern San Diego County*, California, Brian J. Olson, ed.

For more information, visit us online at www.SanDiegoGeologists.org.

South Coast Geological Society PUBLICATIONS

1972 *Geologic Guidebook to the Northern Peninsular Ranges*
1973 *Guidebook to the Tertiary Geology of the Eastern Orange and Los Angeles Counties*
1974 *Guidebook to Selected Features of Palos Verdes and Long Beach*
1976 *Geologic Guidebook to the Southwestern Mojave Desert Region, California*
1977 *Geologic Guidebook to the Ridge Basin, Northwest Los Angeles County, California*
1978 *Geologic Guidebook to the Santa Ana River Basin, Southern California*
1979 *Guidebook to Selected Geologic Features, Coastal Areas of Southern Orange County and Northern San Diego County, California*
1980 *Geology and Mineral Wealth of the California Desert*
1981 *Geology of the San Jacinto Mountains*
1982 *Geology and Mineral Wealth of the California Transverse Ranges*
1983 *Geology of the Northern Elsinore Trough*
1984 *Geology of the Long Valley – Mono Craters – Mammoth Lakes Area*
1985 *Geology of the Santa Catalina Island*
1986 *Geology of the Imperial Valley, California*
1987 *Geology and Mineral Wealth of the Owens Valley Region, California*
1988 *Geology of the Death Valley Region*
1989 *San Andreas Fault – Cajon Creek to Wallace Creek, V1 & V2*
1990 *Lower Colorado River Extensional Terrain and Whipple Mountains*
1991 *Southern Coast Ranges*
1992 *The Regressive Pleistocene Shoreline*
1993 *Geologic Investigations in Baja California, Mexico*
1994 *Mojave Desert*
1995 *The Garlock Fault and El Paso Mountains*
1996 *Sturzstroms and Detachment Faults, Anza-Borrego Desert State Park*
1997 *Southern San Andreas Fault – Whitewater to Bombay Beach – Salton Trough*
1998 *Black Canyon of the Colorado River, Arizona/Nevada and Castle Mountains Gold Mine, San Bernardino County, California*
1999 *The Long Valley Caldera, Mammoth Lakes, and Owens Valley Region, Mono County, California*
2000 *Geology of the Las Vegas Area*
2001 *Geologic Guidebook to the Long Valley-Mono Craters Region of Eastern California by Steve Lipshie*
2001 *Geology of the Owens Valley and Inyo Mountains Region, California*
2002 *Structural Geomorphology, Petroleum and Water Resources – Northwestern Transverse Ranges*
2003 *The Geology of the Elsinore Fault Zone San Diego Region*
2004 *Geology and Tectonics of Santa Catalina Island and the California Continental Borderland*
2006 *Geology of the Orange County Region, Southern California*
2007 *Geology and Tectonic Development of Western San Luis Obispo County, Southern Coast Ranges, California*
2008 *Geology and Hydrogeology of the Big Bear Valley and San Bernardino Mountains, Transverse Ranges, California*
2009 *Geology and Hydrology in the Eastern San Gabriel Mountains*
2010 *Geology and Lore of the Northern Anza-Borrego Desert Region*
2011 *Mammoth Lakes - Volcanoes, Glaciers, Rivers, and More*
2012 *Palms to Pines: Geological and Historical Excursions through the Palm Springs Region*
2013 *Eastern Mojave Desert Geology, Mining and Earth Resources*

For more information, or to order the available publications,
visit our website (www.southcoastgeo.org) *to download our publication order form.*

RELATED PUBLICATIONS

Palm Springs Legends: Creating a Desert Oasis, Greg Niemann
From the Land of the Cahuilla to the Legacy of Indian Canyons, vibrant chapters portray this storied land.

Tortoises through the lens: A Visual Exploration of a Desert Icon, David Lamfrom
Remarkable close-ups and scenes produced by some of today's students who'll drive tomorrow's science.

Gold and Silver in the Mojave: Images of a Last Frontier, Nicholas Clapp
"The glamour, grit, grandeur, and gold of the desert" on Coachella Valley's northern ramparts.

Baja California Plant Field Guide San Diego Natural History Museum, Jon Rebman & Norm Roberts
Full-color detail with over 50% of the species also occurring in the deserts and highlands north of Baja.

Desert Legends and Lore of California's Colorado and Mojave Deserts, Choral Pepper
Lost mines, Indian myths, legendary characters, strange natural features, and fabled treasure ships.

Earth Pigments & Paint of the California Indians: Meaning & Technology, Paul Campbell
Native Californian rock art from the Klamath to Cabo including Baja's world-class giant figures.

Cycling the Palm Springs Region, Nelson Copp & Margaret Gooding
Includes 35 route descriptions and superb maps, many showing fault features, in the Coachella Valley and surrounding mountains.

Afoot & Afield Inland Empire: A Comprehensive Hiking Guide, David and Jennifer Money Harris
Includes 75 Coachella Valley, Santa Rosa, and San Jacinto Mountains National Monument and Joshua Tree National Park trails.

Palm Springs-Style Gardening: Plants & Practices for Dryland Gardens, Maureen Gilmer
Bountiful botanic inspiration from Coachella Valley's oasis communities for arid lands anywhere.

For Tourism and a Good Night's Sleep: J. Win Wilson, Wilson Howell, and the Beginnings of the Pines-to-Palms Highway, Steve Lech.
A history of the circumstances and people who created the Pines to Palms Highway.

These and similar publications are available from
www.SanDiegoGeologists.org
or
www.sunbeltbooks.com

Palms to Pines Guidebook
INTRODUCTION

Between the peaks of Mounts San Gorgonio and San Jacinto, through San Gorgonio Pass into the basins of Coachella Valley lies the San Andreas fault system. This is the transform plate boundary whose effects are responsible for some of the steepest topographic relief in the United States.

Journey along the "Pines to Palms" State Highway 74. Wonder at a 5 mile rock slide at Martinez Mountain. Witness the Whitewater River, a torrent swallowed up by the desert. Imagine fishing in ancient Lake Cahuilla with early Native Americans. Contemplate tectonic forces responsible for this landscape, hot springs, and cool fan palm oases.

Follow the road logs from the Pacific tectonic plate to the North American tectonic plate. Ponder modern water wars, while exploring this desert land where Native Americans have made their homes for millenia. The road logs provide descriptions of specific stops to visit plus information about the area as you drive past points of interest.

Discover neotectonics and faulting, hydrogeology, landslides, history, and archaeology of the Palm Springs region. Expert authors contribute general interest and technical papers on a range of views on these topics.

Hydrogeology is a major topic and was the focus of the San Diego Association of Geologists' 2008 Field Trip. Structural geology of the mid- and upper-crustal zones expressed by well-exposed mylonitic shear structures and associated detachment faulting is explored in detail and was the major topic of the San Diego Association of Geologists' 2009 Field Trip. Neotectonic and groundwater controversies remain unresolved at this active plate boundary and were the main themes of the South Coast Geological Society's 2012 Field Trip. Highlights include the spectacular Santa Rosa and San Jacinto Mountains National Monument, Mission Creek Reserve, Boyd Deep Canyon Research Center, the original Palm Springs, Thousand Palms Oasis, and Morongo Valley.

PLEASE READ:
This book describes driving and hiking situations and other potentially hazardous activities that are the sole responsibility of the reader. Exploring geological features in mixed terrain from remote desert wilderness to urban environments is not without risk. All who engage should be aware of their own capabilities, be knowledgeable, prepared, and alert. When exploring such environments, please abide by laws of the road, respect private property, and don't forget about common sense. The conditions, access, and features described in this book can and do change with time.

"The geologist takes up the history of the earth
at the point where the archaeologist leaves it,
and carries it further back into remote antiquity."
—*Bal Gangadhar Tilak*

Part I

ROAD LOGS

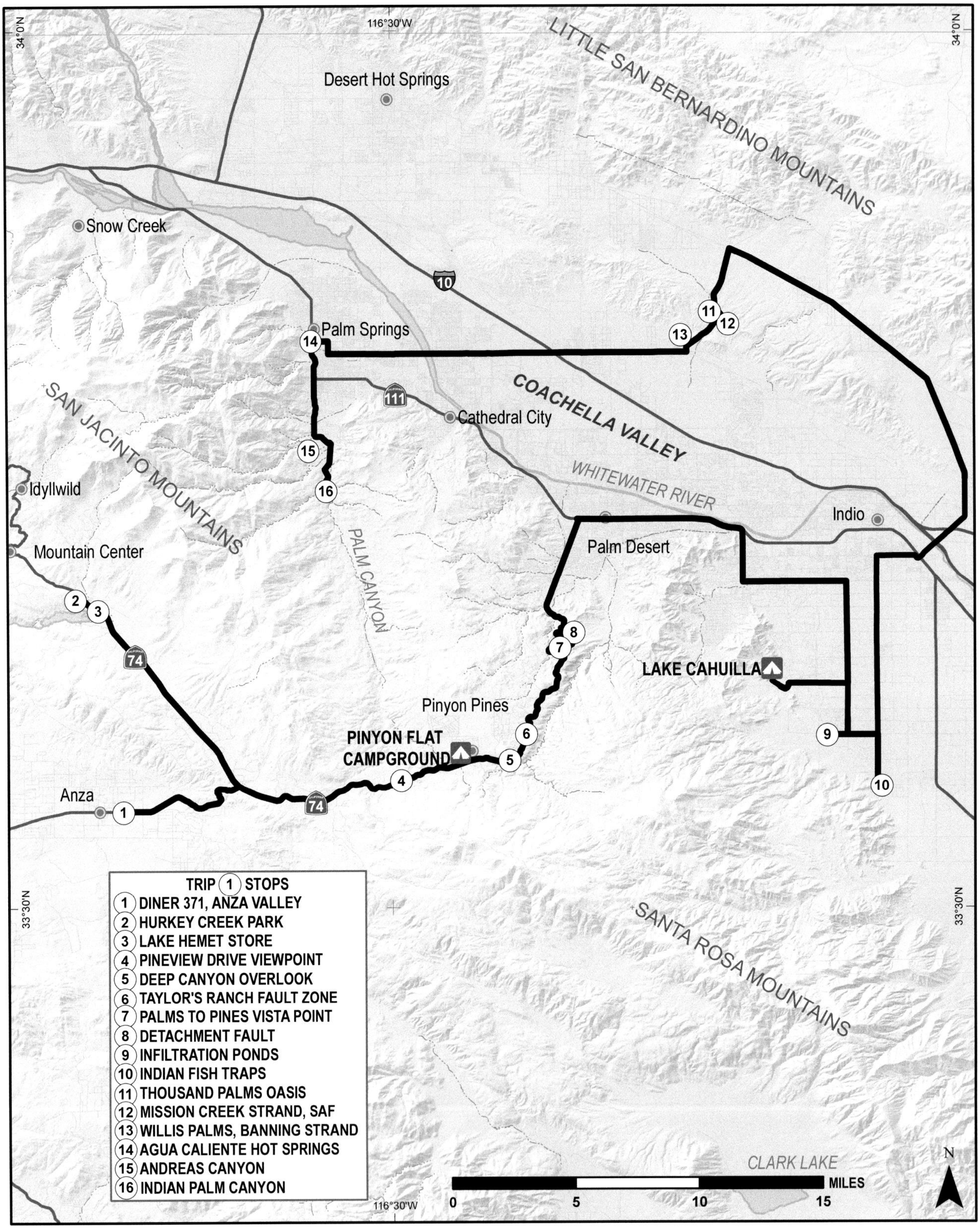

ROUTE MAP, TRIP 1

Previous Page: The Switchbacks. The Pines-to-Palms Highway was built between 1929 and 1932, opening in June 1932. It connected Keen Camp (today's Mountain Center) with the Palm Springs-Indio Highway (today's Highway 111). It was hailed as an engineering marvel and was Riverside County's first scenic highway. Source: Collection of Steve Lech.

TRIP 1
Oasis in the Desert: Coachella Valley Groundwater and Santa Rosa Mylonite

"Splish, Splash, I was taking a bath."

FIELD TRIP LEADERS: Scott Snyder, Monte Marshall, Bradley Erskine, and Bryan Miller-Hicks

INTRODUCTION

The Palm Springs region is a place of extremes. Mount San Jacinto looms over Palm Springs to the southeast onto the valley floor with some of the greatest relief in the United States. The famous San Andreas fault system cuts across the eastern edge of the Coachella Valley. This active transform system controls surface topography and subsurface groundwater systems. The depression at the low point of the valley is currently occupied by the Salton Sea, formed in 1905–1907 when Colorado River flood waters accidentally escaped through an unprotected irrigation cut. Tectonic activity in the Mesozoic Era records a period of compression that left a record of mylonitic textures in the basement rock, now exposed by later crustal extension and transpressional uplift. This road log takes the geologic explorer in a grand loop from the uplands of the Santa Rosa and San Jacinto mountains, down the Pines to Palms Highway to Palm Desert and La Quinta, east across the Coachella Valley to the desert palm oases along the San Andreas fault, then back west to Palm Springs and concludes in Palm Canyon.

Photo 1 (above). Infiltration ponds on the Whitewater River at the base of Mount San Jacinto. Chino Canyon in center of photo. Palm Springs and Santa Rosa Mountains in the distance. (Bruce Perry, Department of Geological Sciences, CSU Long Beach)

MILEAGE *Trip 1 covers 122.0 miles. Three days should be ample time to visit all Stops.*
(cumulative and intervals)

0.0 0.0

STOP 1 SAN JACINTO FAULT ZONE

Speaker: Nate Onderdonk

Parking lot of Diner 371, 57331 State Highway 371, near cross street Hamilton Drive in the Anza Valley, makes a convenient landmark to begin this excursion.

The San Jacinto fault zone is a major structure in southern California and delineates the southwestern side of the San Jacinto and Santa Rosa mountains. The fault zone splays off the San Andreas fault in the eastern San Gabriel Mountains and extends over 200 km southeast to the western Imperial Valley. The total offset on the fault zone has been estimated to be between 24 km and 30 km of right-lateral displacement based on offset metamorphic rocks (Sharp, 1967; Morton and Matti, 1993). The zone here is made up of several major fault strands, including the Casa Loma and Clark faults that run through the Anza area at the foot of Thomas Mountain. In the Anza area, Sharp (1981) estimated between 5.7 km and 8.6 km of offset since 700 ka* along the Casa Loma fault based on Pleistocene gravels offset from their source area on the south side of Thomas Mountain. This relationship yields a slip rate between 8 mm per year (mm/yr) and 12 mm/yr. More recently, extensive trenching studies by Rockwell and others (1990, 2007) at Hog Lake near Anza have yielded slip rates of 6–23 mm/yr and most likely in the range of 13–14 mm/yr.

The San Jacinto fault zone is now receiving a fair amount of attention in an attempt to re-evaluate the distribution of strain across the plate boundary system. Recent studies have shown that Holocene slip rates along the San Jacinto fault may be larger than along the southernmost San Andreas fault (McGill et al., 2007; Fletcher et al., 2006; Kendrick et al., 2002; Morton and Matti, 1993). In addition, geodetic studies indicate that the San Jacinto fault may be accommodating a larger amount of the plate boundary displacement than the southernmost San Andreas fault (Becker et al., 2005; Fialko, 2006) suggesting that the fault zone may be the dominant plate boundary structure in southern California.

Drive east on Highway 371.

2.0 2.0 Road changes course from east to northeast, then southeast. When the road changes to southeast you enter a roughly straight canyon of Hamilton Creek. You are traveling along the trace of the Clark fault, a major 50 mile strand of the San Jacinto fault zone (Figure 1). Behind you, you could trace this fault northwest in a generally straight line to Hog Lake (6 miles). The fault continues northwest another 12 miles toward the town of Hemet where the Clark fault ends; right-lateral strain to the northwest is taken up by the en echelon Casa Loma fault. The Clark fault strand continues to the southeast into the Imperial Valley for about 35 miles.

*ka = kilo annum or a thousand years ago

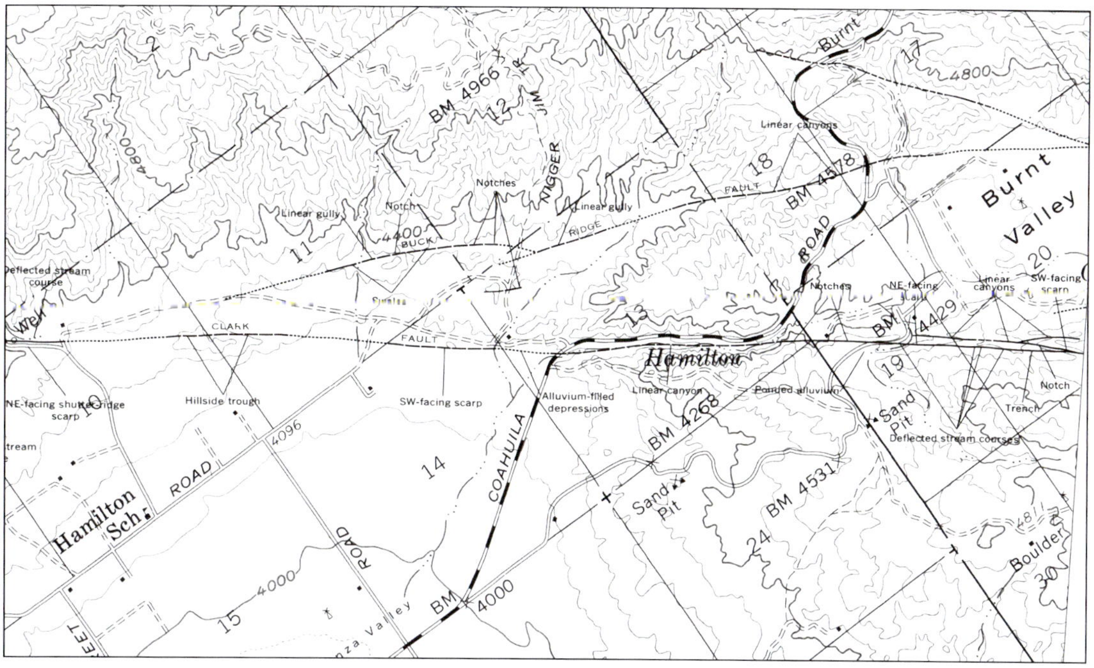

Figure 1. Fault map showing Clark fault through canyon of Hamilton Creek and Buck Ridge fault east of Anza, mile 2.2 to mile 3.7 of road log.

Figure 2. Clark fault on northeast side of Highway 371 in canyon of Hamilton Creek.

2.7 0.7 Exit canyon of Hamilton Creek and Clark fault.

3.5 0.8 Cross Buck Ridge fault, a splay of the Clark fault which trends more easterly than the main Clark fault. The Buck Ridge fault joins the Thomas Mountain fault about 1.5 miles to the east.

4.1 0.6 Cross Thomas Mountain fault, which trends north-northwest. The Thomas Mountain fault will parallel our route along Highway 74. Along this portion of Highway 371, on the northeast side of the canyon, fault gouge may be seen from the road.

5.0 0.9 Turn left at Junction with Highway 74 and head northwest along the base of the San Jacinto Mountains.

14.1 9.1 Turn right into campground parking lot.

STOP 2 HURKEY CREEK PARK AND CAMPGROUND: HOT SPRINGS FAULT OUTCROP

Speaker: Nate Onderdonk

Figure 3 (left). Hot Springs fault at stop 2. Fault dips south placing Cretaceous tonalite against Quaternary fanglomerate.

Figure 4 (right). Quaternary fanglomerate in footwall of Hot Springs fault at Stop 2.

The Hot Springs fault is a major strand of the San Jacinto fault zone that bounds the southwestern edge of the San Jacinto Mountains. The fault is an important structure in this area because of its role in the uplift of the San Jacinto Mountains and the development of the San Jacinto fault zone. It forms a restraining step-over with the Claremont fault at its north end near the town of San Jacinto, and extends 40 km southeast to the Garner Valley area.

At the north end of Garner Valley, the Hot Springs fault is exposed in a roadcut on the southwest side of Highway 74 near Hurkey Creek Park. At this location, the fault juxtaposes Cretaceous tonalite over Quaternary fanglomerate. The fault dips 60° to the south and exhibits southwest-side-up reverse separation. The fault most likely becomes steeper at depth as indicated by the distribution of hypocenters along the fault and abundant outcrops to the northwest that show a near-vertical dip. Along most of the length of the Hot Springs fault, the sense of dip-slip separation is the opposite of that seen here, with the northeast side elevated. Sharp elevation changes across the fault, and vertically offset Plio-Pleistocene sediments indicate that uplift of the San Jacinto Mountains was partly the result of displacement along this structure. The amount of strike-slip displacement was postulated to be 5 km by Hill (1981) based on an offset intrusive contact in this area.

Southeast of this location the fault is concealed beneath recent alluvium of Garner Valley. The southeastern termination of the Hot Springs fault is unknown. The north side of Garner Valley is marked by a linear mountain front of crystalline rock that has been overlapped by Quaternary fan deposits. This old mountain front is on alignment with the fault, and is most likely the result of early displacement along the Hot Springs fault trend.

The distribution of seismic activity in the area suggests that the Hot Springs fault is currently not as active as the parallel Casa Loma fault strand 3 miles to the southwest. However, the fault exhibits abundant geomorphic and geologic evidence of late Quaternary activity such as pronounced scarps, offset Quaternary terrace deposits, fault benches, linear valleys, and deflected streams.

For additional discussion of the Hot Springs fault and its role in uplift and shaping the San Jacinto Mountains and the Santa Rosa Mountains see Onderdonk (this volume).

Turn around and backtrack southeast on Highway 74.

14.7 0.6 Pull over to the right shoulder.

STOP 3 GARNER VALLEY AND LAKE HEMET

Speaker: Nate Onderdonk

Lake Hemet Market, 56571 State Highway 74, Mountain Center (cross street: Lake View Drive).

Garner Valley is a fault-controlled valley that developed in conjunction with slip along the San Jacinto fault zone. Several strands of the fault zone are present in the area and have created the local topography visible from this location. All the topographic features in the area, including the valley itself, trend in a northwest-southeast direction parallel to the San Jacinto fault zone (Hill, 1981). Along the northeast side of the valley, Quaternary fan deposits overlap an older linear mountain front that was most likely formed by slip along the Hot Springs fault. The spine of the southeastern San Jacinto Mountains was elevated along the east side and is composed of Cretaceous tonalite and older metasedimentary rocks (Brown, 1968; Dibblee, 1968).

The south side of the valley is bordered by Thomas Mountain, which is also composed of Cretaceous intrusive rocks with minor screens of older metamorphic rocks. The Thomas Mountain fault lies at the base of the mountain and defines the edge of the valley. The fault truncates Plio-Pleistocene sediments in the southern part of Garner Valley and juxtaposes Quaternary alluvium against tonalite at its northern end, but appears to be overlain by older Quaternary alluvium deposited at the base of Thomas Mountain. At its southeast end, the Thomas Mountain fault merges with the Buck Ridge fault in the area of Burnt Valley. At its north end near Lake Hemet, the fault splits into two strands that merge to the northwest with the Casa Loma fault. The steep canyon that drains Lake Hemet and Garner Valley was eroded into gouge of the northern fault splay. Excellent exposures in the spillway of the Lake Hemet dam show that the Thomas Mountain fault dips south at approximately 50°. The fault most likely merges at depth with the Casa Loma fault located along the opposite side of Thomas Mountain.

The bedrock floor of Garner Valley was continuous with the Anza plateau during deposition of the Plio-Pleistocene terrestrial deposits known as the "Bautista Beds." Later movement on the Thomas Mountain and Casa Loma faults resulted in the uplift of Thomas Mountain and the separation of Garner Valley from the rest of the elevated plateau to the south. Garner Valley was subsequently filled with locally derived sediment and now drains to the northwest via San Jacinto Canyon. Lake Hemet was formed when a dam was built across this drainage in the 1890s to provide water for agriculture in the valley.

23.2 8.5 Pass Highway 371, where we turned earlier.
Reset odometer to zero.

0.0 0.0 Travel along the Pines to Palms Scenic Byway.

1.0 1.0 Pacific Crest Trail crossing. The trail follows the approximate drainage divide between streams that flow west to the Pacific Ocean, and those that drain east into the Coachella Valley.

5.2 4.2 Santa Rosa Mountain Road.

6.2 1.0 Turn left into the viewpoint turnout.

STOP 4 PINEVIEW DRIVE VIEWPOINT. (On left; if you pass Ribbonwood Road, you went too far.)

Speakers: Bradley Erskine and Monte Marshall

Figure 5. View from stop 4, Pineview Drive viewpoint looking north into Palm Canyon (Photo: Greg Farrand).

Walk north a few hundred feet to see a great view down into Palm Canyon, where you will see an excellent east-west cross-section of the major geologic features in this area (Figures 6 and 7). The units and structures from west to east are:

- Basement, composed of plutonic rocks and the metasediments of the Desert Divide Group, first mapped by Dick Brown (1968, 1980).
- Mylonite zone, showing east dipping foliation and north-northeast plunging lineation, in conformable contact with the basement.
- Palm Canyon fault, within the bottom of the canyon, dipping east. It is recognized by its green color and brecciated fabric.
- Paleozoic Palm Canyon complex, with east dipping foliation and north-northeast plunging lineation.
- Asbestos Mountain fault, with a shallow dip to the east.
- Granodiorite of Asbestos Mountain showing mild foliation, dipping east; and north-northeast plunging lineation.

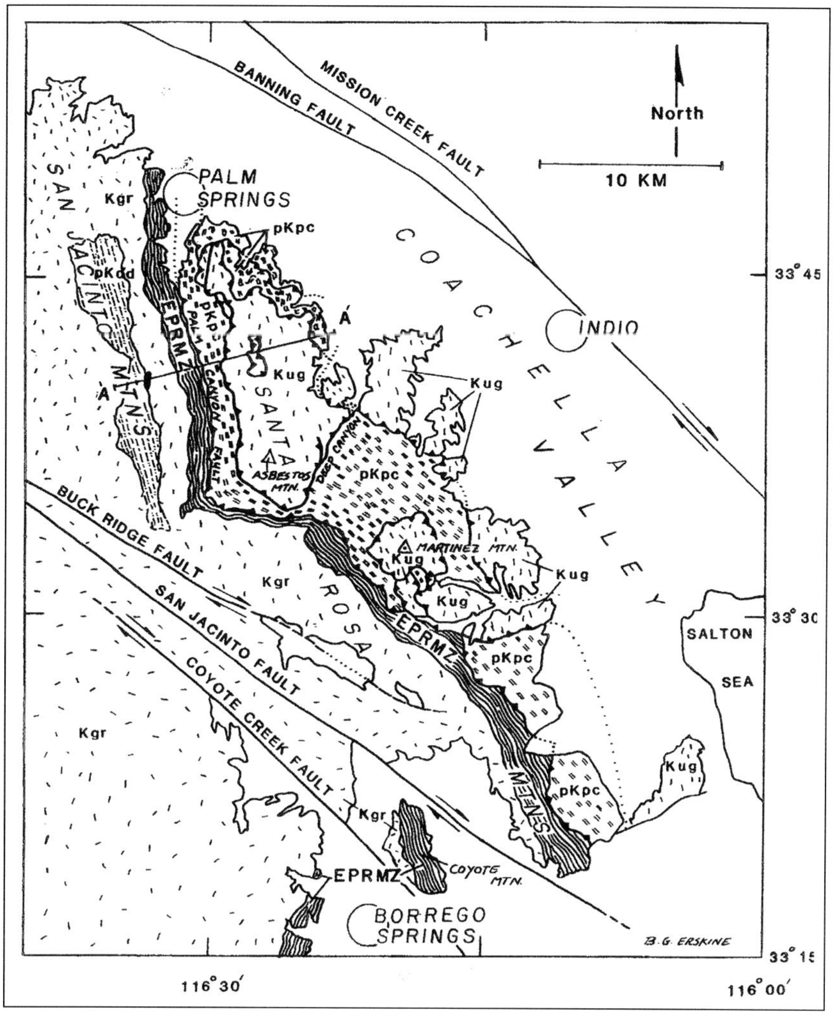

Figure 6. Brad Erskine's generalized geologic map of the northeast San Jacinto Mountains and the Santa Rosa Mountains. See Figure 7 for meaning of lithologic symbols (from Todd et al., 1988).

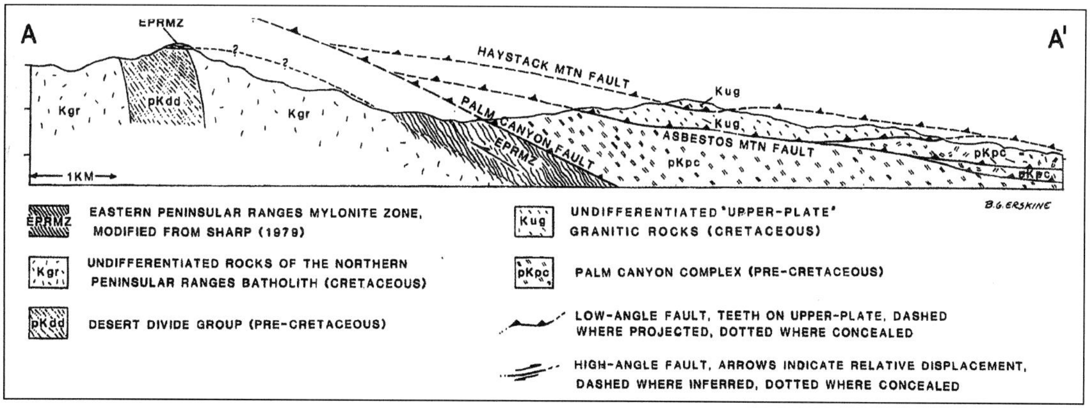

Figure 7. Brad Erskine's east-west cross-section 10 km south of Palm Springs. No vertical exaggeration. See Figure 6 for location (from Todd et al., 1988).

9.5 3.3 Palm Canyon Drive

10.2 0.7 Pinyon Flats Campground and Pinyon Pines.

SIDE TRIP AND HIKE INTO PALM CANYON

Turn here and park at Pinyon Flats Campground 0.2 mile north of Highway 74. Hike from the campground north on Pinyon Flat road about 3.5 miles to its end near an abandoned and sealed tremolite asbestos mine, located on the Asbestos Mountain fault. Hike across the southwest slope of Asbestos Mountain and descend approximately 0.5 mile and 200 vertical feet down the east side of Palm Canyon through a beautifully exposed section of the Paleozoic Palm Canyon complex, composed of quartzites, marbles, sillimanite schists with mylonitic fabrics and evidence of anatexis (partial melting). At the bottom is the Palm Canyon fault and the mylonitized granodiorite. The fault is generally located in the bottom of Palm Canyon and is marked by its chloritic green color and brecciated fabric. The mylonites are on the west flank of the canyon; the metasediments are on the east.

10.7 0.5 CalFire Fire Station and Sugarloaf Café

0.0 0.0 Reset odometer to zero.

For more information on the various aspects of geological and hydrological resources of the Santa Rosa San Jacinto Mountains National Monument refer to Harry Quinn's 1995 introductory paper on this subject, reprinted in this volume. Included at the end of his paper is a geologic road log that takes the reader from the intersection of Highway 74 and Highway 111 in Palm Desert to this point, then turn around back down the Pines to Palms Highway.

1.0 1.0 Pull into turnout on the right.

STOP 5 DEEP CANYON OVERLOOK

Speakers: Harry Quinn and Brian White

From here you can look down into the upper part of Deep Canyon. Looking south across the canyon you can see a couple of hills containing dolomite and to the southwest the old dolomite mine that operated in the early 1950s. Also to the southwest you can see where the canyon has cut into the Pigeon Creek fan. To the south can be seen some remnants of the fan atop the hills with the dolomite. Behind the dolomitic hills can be seen a bench along the base of the mountain. This bench is at about the same elevation as the place where we are standing now and the two areas may have been part of the same flat surface in the past. This suggests that Deep Canyon has been cut into an old flat area that was probably present in Pleistocene time. The late origin of the canyon is further supported by the presence of the Pigeon Creek fan remnants. To the southeast is another flat area, Pinyon Alto Flat, that was probably once part of this old large flat area and there it has been incised into by Horse Thief Canyon, a tributary to Deep Canyon.

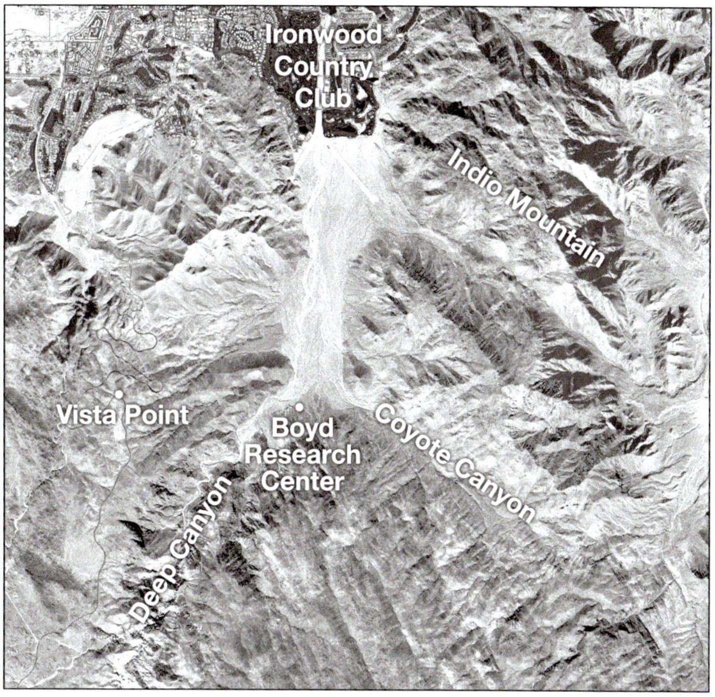

Figure 8. Aerial image of view from 2 miles above Deep Canyon, Palm Desert, California. Image: Google, Inc.

1.3 0.3 Tewanet Overlook for a nice view and some Native American information.

2.3 1.0 Pull into turnout on the right.

STOP 6 TAYLOR'S RANCH FAULT ZONE

Speaker: Harry Quinn

To the west can be seen two nearly parallel quartz-feldspar dikes that abruptly stop at the base of the hill. If you look to the east you can see the continuation of these dikes on the north side of the fault, and they abruptly stop at the base of the hill. Walk north along the highway a short distance to see the fault zone exposed in the road cut. The width of the fault zone is plainly seen by the gouge zone present between the undisturbed rocks to the north and south of the zone.

Figure 9. Taylor Ranch fault zone at Stop 6.

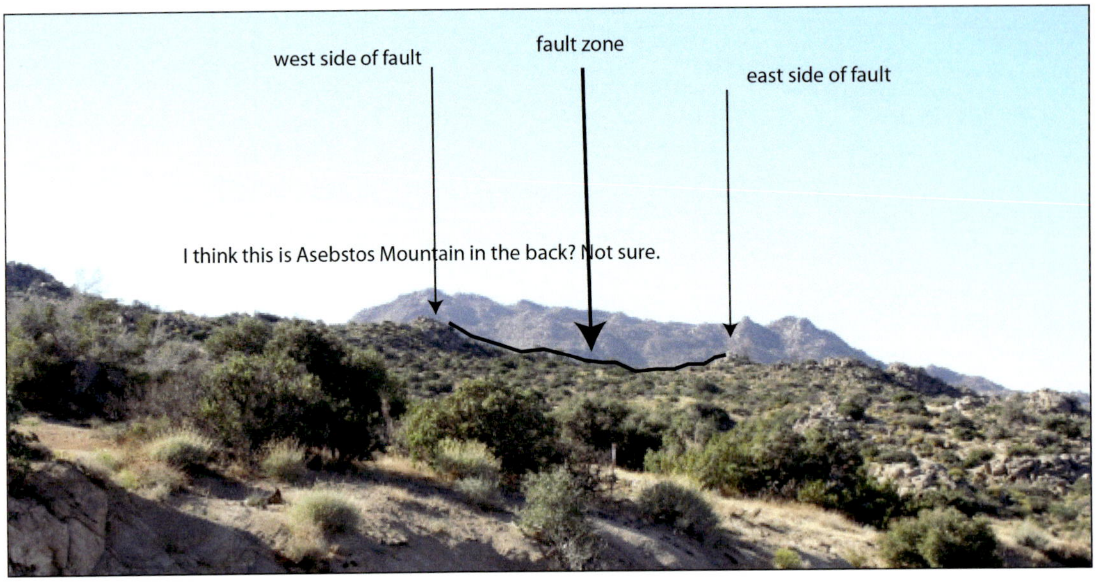

Figure 10. Taylor Ranch fault zone exposed at the crest of a ridge near Stop 6.

Look east to see a topographic bench at about the same elevation as our stop. This may be a remnant of a flat area that probably existed before the canyon formed.

6.6 4.3 Pull into the parking lot on the right.

STOP 7 COACHELLA VALLEY VISTA POINT. Pines to Palms overlook of the Coachella Valley. *Speakers:* Harry Quinn and Bradley Erskine

Figure 11. View to north to northeast from Stop 7, Coachella Valley Vista Point on Highway 74, with snow capped summits of San Jacinto and San Gorgonino in the far distance, left, to Little San Bernardino Mountains forming a backdrop of Coachella Valley with the city of Palm Desert in the middle ground.

From Vista Point, if the air is clear, one can look northward to the Indio Hills and Little San Bernardino Mountains, beyond which is Joshua Tree National Park. To the north you may be able to see the flanks and summit of Mount San Jacinto, the tallest of the Peninsular Ranges mountains at 10,834 feet. Farther in the distance and slightly to the east is Mount San Gorgonio, the tallest mountain in southern California with a summit at 11,499 feet. If you look closely you can see both major strands of the San Andreas fault, the Banning fault along the base of the Indio Hills and the Mission Creek fault cutting through the Indio Hills. These fault zones are accentuated by lines of native palm oases.

To the southeast the Salton Sea can be seen, a feature that formed from accidental flooding from the Colorado River in 1905–07. The Salton Sea is located in a depression with no outlet and would have dried up in this arid location but for irrigation returns which have maintained its elevation since its creation. You can also see two detachment-related fault blocks, the largest being the one that extends all the way out to Point Happy at Highway 111 and Washington Street, and beyond under the valley. In front of us is a third portion of the detachment fault block and if you look carefully you can actually see the fault plane. It is slightly greenish and parallels the lowest curve of the highway visible below us. We'll visit this fault plane at Stop 8.

Figure 12. View to southeast from Stop 7, Vista Point on Pines to Palms Highway, Indio Mountain in the center and Little San Bernadino Mountains in distance.

To the west-northwest Haystack Mountain can be seen. This is a klippe, one of several klippen formed by the low angle Asbestos Mountain fault as it intersects some of the hills and mountains below. It is composed of a rock type not found elsewhere in the area. Several other possible thrust faults and detachment faults can be seen from here. The upper part of the Paleozoic section in Deep Canyon is visible to the south.

For a discussion of the Pinyon Flat landslide, a mega-landslide postulated to be 15 miles long and 27 miles wide on the northeastern flank of Santa Rosa Mountains from Palms Springs to Martinez Canyon, covering about 310 square miles, including Vista Point, refer to Elliott and Scaramella's paper in this volume.

Point of Discussion: Geometry of the stacked low-angle faults.

Resume drive on Highway 74 descending through a series of hairpin curves to the desert floor. Note that during the first few miles of the descent, you are in upper-plate Cretaceous granodiorite of Asbestos Mountain. This unit is characterized by large euhedral sphene and large subhedral, oriented hornblende. Look to the south and you will see Deep Canyon and the Paleozoic metasediments of the Palm Canyon Complex (Figs. 3 and 4) intruded by Cretaceous igneous rocks. The Asbestos Mountain fault that places the granitic rocks on top of the metasediments is exposed in Deep Canyon. The exposures in the roadcuts are fabulous, but should only be enjoyed by passengers as drivers will need to pay attention to traffic and hairpin turns.

9.3 2.7 **View a detachment fault in plain view several feet above your head to the right along the side of the road.** This is probably one of your closest views of the fault from the road.

9.5 0.2 Turn left into the wide dirt turnout.

STOP 8 SALTON SEA DETACHMENT FAULT. Turnout on southbound side of the highway. View the detachment fault up close.
Speaker: Mike Hart

Figure 13. Road cut 0.2 mile north of Stop 8 of detachment fault placing fanglomerate on top of metasediments in lower plate, with a prominent greenish chloritized top.

Park in the unpaved parking area on the west side of the Pines to Palms Highway opposite what San Diego State University Professor Eric Frost referred to as "an exquisite exposure of the Salton Sea Detachment fault." At this location the fault can be observed on the east side of the highway as a gently northerly dipping and very sharp contact between lower plate metamorphic rocks and upper plate massive granitic rocks. A short walk of approximately 15 minutes follows a use trail along this spectacular fault contact for several hundred meters on the northwest side of the highway and terminates at a spectacular highway road cut that exposes the northeasterly dipping fault plane. The upper plate rocks at this location are composed of a coarse chaotic granitic breccia that contains deformed bands of Miocene alluvium entrained in the breccia. The lower plate rocks are composed primarily of light brown metasedimentary rocks and gneiss. According to Eric Frost, the breccia, which attains a thickness of several hundred feet, was primarily formed by collapse of steep headscarps formed during Miocene faulting and represents graben infilling resulting from down-to-the east subsidence of Coachella Valley. Similar thick breccias resembling Quaternary landslide debris crop out along Highway 78 near Yaqui Pass in Anza-Borrego Desert State Park.

9.7 0.2 Fanglomerate along the right hand side of the road.

9.9 0.2 Contact of the same fanglomerate on top of metasediments. This fanglomerate is below the detachment fault.

10.2 0.3 Art Smith Trailhead.

10.3 0.1 San Jacinto Santa Rosa Mountains National Monument.

10.6 0.3 As you continue down Highway 74 notice a church and church school on your left. Look behind the church and you can see a white dolomite outcrop. This may be the same dolomite that you saw at Stop 4.

10.7 0.1 Side trip to Deep Canyon at the Philip L. Boyd Deep Canyon Desert Research Center, part of the University of California Natural Reserve System. Access to the research facilities and 6100-acre reserve is strictly controlled. Researchers, scholars, and others may apply to the University of California, Riverside, for access. Every visitor must abide by the rules of the facility and all rock hammers must be left behind before entering the canyon.

DEEP CANYON HIKE

Hike south from the Reserve research center up a dirt road for about ½ mile to the mouth of the canyon, i.e., where the width of the canyon narrows abruptly and the walls become quite steep and high. Those who have attempted to hike down the canyon have learned that it is well named! Hike up Deep Canyon and make frequent stops to experience a pristine, wild gorge that few people, even geologists, have ever seen. This is the same complex that can be seen across Deep Canyon while driving on Highway 74. This is also presumably the same unit as seen on the earlier hike from Pinyon Flats into Palm Canyon. Notice the difference in fabric and lithology. The river bed consists of sand with scattered gravel, up to boulder size. The ascent is gradual and, assuming no recent rains, you will be walking on dry sand with some short boulder hopping—making hiking boots recommended. You will be enclosed on two sides by high, steep walls of igneous and metamorphic rocks, with signs of pervasive ductile deformation and mylonitization.

Points of Discussion:

- Why are these rocks so different than the same unit in Palm Canyon?
- What were the conditions of metamorphism and deformation?

Figure 14. Dr. Monte Marshall at "Z-fold" at the mouth of Deep Canyon. The Z-fold shows that the sense of shear is top of section moved to the right, or uphill. View looking southeast.

Figure 15. Mylonite texture in a streambed boulder in Deep Canyon

Figure 16. Coyote Wash detachment fault, chloritized breccia top of lower plate on the right, fractured upper plate granodiorites on the left. View to the east in the Boyd Deep Canyon Reserve.

COYOTE WASH DETACHMENT FAULT

From the research center, hike along the entrance road for about ¼ mile. At the fork with the paved road, take the right unimproved dirt road to the entrance of a wash, about ¾ mile. Hike up the wash and examine one of the best exposures of a low-angle fault that is part of a detachment fault system in the Peninsular Ranges. Observe light-green chloritized breccia in the footwall, the band of non-chloritized brown fault gouge and breccia, and the fractured upper-plate granodiorites of Asbestos Mountain.

For more information on mylonitization and associated low-angle faulting in this area, refer to Brad Erskine's paper in this volume. Brian White provides a paper, also in this volume, on an interpretation of the structural development of the mylonite zone.

Points of Discussion:

- The presence of the mineral chlorite and pervasive brittle fractures of the breccia in the rock fabric indicate shallow depth of faulting.
- Is this a normal (extensional) or a reverse (compressive) fault?
- Is this really a "Detachment Fault," and is it related to the mylonites and Palm Canyon fault?

Figure 17. Coyote Mountain detachment fault, non-chloritized brown fault gouge at base of brecciated granodiorites of upper plate

15.1	**4.4**	**Turn right on Highway 111 from Highway 74.**
0.0	**0.0**	**Reset odometer to zero.** If you look to the hills to the west, you may be able to make out a large landslide in granitic rocks. For more information on the Palm Desert landslide, refer to Michael W. Hart's paper in this volume.
2.6	**2.6**	**Desert Maps and Aerial Photo store on your left.** A good source for local and southern California geology references.
5.6	**3.0**	**Right on Washington Street.**
7.7	**2.1**	**Left on Avenue 50.**
11.2	**3.5**	**Right on Monroe.**
15.2	**4.0**	**Right on 58th Avenue.** As you turn right look at the 10 o'clock direction to the hills in the foreground. Look at the bottom of the hill and observe the ancient Lake Cahuilla shoreline, characterized by a horizontal line contrasting brilliant white crust below otherwise dark rock above. That crust is composed largely of the fossil remains of calcic invertebrates that encrusted the substrate below the waterline. In the 8 o'clock direction is the toe of Martinez Mountain rockslide.

Figure 18. The ancient Lake Cahuilla shoreline is marked by sharp contrast between the dark rock and white tufa crust that formed at the waterline. Collection of Arthur Sylvester.

17.2	**2.0**	**The road will bear left.**
17.7	**0.5**	**The road will bear right (Wade St.).** This is the entrance to Lake Cahuilla campground. If you forget to bear right a nice guard for the gated community will direct you to turn around.
18.0	**0.3**	**Lake Cahuilla County Park and Campground.** Look across the lake to PGA West, a well known and long standing resort community. Within that community the land surface has subsided due to groundwater withdrawal, forming fissures and cracks in the land surface damaging roadways, sidewalks, and homes. This has resulted in multiple lawsuits that are pending at this time.
0.0	**0.0**	**Return to Madison Street and 58th Avenue. Reset odometer. Turn south on Madison.**
0.5	**0.5**	**On your right (about 3 o'clock) is Coral Island.**
1.0	**0.5**	**Turn left on 60th Avenue.**
2.0	**1.0**	**Turn right on Monroe Street.**
3.0	**1.0**	**Turn right on 62nd Avenue.**

3.7 0.7 Drive to the end of the road.

STOP 9 WEST END OF 62ND AVENUE. Park here and walk up to the top of the dike.

Speaker: Harry Quinn

Below you and to you right are the pilot infiltration ponds for the recharge of groundwater in the southern portion of the Coachella Valley. To date, the recharge north of Palm Springs at Whitewater has not made it down to the lower portion of the valley.

Figure 19. At Stop 9, view to north of pilot infiltration ponds for the recharge of groundwater in the southern portion of the Coachella Valley.

From the same vantage point, to your left toward the southwest, you can see the Martinez Mountain rock slide. The pile of rock at the base of the mountain came from up near the top of Martinez Mountain. It slid off in an easterly direction, went up the bank of Toro Canyon, and then turned northward out onto the desert. Take note of the high patina on the rocks at the base of the slide and the lack of any patina up near the top of the mountain. Since this is the same rock, one can see the importance of weathering in patina formation.

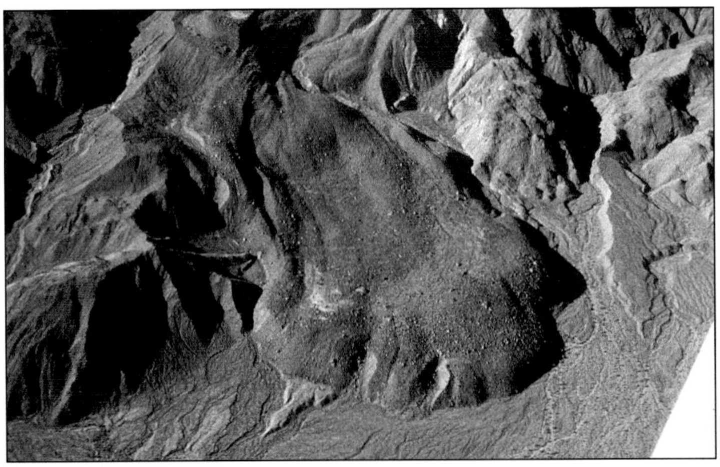

Figure 20. The Martinez Mountain Landslide. This slide occurred in granitic rock on the east side of Martinez Mountain near Palm Springs. Width of the toe is approximately one-half mile. The slide initiated near the top of Martinez Mountain and traveled over five miles to a point just north of the mouth of Toro Canyon (left side of photo). Credit: Mike Hart

Turn around and return the way you came. Drive east on Avenue 60.

5.4 1.7 Turn right on Jackson.

7.4 2.0 Stop before the pavement ends.

STOP 10 INDIAN FISH TRAPS. The farm is private property, the fish traps are not, so you will have to park at the corner of Jackson Street and 66th Avenue and walk around the property. *Speaker:* Harry Quinn

There are two major types of fish traps found here, the rock pit type and the rock wall type. Both appear to have been made in the same way, rocks were taken from the center and piled around the edge. Where the rocks were small, the pit type was produced and where the rocks were large, the rock wall type was made. Just how these traps functioned is debated. The one thing that is apparent is that they were not used here when the lake was full and that each row represents a single year's usage. Because of the method employed to make them, they did not have to destroy the previous year's traps. Down the valley there were once several hundred V-shaped fish traps. Many of these have been destroyed by agricultural and aggregate development. For more information on fish traps, see paper by Harry Quinn (this volume).

Turn around and travel north on Jackson Street for about 3 miles to Avenue 60.

0.0 0.0 Reset odometer at Jackson Street and Avenue 60. Continue north on Jackson Street.

3.0 3.0 Corner of 54th Avenue and Jackson Street where the U.S. Geological Survey and Coachella Valley Water District have detected up to 1 foot of land surface subsidence since 2000 due to groundwater withdrawal. For more information on groundwater withdrawal, see Scott Snyder's paper in this volume.

6.0 3.0 Right on Avenue 48.

7.5 1.5 Avenue 48 becomes Dillon Road.

8.9 1.4 Pass under Interstate 10.

9.9 1.0 Cross over Coachella Canal.

Figure 21. The Coachella Canal delivers water to Coachella Valley from the Colorado River via the All-American Canal. Water deliveries began in 1949. The canal is currently operated by the Coachella Valley Water District. For more information on the history of Coachella Valley Groundwater, see Scott Snyder's paper in this volume.

11.3 1.4 Fargo Canyon Road. We are driving on the Coachella Fan, a broad alluvial fan that has been cut by strands of the San Andreas fault. Miles Kenney and other workers performed detailed trench studies across the San Andreas fault in Fargo Canyon about a mile east of here (see paper by Miles Kenney, this volume).

23.9 12.6 Turn left on Thousand Palms Canyon Road.

25.9 2.0 McCallum Palms. This fan palm oasis occurs due to a splay off the Mission Creek fault.

26.6 0.7 Turn right into the parking area.

STOP 11 THOUSAND PALMS OASIS

Speaker: Harry Quinn

Figure 22. Thousand Palms Oasis, one of the many palm oases that occur along strands of the San Andreas fault in Coachella Valley.

If one looks to the southwest at the cliff face behind the oasis, one can see the old fluvial sediments that make up the hill. Within these old channels can be seen zones with limonite staining. This staining results from the plant material that was growing on the old fan surface. Look closely at the spring and observe limonitic stain in the spring water that has developed from the organic matter.

Figure 23. Monte Murbach at Thousand Palms Oasis along the along the Mission Creek strand of the San Andreas fault in Coachella Valley.

26.8 0.2 Dirt parking lot on the left.

STOP 12 MISSION CREEK STRAND OF THE SAN ANDREAS FAULT

Speaker: Harry Quinn

From the parking lot, walk east along a trail to the road. South of this trail between the palm groves is where the fault trenching took place. The small hill east of the palms has been uplifted along the Mission Creek branch of the San Andreas fault zone. The north face of this hill nearly marks the fault trace. Note the mesquites growing along the fault. Mesquite thrives in areas where the water table is less than 50 feet. Palms can exist where the water table is about 25 feet or less below the surface. For more information on the occurrence of the desert fan palm (*Washingtonia filifera*), see paper by James W. Cornett in this volume.

From the edge of the road, look east to the ridge across the canyon. Notice bedded sediments in the wall along the canyon. Follow the bedding eastward past the steps up to the top of the ridge. The bedding stops at the ridgetop. This area is virtually free of large-clast rocks and is pulverized. This is the gouge area of the fault. Continue looking to the eastward to see bedding reappear in the side of the hill. The fault zone in this area is close to 100 meters wide. For more information on the Indio Hills, refer to an article by Robert M. Norris from 1994 in *California Geology*, reprinted in this volume.

Return to the parking lot.

Figure 24. Water commonly flows on the surface in the Willis Palms area where groundwater reaches the surface due to effects of the San Andreas fault controlling groundwater flow. (Bruce Perry, Department of Geological Sciences, CSU Long Beach)

27.1 0.3 Cross the stream (seasonal). This stream is the result of groundwater reaching the surface at the Willis Palms area and the Paul Wilhelm Grove where the Visitor Center is located. The excess water flows downstream, crosses under or over Thousand Palms Road, and eventually percolates into the alluvium southeast of the road.

Drive down to the trailhead at Willis Palms.

28.1 1.0 Pull into the turnout on the right.

STOP 13 WILLIS PALMS AND BANNING STRAND OF THE SAN ANDREAS FAULT

Speaker: Harry Quinn

This palm oasis is developed along the Banning Branch of the San Andreas fault zone. Of interest here are the outcrops of the Imperial Formation, a marine unit that is of late Miocene to early Pliocene age and was deposited when this area was an extension of the proto-Gulf of California. As we drive away from this stop and turn right onto Ramon Road, look back for a good view of the light yellow Imperial Formation siltstone outcrop just west of the palms.

28.6	**0.5**	**Turn right on Ramon Road.**
33.0	**4.4**	**Cross over Interstate 10.**
40.2	**7.2**	**Turn right onto Sunrise Way.**
40.7	**0.5**	**Turn left onto Taquitz Canyon Way.**

41.2 0.5 Park on the right at the corner of Indian Canyon Drive.

STOP 14 AGUA CALIENTE HOT SPRINGS is located beneath North Indian Canyon Drive and is a source of hot mineral water for the adjacent Spa and Resort. This is the spring for which Palm Springs is named. This location is in the heart of downtown Palm Springs where street parking is limited. In earlier times the area was an amazing oasis that extended for thousands of feet in all directions.

Figure 25. Agua Caliente Hot Springs have been landscaped at the entrance to one of the main facilities of the Agua Caliente Indian Tribe in downtown Palm Springs.

The spring is located along the northerly projection of the Palm Canyon fault. Today the springs are contained with a large structure below the sidewalk, accessed by large steel plates at the surface.

At the intersection of Taquitz Canyon Way and North Indian Canyon Drive, reset odometer to zero.

0.0	**0.0**	**Turn left on Palm Canyon Drive.**
1.4	**1.4**	**Bear right at the Y in the road.** Be sure to continue on South Palm Canyon Drive, not East Palm Canyon Drive (i.e., you are turning off the main road, but generally continuing straight)
3.6	**2.2**	**Pass Tribal building on left.**
4.1	**0.5**	**Reach the entrance of Palm Canyon.**

5.5 1.4 Turn right for Andreas Canyon.

STOP 15 ANDREAS CANYON PICNIC AREA

Speaker: Bradley Erskine

Proceed on the single road for a short distance until you come to a fork in the road. Take the right branch and drive about 1 mile into the Andreas Canyon picnic area parking lot. This peaceful and special site with trees and a flowing river was an Indian acorn grinding site. Note the bowl-shaped morteros ground into the wonderful mylonites. Fantastically exposed on almost mutually perpendicular surfaces of the giant rocks are a spectacular foliation and lineation indicating the strain axes; asymmetric "S-C" structures indicate sense of shear. S-C mylonites are made up of two sets of planar structures, a foliation (S) and shear bands (C) that intersect obliquely. Thin quartzite dikes are folded and sheared into the foliation plane with fold axes parallel to the lineation direction. These document the tremendous strain in these rocks. These deformations are good examples of ductile strain, indicating they formed at depths of 10-15 km or greater, and at temperatures of 300°C to 600°C—at mid-crustal depths. So these rocks are only visible at the earth's surface because of the removal by erosion and faulting of tens of thousands of feet of rocks that formerly covered them. Rarely are the strains of mid-crustal extension, compression, and shear shown so well.

Figure 26. Geology group in 2009 at Andreas Canyon observing mylonites at Stop 15.

For additional information on fluid evolution in the mylonite zone, refer to Gregory J. Holk's paper in this volume. Bradley Erskine and Brian White provide papers in this volume that present their views on mylonitization in this area.

Points of Discussion:

- Is the mylonite zone more of a simple shear zone, that is, deformed like a deck of cards, or a zone of flattening in one direction and extension at right angles with a shear component?
- How much strain is recorded?

Leave Andreas Canyon camp and return to the fork in the road. This time take the right branch and continue driving south for about 2 miles on the narrow road. Go through a very narrow and somewhat low rock tunnel, watching for on-coming cars. Anyone driving a mobile home or large RV should consider parking it before reaching the tunnel. Drive up to the top of a ridge with a parking lot overlooking beautiful Palm Canyon and the Indian Trading Post at the far end. Do not bring alcohol. It is against the law on the reservation.

6.4 0.9 **Park in the lot at the end of the road.**

STOP 16. INDIAN TRADING POST

Speakers: Scott Snyder, Harry Quinn, and Bradley Erskine

Descend the walkway into the canyon to examine the famous mylonitized rocks of the Eastern Peninsular Ranges mylonite zone, formerly called the Santa Rosa mylonite belt. These are the same rocks visited in Andreas Canyon. Continue walking east for several hundred yards to see a beautiful exposure of the Palm Canyon detachment fault with its very well-developed chloritized breccia zone and Paleozoic metasedimentary rocks that have been faulted in Neogene(?) time against the Cretaceous mylonites. This is the Palm Canyon fault. This important and peaceful site is a great place to eat lunch. Follow the fault up a creek to see even better exposures of the Palm Canyon fault.

Figure 27. Palm oasis in Palm Canyon.

The Palm Canyon fault runs along the base of the San Jacinto Mountains. It has been described as a detachment fault, a low angle normal fault, that may be part of the San Andreas fault system. The fault restricts the flow of groundwater which is then forced up to the surface. In the desert, native fan palm oases (*Washingtonia filifera*) and cottonwood-willow riparian areas form where there is surface water, providing critically needed sources of food and water for desert wildlife, and delightful places for humans to visit. These areas provide dramatic contrast to the hot, dry desert environment.

Indian Canyons, located at the base of the San Jacinto Mountains and managed by the Agua Caliente Band of Cahuilla Indians, boast the largest system of native fan palm oases in the United States. Visitors from around the world come to visit these spectacular oases nestled in steep rugged canyons where the force of running water over the millennia has carved the rocks into pools and waterfalls. Higher up the canyon walls, the rocks glisten with the beautiful patina of desert varnish.

Associated with some of the springs along the base of the San Jacinto Mountains is hot mineral water, heated at depth possibly by extensional tectonics in San Andreas fault zone. This hot mineral water is of excellent quality, and is available to visitors at world-class spas throughout the Coachella Valley. Conservation and protection of the watershed which recharges these natural springs is important to maintain the high quality of water in the Coachella Valley, and to continue supporting the tourism industry which uses this resource (U.S. Bureau of Land Management, 2012).

Figure 28. Lake Cahuilla in Coachella Valley.

ACKNOWLEDGMENTS

The field trip guide above is the combination of two field trips sponsored by the San Diego Association of Geologists: one led by Scott Snyder in 2008 and the other led by Bryan Miller-Hicks in 2009.

SCOTT'S ACKNOWLEDGEMENTS. As with any significant undertaking, this field trip and publication would not be possible without the support of more people than I can probably remember:

Rebecca, my wife, who has supported me in (and sometimes marveled at) my love of this art and science since my very first "rocks for jocks" class I took at Penn State in 1991. She has endured far more blathering about this rock and that outcrop than any wife should ever have to, and has put her life in my hands each and every time we pass a road cut or overlook. We've been lucky so far...without her I can truly say I would not be where I am today, physically, mentally, and professionally.

SDAG — Monte and Diane Murbach, Carole, Phil, Lowell, Greg, Dave, Chuck, Mike, you guys and gals are the lifeblood of this great organization. It continues to amaze me how much time and effort people will continue to put in year after year to their profession and obsession. You are all amazing people that are truly special to all of the members of SDAG, whether they know it or not. I have been part of many volunteer organizations, but this one is by far the most giving. Your altruism is admirable.

Margaret Eggers. Thank you for asking me to run for office of the SDAG Board back in 2004, it was the perfect time in my career to do something like this and although I was scared to death, I enjoyed every minute of it.

Contributors. Thanks to Mike Hart, Nate Onderdonk, and Miles Kenney for their original research and papers in this volume. I also would like to thank the authors of the previously published papers that were reprinted here as important and interesting works of the area.

Harry Quinn. It was about 3 months before the 2008 field trip and although I had some good ideas and some of it researched and planned out, I wasn't feeling all that great about the trip. In looking for a contact for the Lake Cahuilla fish traps I was eventually directed to Harry for some info. Boy did I get more than I could have ever expected. All I got was a professional geologist of 50 years that literally grew up in my field trip area! Was I relieved! But then I wasn't sure how much he wanted to be involved, if at all. The second sigh of relief was that he was happy to help me with the trip and be a part of it on the weekend we had planned. That is no small matter for Harry's busy schedule; he's busier in retirement than I am with a full time job. Harry has spent a good part of his life in the Coachella Valley and the Santa Rosa and San Jacinto Mountains up above the valley, and he has the stories to prove it. He knows or knew of all of the springs and streams you could find water in as a kid; he knew the Pines to Palms Highway when it was still just a mountain slope; and he knows of just about every Indian trail and every medicinal or other useful purpose of many of the plants in the region. He is a geologist, paleontologist, anthropologist and any other -ologist you can think of. He has written many papers about the area thereby preserving some of the history. Some of those papers you'll find in here. Thank you, Harry, for your time and invaluable insight into this area. Your work is much appreciated. But he wouldn't tell you that it was work for him. He just does it because he loves it, as do we all.

—Scott Snyder
2008 SDAG Vice-President

BRYAN'S ACKNOWLEDGEMENTS. My heartfelt thanks go to all who pitched in and supported our efforts to make this field trip a success: an informative, educational and, of course, a fun one!

Primary acknowledgment and thanks go to Monte Marshall and Brad Erskine. Monte provided the initial spark and inspiration which launched this field trip. Brad shared with us his deep knowledge of mylonites and the local structure and the best places to see them — knowledge gained by his many years of working in this area. It is due in no small part to the professional advice by Monte, Brad, Chuck Houser, and other colleagues of the San Diego Association of Geologists that I was able to get this trip "off the ground and onto the ground."

In addition to being the spiritual founder and a guide for this field trip, Monte Marshall was my "odometrist" in our planning visits and road log compilation. The detailed road log mileage and narrative are primarily due to Monte's efforts. It's safe to say that the trickiest parts of our scouting trips and mileage logging were those segments which took us through the hazardous streets of Palm Springs and associated desert communities!

Special thanks go to Dr. Allan Muth, who was kind enough to allow a bunch of geologists into his Biological Reserve, the Philip L. Boyd Deep Canyon Desert Research Center, administered by the University of California, Riverside. The center was our lunch stop on Saturday, where we were treated to a presentation of the center's research projects by Dr. Mark Fisher.

Gratitude goes to Diane and Monte Murbach, who are tireless in their efforts to lend a ready hand. They came through with valuable advice and suggestions during the planning phase. We also owe the delicious Boyd Center make-your-own-sandwich lunch to their efforts. Thank you to Scott Snyder for the early morning coffee runs: invaluable!

Thanks to Dave Bloom, our former President, for his planning help and scouting trips into Deep Canyon with Monte Marshall, and his arrangements with Rebecca Rowan, our caterer, for the scrumptious and hearty Saturday dinner and Sunday breakfast.

Thanks to Carole Ziegler and Dave Bloom for all their hard work in editing and putting together this combined volume, Phil Farquharson for his capable work on graphics, and our publication coordinators, Lowell and Diana Lindsay of Sunbelt Publications.

A word of appreciation is also due to the personnel of San Bernardino National Forest who made sure that Pinyon Flats Campground was clean and ready to accommodate a few dozen fun-loving geologists!

And, many thanks to all friends who attended!

—Bryan Miller-Hicks
2009 SDAG Vice-President

REFERENCES

Becker, T., J. Hardebeck, and G. Anderson, 2005, Constraints on fault sliprates of the southern California plate boundary from GPS velocity and stress inversions, Geophysical Journal International,160, 634–650, doi:10.1111/j.1365-246X.2004.02528.x

Brown, A.R., 1968, Geology of a portion of the southeastern San Jacinto Mountains, Riverside County, California. University of California, Riverside, Master of Arts thesis. 95 p. December 11. URI: http://hdl.handle.net/10211.10/167

Brown, A.R., 1980, Limestone deposits of the Desert Divide, San Jacinto Mountains, California, in Fife, D.L., and Brown, A.R., eds., Geology and mineral wealth of the California desert; Dibblee volume: South Coast Geological Society Guidebook, October 11–12, 1980, p. 284–293.

Dibblee, T., 1968, Displacements on the San Andreas fault system in the San Gabriel, San Bernardino, and San Jacinto Mountains, southern California, in Dickinson, W., R., and Grantz, A., eds., Proceedings of conference on geologic problems of San Andreas fault system: Stanford University Publications Geological Sciences, v. 11, p. 260–278.

Fialko, Y., 2006, Interseismic strain accumulation and the earthquake potential on the southern San Andreas fault system. Nature, v. 441, p. 968–971, doi:10.1038/nature04797.

Fletcher, K.E., Johnson, G., Kendrick, K., Hudnut, K., and Sharp, W., 2006, Long-term slip rate on the southern San Andreas Fault determined by Th-230/U dating of pedogenic carbonate. American Geophysical Union, Fall Meeting, Abstract #T13B-0506.

Hill, R. L., 1981, Geology of Garner Valley and vicinity, in Geology of the San Jacinto Mountains, South Coast Geological Society Annual Field Trip Guidebook No. 9.

Kendrick, K. J., D. M. Morton, S. G. Wells, and R. W. Simpson, 2002, Spatial and temporal deformation along the northern San Jacinto fault, southern California: Implications for slip rates, Bulletin of the Seismological Society of America, 92, 2782–2802, doi:10.1785-/0120000615.

Hill, R. L., 1981, Geology of Garner Valley and vicinity, in Geology of the San Jacinto Mountains, South Coast Geological Society Annual Field Trip Guidebook No. 9.

Matti, J. C., and Morton, D. M., 1993, Paleogeographic evolution of the San Andreas Fault in Southern California; a reconstruction based on a new cross-fault correlation, in Powell, R. E., Weldon, R. J., and Matti, J. C., eds. Geological Society of America Memoir 178, p. 107–159.

Morton, D. M., and Matti, J. C., 1993, Extension and contraction within an evolving divergent strike-slip fault complex; the San Andreas and San Jacinto fault zones at their convergence in Southern California, in Powell, R. E., Weldon, R. J., and Matti, J. C., eds. Geological Society of America Memoir 178, p. 217–230.

Rockwell, T., Loughman, C., Merifield, P., 1990, Late Quaternary rate of slip along the San Jacinto fault zone near Anza, Southern California. Journal of Geophysical Research B, v. 95, n. 6, p. 8593–8605.

Rockwell, T. K., and Y. Ben-Zion, 2007, High localization of primary slip zones in large earthquakes from paleoseismic trenches: Observations and implications for earthquake physics, Journal of Geophysical Research, v. 112, p. B10304. doi:10.1029/2006JB004764.

Sharp, R.V, 1967, San Jacinto fault zone in the Peninsular Ranges of Southern California. Geological Society of America Bulletin, v. 78, p. 705–729.

Sharp, R.V., 1981, Variable rates of late Quaternary strike slip on the San Jacinto fault zone, Southern California, Journal of Geophysical Research B, v. 86, p. 1754–1762.

Todd, V.R., B.G. Erskine, and D.M. Morton, 1988, Metamorphic and tectonic evolution of the northern Peninsular Ranges batholith, southern California, in Metamorphism and Crustal Evolution of the Western United States, Rubey vol. VII, edited by W.G. Ernst, p. 894–937, Prentice-Hall, Englewood Cliffs, N. J.

U.S. Bureau of Land Management, 2012, Santa Rosa San Jacinto Mountains National Monument. Accessed September 12, 2012, from http://www.blm.gov/ca/palmsprings/points_of_interest.html. Checked October 25, 2013, from https://web.archive.org/web/20121018104518/http://www.blm.gov/ca/st/en/fo/palmsprings/santarosa/points_of_interest.html

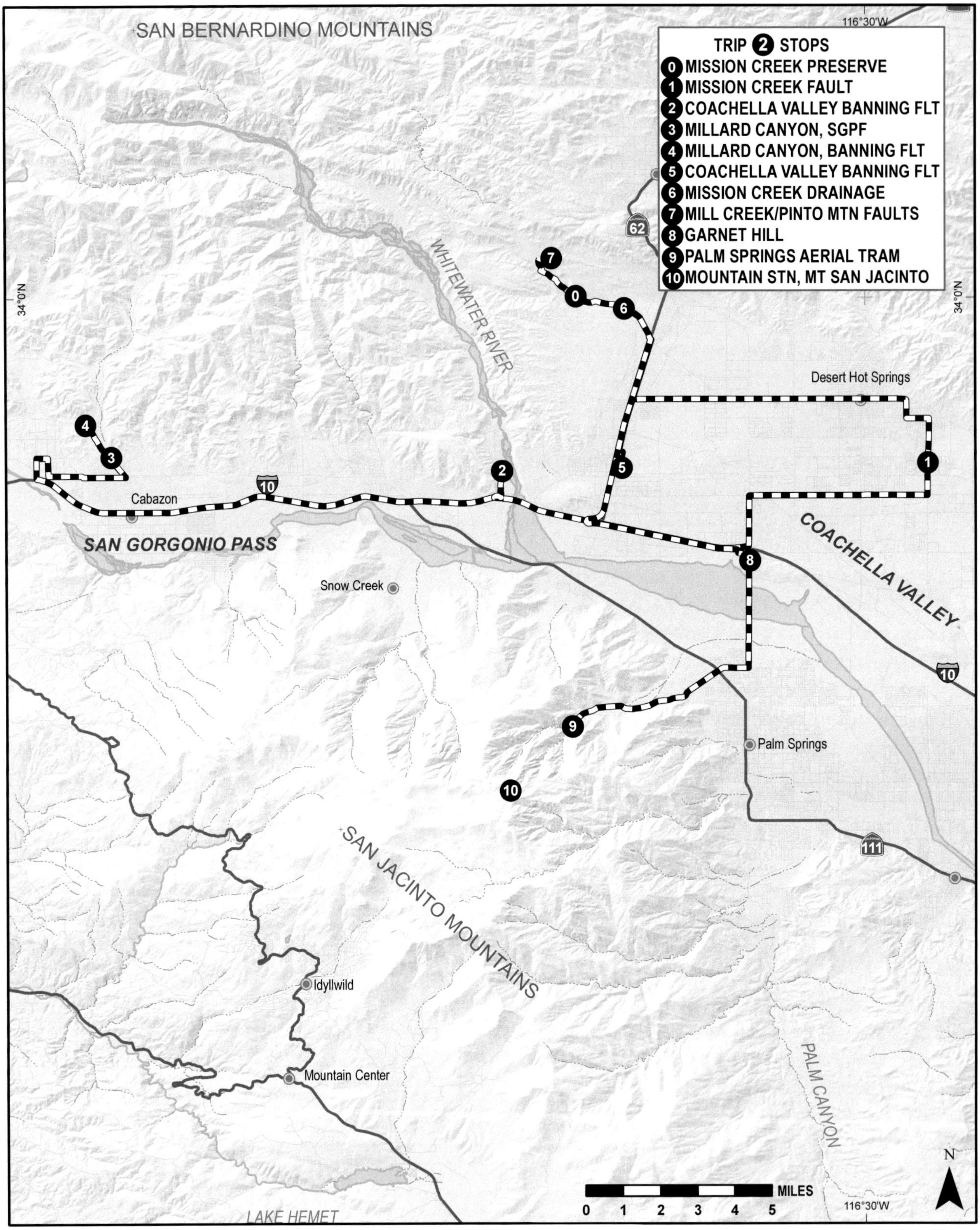

ROUTE MAP, TRIP 2

TRIP 2
Exploring an Active Plate Boundary: San Gorgonio Pass and Northern Coachella Valley, Southern California

"Slip, Sliding Away"

FIELD TRIP LEADER: Jeffrey Miller

INTRODUCTION

You don't have to be the world's greatest geologist to know there is something special going on in the vicinity of San Gorgonio Pass. From the highs of Mount San Gorgonio and Mount San Jacinto to the lows of San Gorgonio Pass and the Coachella Valley and the San Andreas fault somewhere in between, there's a lot of geological explaining to do! We hope you find this field guide useful in developing and answering questions regarding this field laboratory. So dust off your boots, grab a gallon of water and set off on this fun adventure!

Figure 1 (above). Whitewater River, Mission Creek, and San Bernardino Mountains (Bruce Perry, Department of Geological Sciences, CSU Long Beach)

MILEAGE *Trip 2 covers 82.5 miles. Two days should be ample time to visit all Stops.*
(cumulative and intervals)

0.0 0.0 Meet at Mission Creek Preserve Gate, located at 60550 Mission Creek Road, Desert Hot Springs, California. The Mission Creek Preserve is operated by the Wildlands Conservancy and is also the location of our group campground.

"Located in a transition zone between the Sonoran and Mojave deserts, the 4760-acre Mission Creek Preserve has a rich species composition, with flora and fauna representing both deserts. Somewhat surprisingly, at the center of the preserve is a lush wetland, juxtaposed on extensively eroded painted hills with Mount San Gorgonio looming in the background. The wetlands and perennial stream are important habitat for the endangered least Bell's vireo and southwestern willow flycatcher. The preserve hosts deer, bear, big horn sheep and mountain lions and has spectacular displays of spring wildflowers in wet years. The preserve trail system leads onto the Pacific Crest Trail which is two miles from the 'stone house' group campground (Mission Creek Preserve Website, 2012).

From gate, head southeast on Mission Creek Road toward California Highway 62.

2.2 2.2 Turn right on California Highway 62 heading south.

3.8 1.6 Merge into left turn lane, turn left and cross northbound California 62 onto eastbound Pierson Boulevard.

Desert Hot Springs is located on the southern flank of the Little San Bernardino Mountains in the northern Coachella Valley. The city may owe its very existence to the Mission Creek fault. Not surprisingly, the city is the benefactor of underlying hot springs resulting from groundwater fault interaction. Cabot Yerxa discovered hot water flowing from "Miracle Hill" in 1913, beginning the first movement to settle Desert Hot Springs. Today, travelers can visit Cabot's Pueblo Museum, designated a State historical site following his passing in 1965. Since then the town has undergone periods of intermittent growth and is now a city with current population of approximately 26,000 (Wikipedia, 2012).

Located in the northwestern most portion of the Colorado Desert Geomorphic Province, Desert Hot Springs is affected by both the Coachella Banning Fault and Mission Creek Faults of the San Andreas fault system (SAF). Here in the northern extent of the Coachella Valley we see the first evidence of the complicated geometry of the SAF. From the southern terminus of the SAF, approximately 50 miles southeast of Desert Hot Springs, northward through the Coachella Valley, the SAF is primarily composed of vertically dipping strands with right lateral displacement. Looking to the southeast you will be able to observe the Indio Hills in the distance, resultant from north-south shortening between the Banning and Mission Creek faults (Proctor, 1968).

9.9 6.1 Pierson Boulevard veers right toward the south and becomes Miracle Hill Road. Continue on Miracle Hill Road.

10.4 0.5 Turn left onto Hacienda Avenue heading east.

10.9 0.5 Turn right into Mountain View Road heading south.

12.0 1.1 Mission Creek Fault (destination on right unmarked). Here we stand on the scarp of the Mission Creek Fault underlying Mountain View Road.

STOP 1 THE EFFECT OF THE MISSION CREEK FAULT ON GROUNDWATER FLOW: ESTIMATION OF FAULT ZONE HYDRAULIC CONDUCTIVITY USING DIRECT MEASUREMENTS AND GROUNDWATER MODEL CALIBRATION.

Speakers: Jan Gillespie, California State University at Bakersfield, Alex Mayer, Michigan Technological University

In the northern Coachella Valley, the San Andreas fault splits into two strands. The Mission Creek fault at Desert Hot Springs, California is the north-eastern strand of the San Andreas, and the Banning fault is the south-western strand. A recent study by Catchings et al. (2009) presents seismological and seismic data that suggest that these faults merge to form a single strand at depths of about 10–12 kilometers.

At Stop 1, a poorly consolidated outcrop of Pleistocene alluvium (probably the Ocotillo Conglomerate) occurs between the Mission Creek fault on the south end of the outcrop and the Miracle Hill fault on the north side. The Mission Creek fault is marked by a conspicuous northwest-trending scarp which is easily seen. The scarp is marked by numerous springs and seeps.

Slip on the fault is right lateral oblique and the fault is nearly vertical. The north-eastern block of the fault is the upthrown block. Seismic data interpreted by Catchings et al. (2009) suggest that reflectors representing more consolidated rocks are offset 150 meters across the Mission Creek fault.

The outcrop at Stop 1 lies on the upthrown block of the Mission Creek fault and is composed of north-dipping layers of sandstone and conglomerate. The outcrop is cut by numerous predominantly vertical to south-dipping faults. Many of the faults are filled with white, calcareous precipitates which are more resistant than the surrounding sandy matrix. These are easily seen on the north end of the outcrop both to the right and left of the power pole.

Faults that act as barriers can affect groundwater flow at the basin or regional scale. The quality and availability of groundwater resources in faulted, alluvial basins depends on the hydraulic properties of the faults. The magnitude of the impacts depends on the materials found in the fault zone, the properties of adjacent aquifer materials, and the externally imposed flow regime.

12.0 0.0 Return to vehicles. Begin heading south on Mountain View Road.

13.0 1.0 Turn right onto Dillon Road heading west.

Note the view to the west of San Gorgonio Pass bounded to the north by Mount San Gorgonio and to the south by Mount San Jacinto. San Gorgonio is also known as "Old Greyback," the highest point in southern California at elevation 11,503 feet above mean sea level. On the opposing side of the narrow alluvial valley, to the south, rises San Jacinto Peak which is the second highest southern California peak at 10,834 feet. According to Wikipedia the north face of this escarpment, above Snow Creek, boasts the largest elevation gain [9,000 feet] over a horizontal distance [4 miles] in the contiguous United States.

17.0 4.0 Turn left onto North Indian Canyon Road heading south towards Palm Springs.

18.3 1.3 Turn right onto Interstate 10 westbound, merge and continue.

Believe it or not, rising before you in the distance are over 3,000 wind turbines. Rising from their bases up to 160 feet above ground surface, these turbines collectively generate a potential of over 615 megawatts of electricity (Wikipedia, 2012)! Keep your cameras ready because on the drive to the upcoming stop you will be as up close and personal to the turbines as you can get without a special tour.

24.2 5.9 **Take exit 114 towards Whitewater.**

24.4 0.2 **Turn right onto Tipton Road, continue onto Whitewater Cutoff.**

The Whitewater Rock and Supply Company turned 50 in 2012. This popular rock supply company owes its existence to Al Bankus who left Oregon in 1962 to search for gold and discovered a different kind of gold northwest of Palm Springs. Though not gold bearing, his discovery of sericite schist has led to the founding of the Painted Hills Quarry. The eagle eyed geologist will no-doubt recognize that this quarry is located just north of the Banning strand of the San Andreas fault.

The Painted Hills Mine (CA Mine ID#91-33-0003) is approximately 45 acres in size and produces a rock referred to as "Palm Springs Gold" or "Whitewater Stone." This popular building material and decorative rock is the result of hydrothermal alteration of gneiss to sericite schist with red hematite stained micas and porphyroblasts of black tourmaline (NAGT, 2007).

24.6 0.2 **Turn left onto Whitewater Canyon Road.**

25.1 0.5 **Pull off in open area to left of Whitewater Canyon Road.**

(Note: On non-mine operational days it is possible to continue on Whitewater Cutoff Road to an alternative lookout stop that overlooks the canyon to the west, however this should not be attempted if large trucks are present on the road.)

STOP 2A. COACHELLA VALLEY BANNING FAULT AS AN OBLIQUE-SLIP FAULT.

Speaker: Doug Yule, California State University Northridge.

The view to the northwest from the parking area reveals thick vegetation upstream from a distinct lineament across Whitewater Canyon. Here the Coachella Valley Banning fault is an aquitard and raises groundwater to shallow levels to allow an oasis of lush vegetation to grow year round on the upstream side of the fault. Vegetation is scarce on the downstream side of the fault where ground water levels drop rapidly in porous gravel.

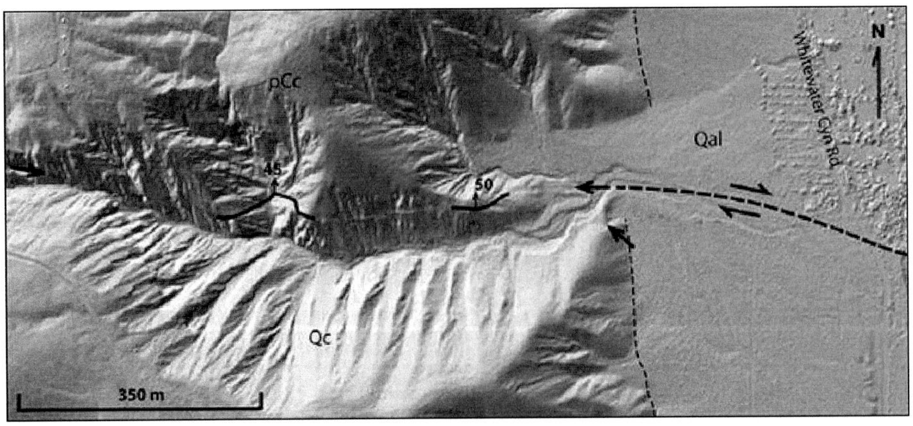

Figure 2. Trace of the Coachella Valley Banning Fault at Whitewater Canyon.

The vegetation lineament trends directly into the mouth of the side canyon at this stop. The fault trace can be followed to the west as it cuts across the side of a steep, south-facing slope. The fault strikes E-W and dips North at 45–50° (Figure 2) and contains heavily fractured and faulted granite and gneiss in its hanging wall and Pleistocene alluvial fan gravels of the Cabazon Formation in its footwall. The fault plane is exposed ~0.75 kilometers west of the stop where it cuts across a tributary of the side canyon. Thus, the Coachella Valley Banning fault at this location is an oblique-reverse slip fault (Yule, 2010), unlike locations to the east-southeast where it is almost purely strike slip.

STOP 2B. THE NOT SO SIMPLE WATER STORY OF THE NORTHERN COACHELLA VALLEY.

Speaker: Marina West, Bighorn Desert View Water Agency

At the intersection of the I-10 Freeway and Whitewater River crossing begins the desert journey of native flows collected from tributaries on the backside (north side) of Mt. San Gorgonio wilderness. The Whitewater River might appear rather formidable given its constant and seemingly high flow rates even during the hot dry summer seasons, but in this case appearances are deceiving. The explanation lies in the fact that the Whitewater River channel houses a turnout from the Colorado River Aqueduct which crosses the valley floor and via a tunnel passes through the San Jacinto Mountains headed towards Lake Perris, California.

This field trip is focused on the northern Coachella Valley, California. In hydrogeologic terms, the "northern" Coachella Valley geographically approximates the "upper" Coachella Valley Groundwater Basin, a 250-square mile area extending from the east end of San Gorgonio Pass to Indio; bordered on the north and east by the San Bernardino and Little San Bernardino Mountains and on the southwest by the San Jacinto and Santa Rosa Mountains (Figure 3).

As shown on Figure 3, the Upper Coachella Valley Groundwater Basin (GB) is divided into four subbasins: Whitewater River, Garnet Hill, Mission Creek and Desert Hot Springs. Most of the groundwater inflow to the Upper Coachella Valley Groundwater Basin is from the San Gorgonio Pass area with the Whitewater River channel contribution being the greatest native base flow contributor.

The long-term average annual stream flow (gaged) of the Whitewater River is estimated to be 12,000 acre feet per year (AF/yr). While the total natural tributary inflow (gaged and ungaged) to the Upper Coachella Valley Groundwater Basin is estimated to be 36,000 AF/yr. The accepted annual outflow from the basin is estimated at 7,000 AF/yr as derived from computer modeling (Reichard & Meadows, 1992, 9). In total the Upper Coachella Valley GB receives native inflow from drainages of the San Gorgonio Mountains, Whitewater River, Mission Creek, and Morongo Valley. Additional inflows into the valley are from Snow Creek, Falls Creek, and Chino Creek originating in the San Jacinto Mountains.

Historically, flows from Snow and Falls Creeks percolate into the Whitewater River Subbasin along with flows from the Whitewater River. However, since 1947, the Palm Springs Water Company (now part of Desert Water Agency) has captured, filtered and diverted these flows, along with flows from Chino Creek, directly into the

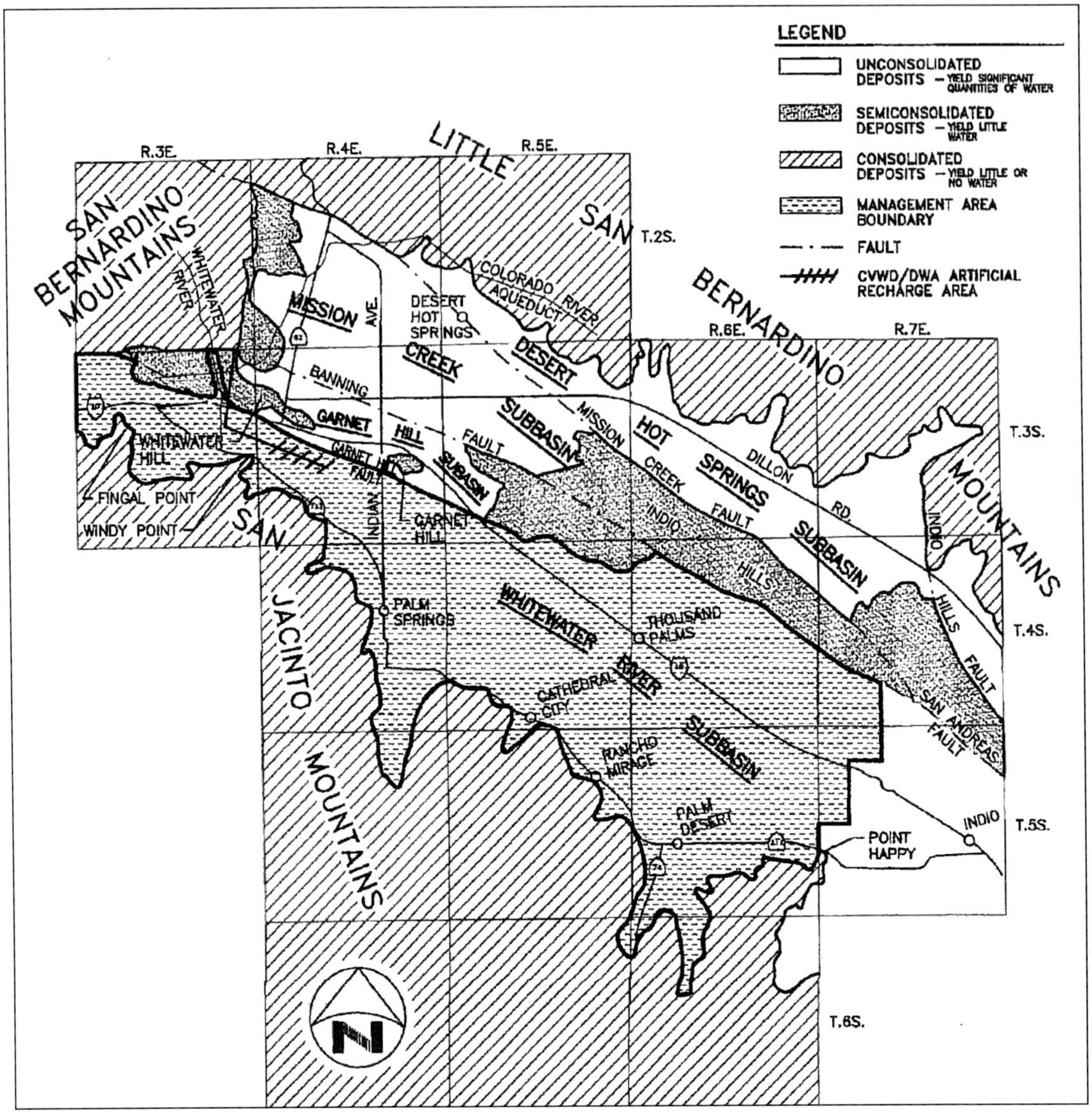

Figure 3. Whitewater River Subbasin

Palm Springs area water system. According to a Huell Howser video, posted on the website dwa.org, approximately 3,400 AF/yr is diverted annually from these streams. However, the annual Engineer's Report on Groundwater Replenishment for the Whitewater River Subbasin states that diversions have been as high as 8,530 in 1978, and have steadily declined. In 2011, total diversions are reported to be 1724 AF. (Scriven, 2012, Exhibit 1). Upon further inquiry it was discovered that increasingly stringent health regulations have essentially prohibited the direct diversion of stream flows into the drinking water system.

The geohydrology of the Upper Coachella Valley area has been studied, explained, and published in previous reports (Tyley, 1974, 7) and will be liberally summarized herein to give the reader a quick perspective of the underground aquifer system:

The basement complex of the Coachella Valley is comprised of consolidated undifferentiated granitic intrusive and metamorphic rocks of Precambrian to Tertiary age. They transmit no measurable water and are considered a no-flow groundwater boundary.

The main water-bearing unit is the unconsolidated Pleistocene and Holocene age valley fill. Thickness of water bearing deposits is estimated to be 3,000 feet. However, according to Reichard, groundwater wells in the basin only reach a maximum depth of 1,200 feet bgs (1992, 5) causing speculation that the basement is much shallower than previously summarized. The San Andreas fault system affects the movement of groundwater in and through the valley fill and groundwater movement is also affected by folding as a result of compression and drag associated with displacement along these faults. Those faults representing groundwater barriers are mapped as the 1) Mission Creek, 2) Banning, 3) Garnet Hill and 4) Indio Hills faults and their associated folds. Topographic highs in the valley represented by such folding are known as the Indio Hills, Garnet Hill and Whitewater Hill (Figure 3). In these areas the permeability and storage capacity have been reduced and they assist in defining the compartmentalization of the Upper Coachella Valley into the four groundwater subbasins noted earlier.

The faults and folds act as groundwater barriers evidenced by offset water levels, fault springs and changes in the vegetation. Water level offsets, across individual subbasin boundaries, range from 30 to 250 feet.

The growth of agriculture, and beginning in the early fifties, tourism have drawn heavily on the groundwater supply and groundwater levels have declined as annual extractions have increased more than tenfold during the period 1936–67 (Tyley, 1974, 4). During this period, cumulative pumping of about 740,000 AF led to a depletion of groundwater in storage of about 600,000 AF.

In response to diminishing groundwater supplies the Desert Water Agency (DWA) and Coachella Valley Water District (CVWD) began to purchase imported water in 1972. However, these purchases were from the State Water Project (SWP) and with no delivery system in place (no access to the SWP water they had purchased) the agencies concocted an Exchange Agreement with Metropolitan Water District of Southern California (MWD) to swap equal shares of SWP for Colorado River Water which could then be artificially recharged into the Upper Coachella Valley groundwater basin.

Between 1973 and 2011, CVWD and DWA have replenished the Whitewater River Subbasin with approximately 2,266,228 AF of SWP water while MWD pre-stored approximately 466,000 AF of SWP water. According to Scriven, groundwater overdraft in the Whitewater River Subbasin has averaged 99,000 AF/yr (excluding artificial recharge) during the last five years. Cumulative overdraft offset by artificial recharge is currently estimated at 736,000 AF (1992, II-2). Although artificial recharge has met increasing average annual groundwater demands during the past 30 years, it has not reduced the cumulative groundwater overdraft that existed prior to initiation of recharge activities, but the trend has been arrested (Scriven, 1992, III-9). In summary, the Whitewater River Subbasin is in a cumulative overdraft condition and will remain so until higher proportions of State Water Project water become available for artificial recharge.

25.1 0.0 Return to vehicles. Begin heading south on Whitewater Canyon Road.

25.6 0.5 Turn right onto Whitewater Cutoff Road.

25.8 0.2 Turn right onto Interstate 10 westbound.

26.8 1.0 Take the exit for the Rest Area.

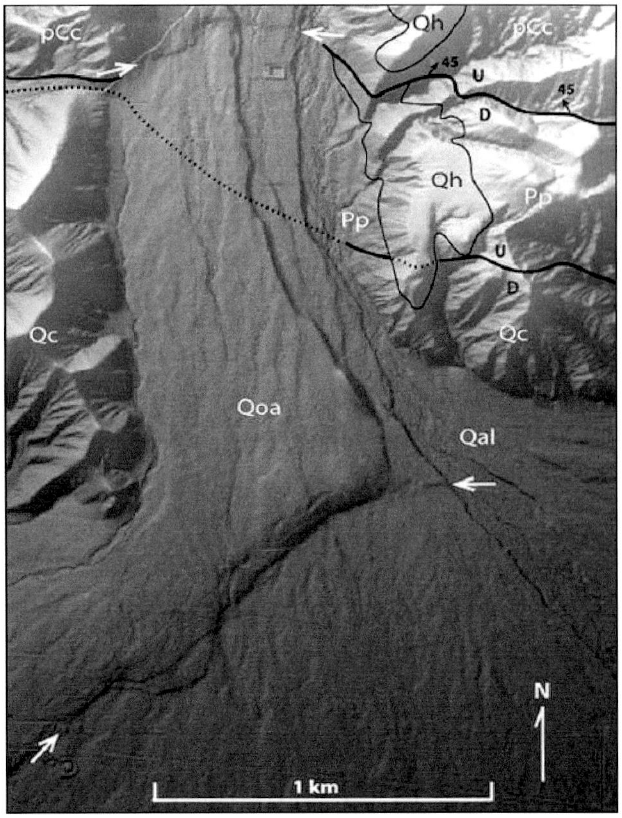

Figure 4. Active and Inactive Fault Traces in Millard Canyon

"Upon approaching the 27-story Casino Morongo and Hotel, look to the north and notice a scarp that cuts at a high angle across the mouth of Millard Canyon. The base of this scarp marks the surface trace of the San Gorgonio Pass thrust fault. We will visit this feature at Stop 3. The scarp at Millard Canyon continues southwest where it almost intersects Interstate 10 (Figure 4). Here, cutting through the parking lots behind a large factory-outlet shopping mall, the fault trace bends northwest toward the mountain front along a series of strike-slip tear faults" (Yule and Sieh, 2003).

After entering the Morongo Reservation, look to the south while driving east on Martin Road and notice that the alluvial surface projects above the stores of the outlet mall. This surface has been uplifted in the hanging wall of the San Gorgonio Pass fault zone" (Yule, 2010).

36.3 9.5 Take Exit 103 for Malki Road.

36.5 0.2 Continue on to Malki Road. Stop at gate for access to Morongo Band of Mission Indians Lands. Permission to access tribal lands must be obtained in advance of any field trip from the Tribe's Environmental Office (website: www.morongonation.org and click on "Environmental Department"; phone: 951-849-4697).

36.5 0.0 Continue through gate to the north.

36.9 0.4 Turn right at Sullivan Road.

37.2 0.3 Turn right at Ramon Road.

37.7 0.5 Turn left at Martin Road.

38.4 0.7 Continue to end of pavement and take left onto dirt road.

39.1 0.7 Reach intersection with small dirt road and park.

STOP 3 MILLARD CANYON: THRUST SCARP OF SAN GORGONIO PASS FAULT ZONE

Speakers: Doug Yule, California State University at Northridge and Eldon Gath, Earth Consultants International

A fluvial terrace riser is evident approximately 150 meters east of the parking area. Climb the riser and walk east to the highest point on the scarp, a height of 12.5 meters above the base of the scarp (Figure 4). Walk down the scarp and notice that the slope steepens to a maximum at the base of the scarp. This steepening down the slope defines an asymmetric fold and suggests that little, if any, diffusion has occurred across the scarp. Shortening at the surface may be accommodated entirely by folding above a blind thrust. If the thrust has reached the surface here it must have a slip vector of ~S40°E parallel to a small fluvial terrace riser that can be followed from the hanging wall block to the footwall block with no lateral offset.

Walk east from the high scarp and climb down another fluvial terrace riser onto a much younger alluvial surface. Look for a 1.5 meter scarp on this lower surface to the east of a dirt road. The west wall of the modern arroyo exposes gravel and sand layers cut by a 24° north-dipping thrust.

Ages of the high and low scarps are poorly constrained. Detrital charcoal from gravel layers beneath the lower surface yields an age of 2850-3600 radiocarbon years (J.C. Tinsley and J.C. Matti in Matti et al., 1992a, p. 26). The poor degree of soil development on the upper surface suggests that it is no older than latest Pleistocene in age. The preliminary age data from these surfaces constrain uplift and north-south shortening each to be about 1–2 millimeters per year over the past 10,000 years (Yule, 2010).

39.1 0.0 Get back in vehicles and drive north.

39.8 0.7 Reach 4-way dirt road intersection and park.

STOP 4 MILLARD CANYON: THRUST SCARP OF BANNING FAULT

Speakers: Doug Yule, California State University at Northridge and Eldon Gath, Earth Consultants International

Much of the Banning fault in San Gorgonio Pass is inactive. However, a 3-kilometer long segment of the northern strand is active. About 100 meters to the north of the parking area, a 5-meter-high scarp and a 2.5-meter-high scarp occur in older and younger alluvial surfaces. The fluvial terrace riser between these two surfaces appears to be offset right-laterally approximately 4.5 meters. Thus the dextral offset across the fault appears to be ~80% greater than the vertical component.

To the east, this active portion of the Banning fault also offsets the late Pleistocene Heights fanglomerate several hundred meters. On the eastern wall of the canyon, the fault trace rises from the canyon floor at a dip of ~45° N and juxtaposes crystalline rocks over the fanglomerate. Farther up the canyon wall the fault flattens to a nearly horizontal dip and overrides the surface of the Heights fanglomerate. Exposures in the canyon wall reveal a buttress unconformity between the fanglomerate and an older eastern wall of the canyon. The base of the buttress unconformity is offset obliquely across the Banning fault with a strike-slip:dip-slip ratio of ~2.5:1, greater than that derived from the younger scar on the valley floor (Yule and Sieh, 2003). Nonetheless, both observations show that strike-slip offset exceeds dip-slip offset on this active section of the Banning fault (Yule, 2010).

39.8	**0.0**	**Return to vehicles and head south.**
41.2	**1.4**	**Veer right at dirt road to paved road.**
41.9	**0.7**	**Turn right at Ramon Road.**
42.4	**0.5**	**Turn left at Sullivan Road.**
42.7	**0.3**	**Turn left at Malki Road.**
43.1	**0.4**	**Reach Gate.**
43.2	**0.1**	**Turn left onto Seminole Drive.**
43.4	**0.2**	**Turn right onto Johnson Lane.**
43.5	**0.1**	**Turn left and merge onto eastbound Interstate 10.**

While driving back, consider that we have seen features resultant from strike slip, oblique slip and reverse and thrust faulting. How do these features fit into the typical San Andreas fault model we have all come to know? How would this complexity impact the possibilities for a large through-going earthquake on the San Andreas fault? As discussed by Doug Yule at the October 2012 SCGS meeting, there may be two models for rupture along this stretch of the San Andreas. Perhaps there are larger magnitude 8 quakes that rupture large stretches of the southern San Andreas through San Gorgonio Pass that are less frequent than the smaller magnitude 7 quakes that are stymied at the San Gorgonio Pass.

55.9	**12.4**	**Take Exit 117 toward California Highway 62 northbound.**
57.4	**1.5**	**Turn right onto Dillon Road shortly after merging onto northbound Highway 62.**
57.6	**0.2**	**Turn left onto Worsley Road.**
58.3	**0.7**	**Stop on east side of road.**

STOP 5 COACHELLA VALLEY BANNING FAULT AS A STRIKE-SLIP FAULT

Speaker: Doug Yule, California State University at Northridge

Walk about 150 meters to the east of the parking area to view some classic geomorphic features of strike-slip faulting, including shutter ridges, a set of parallel fault splays, beheaded streams, and scarps with opposing facing directions (Figure 5). On a clear day, Garnet Hill (Stop 8) is visible approximately 5 kilometers to the southeast. The shutter ridge here is approximately 750 meters long and 200 meters wide and is comprised of older alluvium. It consists of an unusually wide zone of faulting with at least four subparallel fault traces striking ~N70ºW (small arrows on figure) but the main faulting appears to cluster at the side of the shutter ridge. Scarps face both to the north and to the south and some scarps change facing direction along strike. Antecedent streams cut across the shutter ridge — a large one at Worsley Road and a smaller one at the east end of the ridge — and indicate that the uplift rate of the ridge is less than the stream incision rate. Several small scarps can be seen cutting the fill in the gap at Worsely Road where a recent stream has not managed to remove evidence of the most recent earthquake(s).

The eastern end of the shutter ridge impinges on an old alluvial fan remnant (Qoa) that forms an elevated terrace to the south of the fault. Numerous beheaded gullies can be viewed on the old fan surface. The terrace level is 7–10 meters above the modern alluvial fan surface to the north of the fault (Qal). It is intriguing to consider the whereabouts of the upstream portion of this older fan. One possible correlation is the old alluvial fan surface to the east near Dillon Road. This possible correlation suggests approximately 1.6 and 2.0 kilometers of dextral slip since the abandonment of this older fan (Yule, 2010).

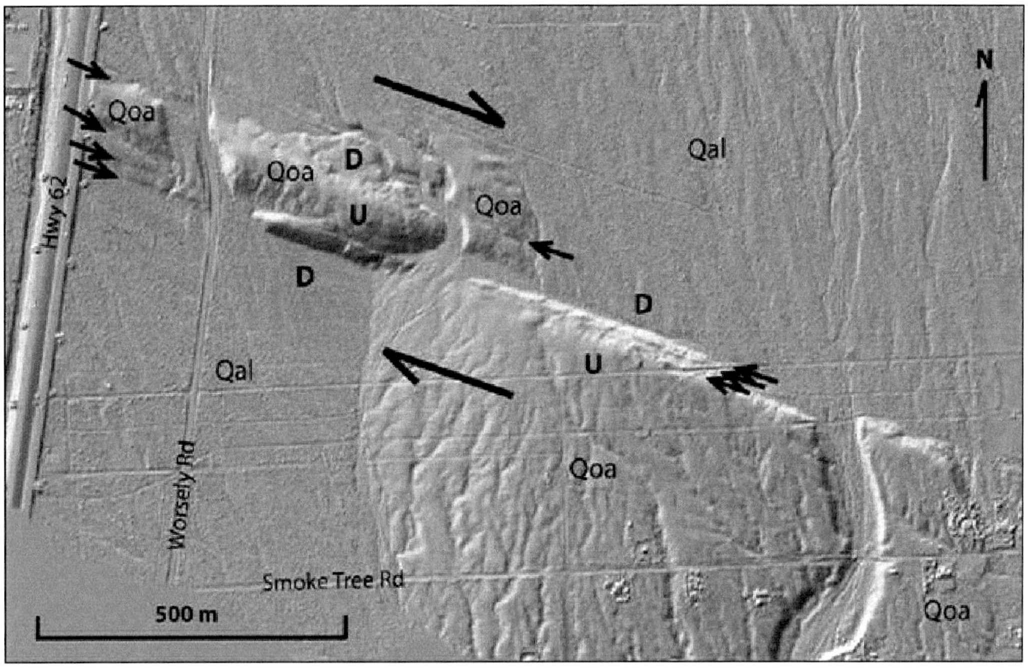

Figure 5. Shutter Ridge and truncated Old Alluvial fan along Coachella Banning fault

The epicenter of the 1986 ML 5.9 North Palm Springs earthquake is located approximately 7.5 kilometers to the north of this stop. The rupture is interpreted to have occurred along a plane oriented approximately N70°W, 45°NE with a first-motion rake of 180°, indicating pure strike-slip faulting (Jones et al., 1986). First motion solutions for aftershocks indicate predominantly oblique motion (dextral reverse). Surface cracks were observed along the trace of the Coachella Valley Banning fault from Devers Hill to Whitewater Canyon. Cracks were also observed in the fault scarp of the Garnet Hill fault at the mouth of Whitewater Canyon (Sharp et al., 1986). Displacements across these cracks were determined to be negligible and interpreted to have resulted from ground shaking, not from fault slip at the surface.

58.3 0.0 Return to vehicles and head north.
60.2 1.9 Turn left on Pierson Boulevard.
60.5 0.3 Turn right onto California Highway 62 northbound.
62.1 1.6 Turn left onto Mission Creek Road heading back towards the campground.
63.2 1.1 Pull off in dirt parking area.

STOP 6 MISSION CREEK DRAINAGE

Speaker: Katherine Kendrick, United States Geological Survey (USGS)

A nested alluvial complex within the Mission Creek drainage records fault-activity histories of the Mission Creek and Mill Creek strands of the San Andreas fault (SAF) in the San Gorgonio Pass region (see Figure 6). USGS is investigating this complex using detailed geologic mapping, morphometric and stratigraphic analysis, geochronology of fills using infrared stimulated luminescence and surfaces using cosmogenic-nuclides, and pedogenic analysis. Surfaces are correlated by morphometric analysis, including longitudinal profiles, surface texture and drainage density. The degree of soil development, compared to independently-dated surfaces regionally, provides temporal boundaries for the tectonic evolution. The data has been used to reconstruct a preliminary movement history for the SAF that includes the following: (1) The fill associated with the youngest pedon is positioned directly downstream from the canyon mouth of Mission Creek, suggesting that the Mill Creek and Mission Creek strands of SAF along this reach have not generated significant slip in approximately the last 100,000 years. (2) The deposits associated with the two older pedons have been displaced by the SAF no farther northwest than about 8–10 km. This displacement is similar to slip estimates for the Mill Creek strand of the SAF north of San Gorgonio Pass (Matti and Morton, 1993). (3) All alluvial fills in the Mission Creek area are within 10 km of the source drainage, suggesting that the Mission Creek strand of the SAF has not generated significant dextral slip in at least the last 400,000 years, and probably longer. Summing total lateral slip accumulated on the fault strands in the eastern San Gorgonio Pass region, including the Mission Creek (0 km), Mill Creek (8 km), Banning and Garnet Hill (2–3 km), during the past approximately 500,000 years results in a maximum long-term slip rate of 20–22 mm per year. This agrees well with some models of tectonic evolution, but not others (Kendrick, 2010 and 2011).

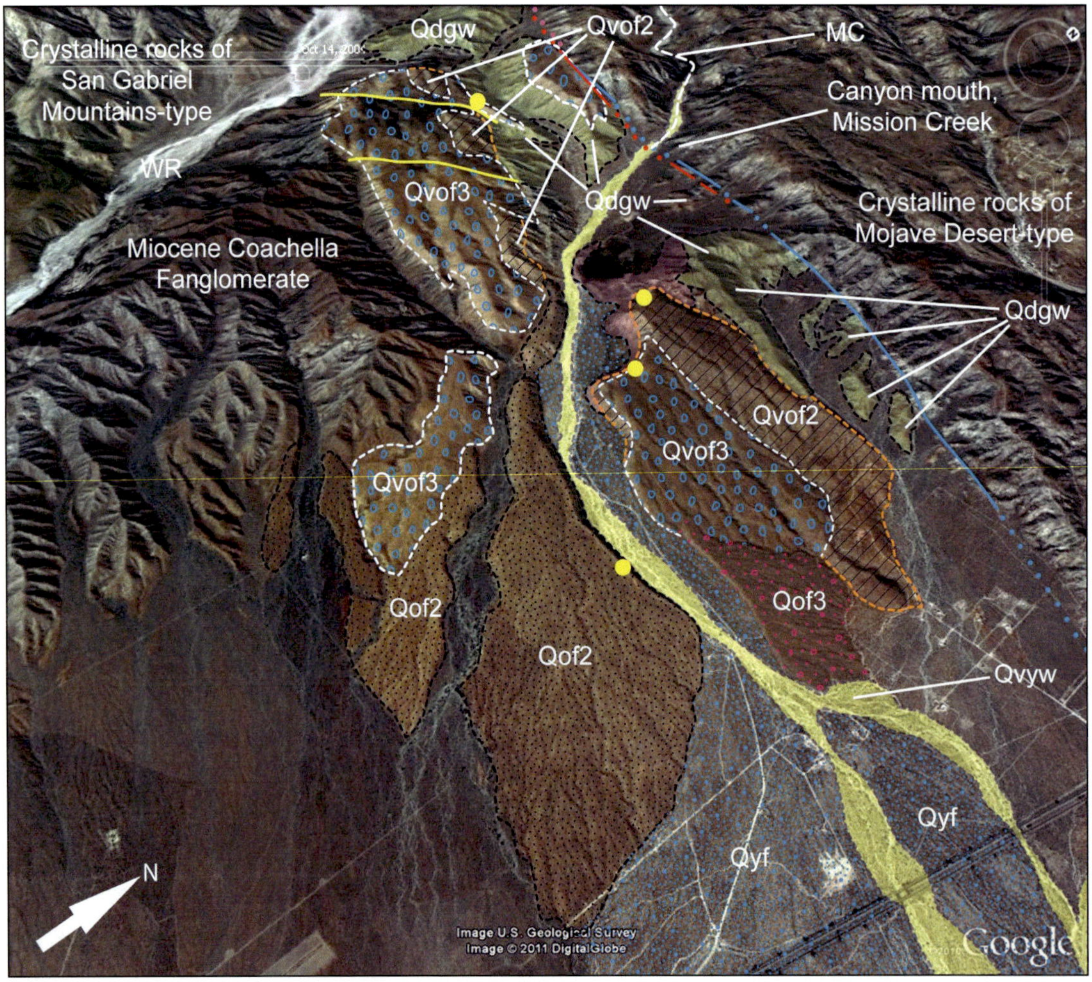

Figure 6. Nested Quaternary Alluvial Complex of Mission Creek Drainage. Note: This figure serves as an excellent "tilted" geologic map for exploring our campground and the volcanic units are marked in magenta.

63.2 0.0 Return to vehicles and continue northwest on dirt road toward campground.
66.0 2.8 Arrive at Mission Creek Preserve Group Campground.

STOP 7 GOOGLE EARTH AND THE MISSION CREEK/MILL CREEK/PINTO MOUNTAIN FAULTS

Speaker: Jeff Miller

Pull on your boots and come along for a pre-dinner hike as we explore the Mission Creek, Mill Creek and Pinto Mountain Fault exposures within reach of our campground. Figure 7 was generated using the USGS Quaternary Fault and Fold Database of the United States (http://earthquake.usgs.gov/hazards/qfaults/).

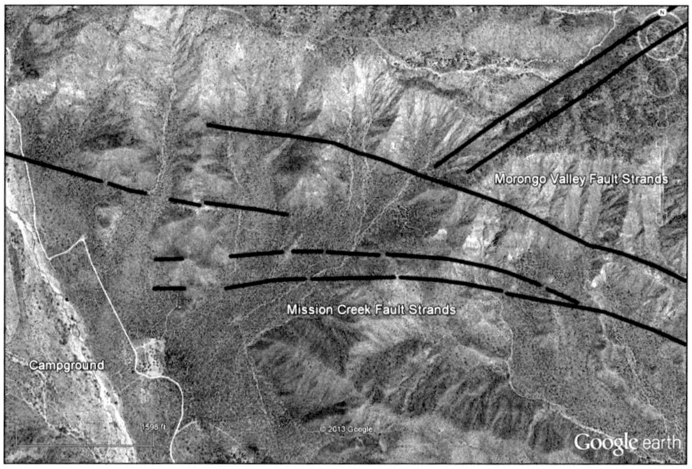

Figure 7. Mapped Faults in the Vicinity of the Mission Creek Preserve Campground

0.0 0.0 Begin at Mission Creek Preserve Gate. Reset odometer to zero. Head southeast along dirt road toward California Highway 62.
2.2 2.2 Turn right onto California Highway 62 southbound.
4.8 2.6 Keep right at the fork in the highway and head towards eastbound Interstate 10.
8.5 3.7 Take exit 120 North Indian Canyon Drive toward Palm Springs (southbound).
8.8 0.3 Turn left onto Garnet Avenue.
9.0 0.2 Turn right onto North Indian Canyon Drive.
9.2 0.2 Turn left into dirt lot and park.

STOP 8 GARNET HILL—THRUST SCARP OF SAN GORGONIO PASS FAULT ZONE

Speaker: Doug Yule, California State University at Northridge

Garnet Hill is a low-relief hill rising from a broad alluvial plain that fills the northern Coachella Valley. Garnet Hill is a growing fold created by dextral-reverse slip across a left stepover in the Garnet Hill fault. To the northwest, Hugo Hill and East Whitewater Hill are also growing folds at similar stepovers in the fault. To the east and west the Garnet Hill fault transfers slip via folding at compressive stepovers onto the Coachella Valley Banning fault and San Gorgonio Pass fault zone, respectively.

Some of the world's best ventifacts can be viewed at Garnet Hill. Westerly winds funnel through the San Gorgonio Pass and focus on Garnet Hill. Here, wind-driven sand from the flood plain of the Whitewater River has sculpted many of the cobbles and boulders. Look for boulders with polished pits, beveled surfaces, and east-west trending grooves. An increasing number of wind turbines in the area are a further testament to the power of the wind here. Marine fossils have been reported from outcrops of marine strata of the lower Pliocene Imperial Formation in canyons eroded into the south side of Garnet Hill (Yule, 2010).

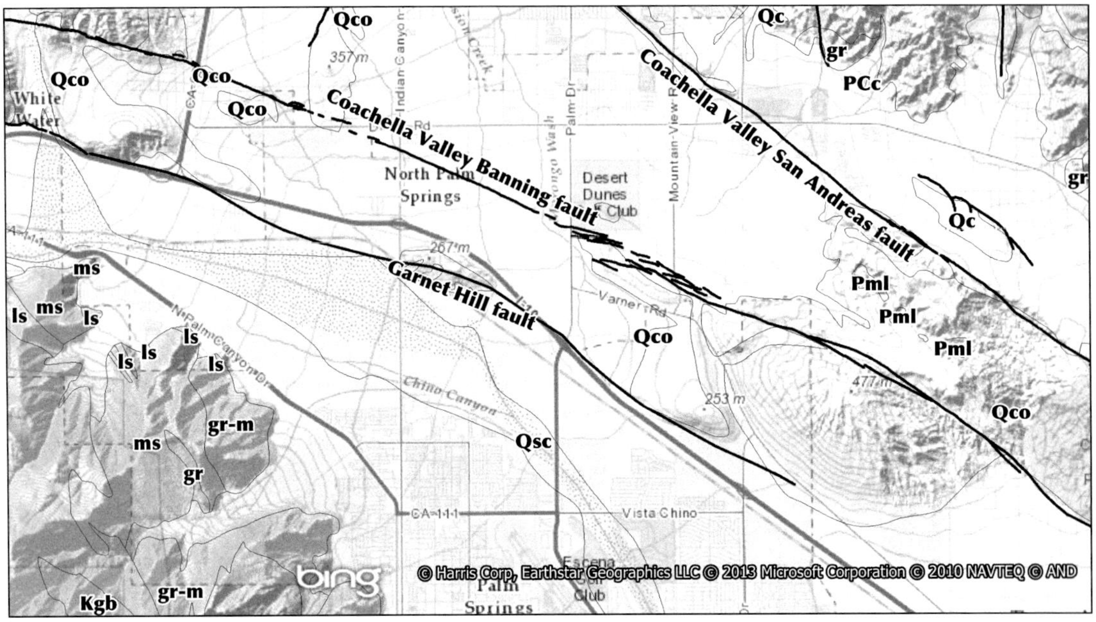

Figure 8. Fault Strands in the northern Coachella Valley

9.2	**0.0**	**Return to vehicles. Turn left onto North Indian Canyon Drive heading south.**
12.0	**2.8**	**Turn right onto West San Rafael Drive.**
12.7	**0.7**	**San Rafael Drive becomes Tram Way.**
16.5	**3.8**	**Arrive at Palm Springs Aerial Tram. Park as directed by attendants.**

STOP 9 PALM SPRINGS AERIAL TRAMWAY AND MOUNT SAN JACINTO DIKES

Speaker: Jeff Miller, E2ManageTech, Inc.

The Palm Springs Aerial Tramway was first proposed in 1935 by an electrical engineer names Francis F. Crocker. Due to the timing coincident with the onset of World War II and the Korean War, the project was delayed. Construction of the Tramway began in 1960 and included using helicopters to aid in building infrastructure in the precarious Chino Canyon. With an elevation gain of 6,000 feet above the base elevation of 2,500 feet, this was no simple task, yet amazingly there were no fatalities and very few serious injuries during construction. Within the San Jacinto block, "the tramway was built across some of the most difficult terrain over which any major construction project was ever attempted. The lower portion of the Tramway is founded in Paleozoic, or older, metamorphic rocks which consist of mica schist, hornblende schist, and limestone and quartzite. The upper portion of the Tramway is founded on granite gneiss with typically blocky jointing. Vertical slopes of up to 500 feet high are very common along the alignment of the Tramway" (Hood). Designed and operated by a Swiss firm, the project stands as a modern marvel. Four major geological concerns were taken into account: 1) Protection against rockfall and avalanche, 2) protection from frost,

3) proper seismic design based on earthquakes ranging from magnitude 7 to 9 from nearby faults, and 4) scour of the steep canyon (Hood).

As if they hadn't already done enough, the Tramway undertook serious renovations in 2001. Now the aerial tram features "cars" that rotate a full 360 degrees allowing for unparalleled views within Chino Canyon, and beyond to Palm Springs and the northern Coachella Valley, and for getting up close and personal with San Jacinto basement rocks. San Jacinto basement includes gneisses and schists derived from sedimentary rocks and Peninsular Ranges batholithic rocks.

DESCRIPTIONS OF PLUTONIC IGNEOUS ROCKS	
Rock Type	**Description**
Pegmatite Dikes	Coarse-grained dikes. Principally consist of abundant quartz, microcline, sodic plagioclase, and very large biotite crystals. A large pegmatite dike swarm occurs in the upper Chino Canyon area.
Granodiorite	Generally coarse-grained, homogenous, and without mafic inclusions. Principally oligoclase-andesine (plagioclase feldspar) (56%), K-feldspar (2%), quartz (30%), biotite (10%), hornblende (1%), minor sphene (1%), and accessory zircon. Widespread unit that forms the central portion of the San Jacinto pluton.
Quartz Diorite (tonalite)	Variable fine- to coarse-grained unit consisting of andesine (55–60%), K-feldspar (<5%), quartz (25–35%), hornblende (5–20%), and biotite (5–15%). Often hybrid-appearing with mafic xenoliths. Grades locally into granodiorite. Widespread unit along the border of the pluton where it is usually foliated.
Diorite	Principally andesine and hornblende with minor biotite. Occurs as small satellitic bodies in the border of the pluton and as dikes in the Chino acid-basic rock complex.
Gabbro	Pyroxene gabbro and olivine norite with coarse-grained orthocumulate texture. Occurs in a small unit in upper Chino Canyon between the first and second tramway towers. May be associated with nearby hornblende lamprophyres.

Reference: University of California, Irvine, (2005), p. 4, Table 1.1.

The metasedimentary rocks forming the eastern escarpment of the San Jacinto Mountains were named the Desert Divide Group by Dick Brown. The age of metamorphism predates the batholithic intrusion of the Peninsular Ranges and as reported in UCI (2005), "The minerals present in the metamorphic rocks indicate metamorphism at a depth of 10–17 kilometers in the range of 3–5 kilobars of pressure and temperatures of 550° to 650° C. This group of rocks is continuous along the eastern escarpment of the Peninsular Ranges down to at least the Mexican border and was originally deposited as marine sediments on continental crust. Brown (1981) measured the sequence to be about 2,400 m thick and assigned it an age of probably Paleozoic based on its similarity to rocks mapped by Vaughn (1922) north of Big Bear Lake. These rocks are present across Chino Canyon and can be seen as isolated pendants northwest of San Jacinto Peak.

"Brown (1968, 1981) separates the rock sequence of the Desert Divide Group into the Ken Quartzite and Bull Canyon Formation. These formations are assumed to be in correct stratigraphic superposition by Brown (1981) and therefore in stratigraphic sequence, with the Ken Quartzite younger than the Bull Canyon Formation. Brown's work was conducted southeast of Idyllwild near Garner Valley. Dibblee (1981), using Sydnor's (1975) thesis work as a basis, extends the metasedimentary geology into the Chino and Snow Canyon areas. Sydnor's cross section at Chino Canyon clearly shows the relationship of the metasedimentary rocks to the intruding tonalite of Mt. San Jacinto. These rocks are described as mostly schists and marbles and are most likely the northern equivalents of Brown's (1981) Bull Canyon Formation" (UCI, 2005, p. 3).

Arrive at Mountain Center.

STOP 10: A DUAL PRESENTATION AND DISCUSSION OF WHAT WE CAN SEE OR ONLY IMAGINE FROM THE TOP OF THE TRAM HIGH ON THE EAST SLOPE OF MOUNT SAN JACINTO

Speakers: Monte Marshall, San Diego State University, Dave Bloom, Tetra Tech

Starting with the line of palm trees marking the traces of the San Andreas fault (SAF), we will discuss the origin of the fault by jumping from the continental borderland and the opening of the Gulf of California. We will look at gravity data that show its dip and depth of sediments in the Coachella Valley at our feet. Looking northerly we will talk about the transpression associated with the eastern part of the Big Bend, and wonder if there isn't an unmapped, south-dipping thrust fault under Mt. San Jacinto that has uplifted it, like its conjugate partner, the Cucamonga fault, has pushed up Mt. Baldy. This will raise the question of whether most of the relief on southern California's two highest mountains started with the jump of the SAF to its present position at only about 6 million years.

Looking farther east we will find that the basement on the other side of the SAF is ten times older, about 1.8 billion years, than the Cretaceous magmatic arc granitic rocks that we are standing on. That will lead us to a discussion of the origin of the basement on both sides. The Peninsular Ranges Batholith extends from where we stand more than 700 kilometers (500 miles) to the south. It is just one of an almost continuous line of Mesozoic (mostly Cretaceous), subduction-related batholiths that dot western North America from British Columbia south to almost Guatemala. Since the San Jacinto—Santa Rosa mountain ranges lack high, positive aeromagnetic anomalies, they probably are on the east side of Brad Erskine's magnetite-ilmenite line. And, like the eastern half of the batholith in San Diego County, were probably formed by magma intruding the continental shelf of the precambrian shoreline south of present-day Yuma.

ACKNOWLEDGMENTS

I would like to sincerely thank the following people and organizations for their support in making this field trip and publication a reality.

Huge thanks to the Morongo Band of Mission Indians for granting us access to their lands and the folks from the Wildlands Conservancy for being such gracious campground hosts.

Thanks to the speakers! Eldon, Katherine, Marina, Monte, Dave, Jan and Alex — Thanks for giving your insight along the way.

Additional thanks go to the members of the field trip committee who have dedicated untold hours in helping get things ready for an awesome trip. Diane, Monte, Carole, Mike, Lavon, Jo'l, and Dan — thanks for helping me pull things together and supporting the trip. Eldon — thanks for the last minute trip to explore a tricky spot and making sure I "got it." Doug — thanks for being so willing to share your extensive work and helping us gain access to all of the stops — you're a real South Coast'er! Marina — there would be no trip without you. Your support has been inspiring and I am so glad to have you at the top of my speed dial!

Thanks very much to this year's corporate sponsors who allow us to offer this trip at a steep discount. A special thank you to Earth Consultants International for our large form graphics and E2ManageTech for adjusting to my crazy schedule and supporting this endeavour, including an obscene amount of printing.

To the current South Coast officers, great job. We've seen membership and meeting attendance explode in the last year and your dedication is repeatedly praised by the membership.

Thanks to members of the South Coast Geological Society. Many of you have made time to help me personally and professionally. I hope to work to repay your kindness and hope you find this field trip a worthy South Coast romp!

For my parents, Karen and Gary, who advised me to find something I loved doing and pursue it. You are the most genuine and generous people I know.

For Haley and Derek — I hope you find something you love doing and pursue it!

—*Jeff Miller*, SCGS President, 2012

REFERENCES:

Allen, C.R., (1957), San Andreas Fault Zone in San Gorgonio Pass, Southern California, Bulletin of the Geological Society of America. March.

Hood, R.G., (unknown), Palm Springs Aerial Tramway, AEG Report.

Kendrick, K. J., (2010), Pedogenic Constraints on Slip Along the San Andreas Fault, Southern California, Abstract.

Kendrick, K. J., (2011), Depositional constraints on slip along the San Andreas Fault within the eastern San Gorgonio Pass region, southern California, Abstract.

Matti, J.C. and Morton, D.M. 1993. Paleogeographic evolution of the San Andreas fault in southern California: A reconstruction based on a new cross-fault correlation, GSA Memoir 178, Chapter 2, 53 p.

Matti, J.C. and Morton, D.M. 2000. Geologic setting, San Bernardino National Forest. http://www.fs.usda.gov/Internet/FSE_DOCUMENTS/stelprdb5207093.pdf. February 9.

Mission Creek Preserve Website, (2012), http://www.wildlandsconservancy.org/preserve_missioncreek.html.

National Association of Geoscience Teachers (NAGT), (2007), Geology of the Coachella Valley Region, 93 p.

Proctor, R. J., (1968), Geology of the Desert Hot Springs-Upper Coachella Valley area, California, CDMG Special Report 94, 50 p.

Reichard, E. G. and Meadows, J. K., (1992), Evaluation of a groundwater flow and transport model of the upper Coachella Valley, California, US Geological Survey Water-Resources Investigation Report 91-4142, 101 p.

Scriven, D. F., (2012), Engineer's report groundwater replenishment and assessment program for the Whitewater River Subbasin, Desert Water Agency, Palm Springs, California, 56 p.

Tyley, S. J., (1974), Analog model study of the groundwater basin of the upper Coachella Valley, California,: US Geological Survey Water-Supply Paper 2027, 77 p

University of California, Irvine (UCI), 2005, Deep Underground Science & Engineering Lab at Mt. San Jacinto. Detailed Site Information Solicitation No. 2, Proposal. April 20. Accessed October 27, 2013, from http://www.ps.uci.edu/~SJNUS

Wikipedia, (2012), http://www.wikipedia.org/.

Yule, D. and Sieh, K., (2003), Complexities of the San Andreas Fault, Journal of Geophysical Research, November.

Yule, D. and Spotila, J., (2010), Quaternary Geology of the San Bernardino Mountains and their tectonic margins, 24 p.

We crack the rocks and make them ring,
And many a heavy pack we sling;
We run our lines and tie them in,
We measure strata thick and thin,
And Sunday work is never sin,
By thought and dint of hammering.

—*Andrew Lawson*

Part II

PAPERS

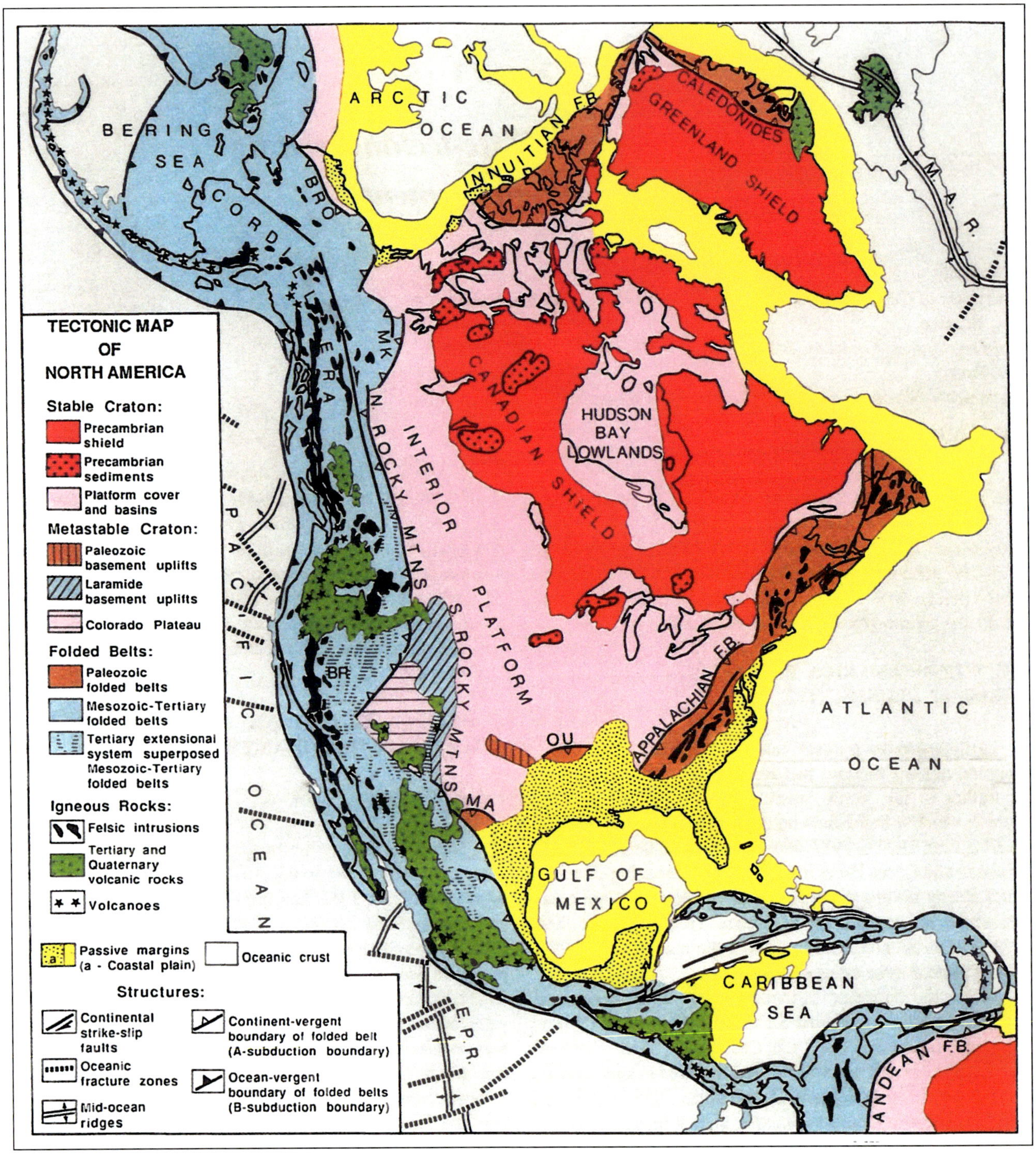

Figure 1. Simplified Tectonic Map of North America. BR-Basin and Range; BRO-Brooks Range; MK-Mackenzie Mountains; MA-Marathon Uplift; OU-Ouachita Mountains; M.A.R.-Mid-Atlantic Ridge; E.P.R.-East Pacific Rise. Note that the part of western North America shown in blue corresponds closely to the part in Figure 2 that is blank. Note also that the Grenville Orogen on Figure 2 is included here in the Appalachian Fold Belt. The three black teeth on the subduction zone off southern and Baja California are incorrect because this section of circum-Pacific subduction is no longer active. (From Bally et al., 1989)

Previous Page: Coachella Valley to the left of Salton Sea, center, and southern California and northern Baja California coastline from Los Angeles to Ensenada, bottom, shown in this photograph taken from an altitude of 396 km by an astronaut aboard the International Space Station in 2002. Credit: Image Science and Analysis Laboratory, NASA-Johnson Space Center. "The Gateway to Astronaut Photography of Earth." <http://eol.jsc.nasa.gov/scripts/sseop/QuickView.pl?directory=ESC&ID=ISS004-E-6119>11/06/2013 21:08:25.

The Unappreciated Role that the Amalgamation and Break-Up of Supercontinents has Played in the Geologic Evolution of the Western Margin of North America

Monte Marshall Professor Emeritus of Geology and Geophysics
Department of Geological Sciences, SDSU

INTRODUCTION

The purpose of this article is to provide a brief overview of how important the amalgamation and break-up of the three known supercontinents in the earth's history have been in the geologic evolution of the westernmost states and provinces of North America. The focus will be on the geologic evolution of southern California because this is the area visited by the 2008 and 2009 field trips of the SDAG and the 2012 field trip of the South Coast Geological Society.

Only in the last few decades have geologic researchers become aware of the existence and paleogeography of the two supercontinents that formed during the two billion years preceding Pangea. The evolution of these three supercontinents consisted of three stages—their aggregation over several hundred My (million years) of most, if not all, of the earth's continents, their existence as a single, and by far the largest, block of continental crust on earth for one hundred to several hundred My, and finally their break-up over 100 to 200 My, as various crustal blocks rifted and spread apart. The forces that caused the continental pieces to come together and then spread apart are unknown. As you will see, the geographic arrangement of the continental blocks, i.e., the paleogeography, of especially the two older supercontinents is currently hotly debated. The actual cause(s) of even current plate motion is a topic of debate! The word "supercontinent" is used somewhat loosely and differently by different authors—and for the same reason that the word "continent" is vague. "Continent" in both Webster's dictionary and the AGI "Glossary of Geology" is defined as "a large land mass." But, how large does a land mass have to be to qualify? Catalina Island certainly wouldn't, but how about Greenland or New Zealand? In the geologic past there have been aggregations of several continents, like that of Gondwana/Gondwanaland. Some authors call such landmasses "supercontinents." But in this article that term will be used only for aggregations of all or most of the world's continental crust.

Although the discussion will focus on the geologic evolution of southwestern North America, and especially that of southern California, the reader should appreciate that the behavior of supercontinents has probably affected almost every area on the earth's surface. Finally, since this paper evolved from a talk given to geotechnical geologists, I have tried to incorporate more background and explanation of recent geotectonic terms and concepts than would be found in a typical journal article. The reader will also find a lot of plate tectonic history. This is partially because the author is one of the rapidly dwindling number of geologists whose university education overlapped the discovery of plate tectonics. And he was personally involved in the continental drift controversy and its resolution. As an undergraduate, he was taught that the idea of continents moving around the globe was "stupid and impossible"! And, since distinguished professors always tell the truth, he whole-heartedly believed it! But, three years later, when he began his PhD thesis, his research ended up showing what rocks on the seafloor (the pillow basalt layer) were largely responsible for generating the marine linear magnetic anomalies (Marshall and Cox, 1971). These oceanic magnetic anomalies, as will be seen shortly, were the final, convincing proof of seafloor spreading and plate tectonics. Hopefully, many readers will enjoy learning parts of this great chapter in the history of science. In order to set the stage for much of what follows and put the geology of California in a meaningful context, we will first take a brief look at the general geologic makeup of our continent.

A BRIEF LOOK AT THE GEOLOGY OF NORTH AMERICA

Figure 1 is the first figure in the first book of the ground-breaking series of books on the geology of North America, called the Decade of North American Geology, or DNAG Series, that was published by the GSA in the late nineteen eighties and early nineties. This figure shows the generalized age and tectonic subdivisions of all of North and Central America. Our continent is special in several respects. Its geologic map is the most symmetrical of all continents, with an Archean and Proterozoic core flanked by the Paleozoic Appalachian orogen on the east and the Mesozoic-Cenozoic Cordilleran orogen on the west. The world's oldest rocks yet dated are 4.0 Ga (billions of years old) granitic gneisses in the Canadian northwest. They are part of the largest surface exposure of pre-Cambrian crust on earth, the Canadian Shield, and form a fairly flat area surrounding Hudson Bay. This mostly granitic and metamorphic crust is rich in diamond-bearing kimberlites and gold and copper deposits. On almost all sides, the shield dips gently below mostly flat-lying Paleozoic and younger sedimentary rocks to form a part of North America (NA) known as the Interior Platform. Together, the shield and platform compose what is called a craton. All the other continents have at least two cratons. These very old parts of continents are characterized by long periods of relative stability and little geologically-recent deformation.

Figure 2 shows the more exact age and nature of the rocks in the craton. Geologists from the Canadian Geological Survey, especially Paul Hoffman (who was later a professor at Harvard), found that the Canadian Shield consists of an intricate patchwork of seven areas of Archean crust, called by various authors "provinces," "microcontinents," or "small cratons," that are welded together by early Proterozoic collisional orogens (Hoffman, 1989).

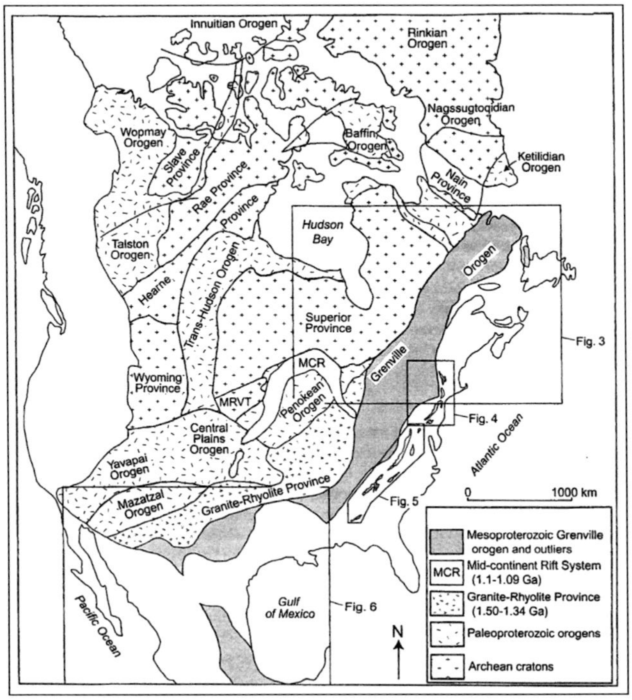

Figure 2. Map of North America Showing the Archean Provinces of the Canadian Shield and the Archean and Proterozoic Provinces and Orogens Buried under the Sedimentary Rocks of the Interior Platform of Figure 1. Map shows the location of the Grenville Orogen, including both its subsurface extent and the location of exposed outliers in the Appalachians and its continuation into Texas and Mexico. Areas of rocks younger than 1.0 billion years are shown with no pattern. MCR, the Mid-Continent Rift System; MRVT, Minnesota River Valley Terrane. The Minnesota River Valley Terrane is the southernmost exposure of the Superior Province and is famous for containing the Morton Gneiss, which has been dated at 3.5 billion years, making it the oldest rock found so far in the United States. Note the absence of the western tier of states from Alaska to California, and the proximity of the 1.8 billion year old Yavapai and Mazatzal orogens to the present Coachella and Imperial valleys. (From Tollo et al., 2004).

Hoffman speculated that even the Archean provinces themselves may be an aggregate of island arcs. Over the past few decades geologists have come to learn that the formation of continental crust is a complicated process of aggregation—the collision of continents and island arcs, addition of subduction zone material, crustal underplating, etc. And, the way continental crust grew in the Archean may well not be the same as today.

Note that the area where the western tier of states and provinces are currently located is blank. This is because the map is designed to show only rocks older than about 1 Ga; most of the rocks currently occupying this N-S coastal strip formed less than 200 Ma (million years ago). The youth of the igneous and metamorphic, or "crystalline," rocks that form the basement in this strip make it the youngest crust in NA! As you will see shortly, this area was occupied by sea water and oceanic crust from about 700 to 200 Ma. How this 'blank' strip was filled in with the rocks that we have come to know and love is one of the principal goals of this paper.

THE THREE SUPERCONTINENTS

In this section we will analyze each of the supercontinents in their order of discovery, which, like most things geological, is in reverse chronological order. There are a lot more young rocks and they are less deformed and metamorphosed than the really old rocks, and are therefore easier to study and understand. Today, even non-geologists are familiar with the fact that most all of the present day continents and large islands were aggregated into one, large supercontinent that lasted from about 300 to 200 Ma. Alfred Wegener, the 'father' of continental drift, adopted the name *Pangea* (Greek for "all earth") for this supercontinent. The continents currently in the northern hemisphere, i.e., North America, Europe, and Asia formed the part of Pangea that is called Laurasia. The "laur" part of that word refers to North America, including Greenland, which is called Laurentia in paleographic terms. South America, Africa, Australia, Antarctica, and India, landmasses that are now mostly in the southern hemisphere, with the exception of the Indian subcontinent, comprised the part of Pangea called Gondwana or Gondwanaland.

PANGEA, PART 1: THE CONTROVERSY AND ITS RESOLUTION WITH THE DISCOVERY OF PLATE TECTONICS

Today's younger geologists cannot appreciate how vigorous, even angry and career-threatening, were the arguments over whether or not the present continents were all together in the past; and, if so, how they got to their present positions. The jig-saw fit of the continents bordering the Atlantic Ocean was certainly suggestive, as had been pointed out by Alfred Wegener in the early twentieth century—and by others even before him. In fact, Sir Edward Bullard used computer modeling to show that if you used a physically more accurate edge of the continents, i.e., the continental shelf-slope break, instead of the shorelines, the fit was almost perfect. But Wegener's hypothesis, while true in many respects, had a fatal flaw that caused decades of controversy in the earth science community. Probably because he was primarily a meteorologist and not a geologist, he didn't really understand what the geophysicists, especially seismologists, were discovering about the difference in the composition and thickness of continental versus oceanic crust. Specifically, the mechanism he proposed for separating the continents, i.e., having them plow through oceanic crust that bordered Pangea, "continental drift" as he called it, was shown by geophysicists to be impossible. Since the oceanic crust is only 7 km thick and the average continental crust is about 35 km thick, Wegener's drift would be totally like a ship plowing thru an ice sheet—you would expect to see up-thrust slabs of oceanic crust in the direction the continents were moving, and deep holes in the oceanic crust behind them. No such deformation or bathymetry of the seafloor was observed by marine geophysicists; and the force necessary to push continents thru oceanic crust and upper mantle would be tremendous.

This Gordian knot was cut by the realization in the late sixties that two layers of the earth were moving—both continental and oceanic crust and the upper mantle, i.e., everything down to a depth of about 100 km. And, because the bottom surfaces of these slabs are essentially flat, and the 100 to 200 km of mantle underlying them was found to probably be partially molten, the lateral motion was much easier to accept and understand. The upper 100 km was then named the *lithosphere*, and pieces of the lithosphere (usually containing both oceanic and continental crust) were named *plates*. Because of its weakness, the layer underlying the lithosphere was called the *asthenosphere*. Geoscientists realized that much geological activity, like mountain-building, volcanoes, and earthquakes, was due to and could be elegantly explained by the motion of the plates relative to one another. So, by 1970, the term "plate tectonics" was coined and these concepts were quickly accepted by most of the world's geologists, except in the Soviet Union.

Because of the role that the oceanic linear magnetic anomalies played in the discovery of plate tectonics and, a few years later, in Tanya Atwater's connection between the subduction of the oceanic plates offshore of western North America and its geologic history, a brief explanation of how the anomalies are caused and used would be useful here. If one looks at a magnetic anomaly map of the ocean basins, made by plane or ship-towed magnetometers, you will see long, linear bands of positive and negative magnetic anomalies on both sides of and running parallel to seafloor ridges and rises. This kind of a zebra-striped pattern is never seen on land. To a geophysicist, one explanation of this curious pattern would be that strips of the seafloor might alternate between being magnetized parallel to and anti-parallel to the earth's present magnetic field. When the magnetometer passes over a strip of crust that is magnetized parallel to the ambient field, the magnetic field of the crust adds to the earth's field and produces a positive magnetic anomaly, i.e., a field stronger than expected. And, just the opposite would be true over a strip of reversely magnetized crust. It's an interesting piece of science history that a 23 year old grad student at Cambridge University, Fred Vine, was one of the first people to realize that if the oceanic ridges and rises really were spreading centers, as had already been proposed by a number of marine geologists, then, as the basaltic lava cooled at the spreading center, it should get magnetized in whatever direction the geomagnetic field had at the time. And, he also knew that paleomagnetists had recently discovered that the geomagnetic field really does reverse itself from time to time. He had read the recent articles, especially those published by two USGS paleomagnetists, Allan Cox and Dick Doell, who were trying to establish a geomagnetic reversal time scale (GRTS). So he realized that if oceanic spreading centers really existed, they would not only create oceanic crust, but would act like double-headed tape recorders and produce the same anomaly pattern on both sides of the ridges or rises. At that time, the Cambridge geology department had many progressive and brilliant thinkers, like Sir Edward Bullard, Harry Hess, Fred Vine and his PhD advisor, Drum Mathews. And so it was that Vine and Mathews were the first to publish a paper (in the prestigious British journal, *Nature*, in 1963) that suggested that the linear oceanic magnetic anomalies, like those that they had measured on an oceanographic cruise across the Indian Ocean Ridge, were caused by a combination of geomagnetic reversals and sea-floor spreading.

As Cox and Doell and other paleomagnetists continued to refine the geomagnetic reversal time scale, they found that the reversals were not at all cyclic or periodic, as they had first believed and had published at the time of Fred Vine's paper in 1963. Instead, the times of constant polarity, whether normal or reverse, were found to have random lengths. Some periods lasted only a few tens of thousands of years, whereas others lasted millions of years. This non-periodicity produced strips of oceanic seafloor that not only alternated between normally and reversely magnetized, but also varied in width. Therefore, a magnetometer towed by a ship or plane across a real seafloor spreading center should record bi-laterally symmetric anomalies, i.e., the complex sequence of variable width, positive and negative magnetic anomalies found on one side should be the mirror image of the sequence measured on the opposite flank. And, the second requirement for true spreading centers is that the sequence of broad and narrow, positive and negative magnetic anomalies should exactly match the sequence of long and short, normal and reversed times of constant geomagnetic polarity, respectively. Or, more simply, the magnetic anomaly pattern observed over oceanic ridges should be the same as the pattern on Cox and Doell's GRTS! Although the Vine and Mathews paper was published in 1963, their Indian Ocean magnetic anomalies were poorly defined and they had unfortunately tried to match them to an early, incorrect GRTS. So the earth science community remained skeptical about the reality of seafloor spreading for five more years—until the big annual meeting of the AGU (American Geophysical Union) in San Francisco in December, 1968. In a special session, the marine geophysicists from Lamont (Columbia University) and Scripps Institution of Oceanography (UCSD) presented slides showing the marine magnetic anomalies they had measured over oceanic ridges and rises in all the world's oceans. When we saw that all the profiles had the same, bilaterally symmetric anomaly pattern, and that the pattern almost exactly matched the theoretical anomaly pattern computed using the newer and more accurate GRTS, we all sat in the hall stunned! Everyone knew immediately that something momentous in geologic history had just happened—that the oceanographers had proved that seafloor spreading as well as some kind of continental drift was real.

When plotted on a map using black, for example, for the strips of seafloor that have positive magnetic anomalies, and leaving the strips with negative magnetic anomalies white, the anomaly pattern looks just like the barcodes on merchandise! And they can contain a lot of information. The random spacing of reversals makes the magnetic barcode anomaly pattern created during each geologic epoch unique to that epoch. The Pliocene barcode doesn't look like that of the Miocene, etc. So, once oceanographers had measured the anomaly pattern from the Mid-Atlantic Ridge to the continental margins of North America and Africa, and once the age of the Atlantic oceanic crust was determined paleontologically by a series of cores drilled across the ocean by the drilling ship *Glomar Challenger*, the age of each unique section of the bar code was known. And Cox and Doell's land-based GRTS, which only went back to 4 Ma, was thereby extended back to 160 Ma.Thereafter, oceanographers could measure the ages of the oceanic crust in many places simply by using a ship-towed magnetometer. As you will see, this oceanic magnetic bar code has proven to be very useful in deciphering the tectonics of not only the oceanic crust, but even some of the tectonic history of the adjacent continental crust. Now you can understand why restoring the current position of the continents to their place in Pangea was relatively easy. When the continental blocks split-up, oceanic crust was created in their wake; and this oceanic crust has a bar code that reveals when and how fast the fragments separated.

The intellectual ferment at Cambridge in the sixties produced other discoveries relevant to this article. At that time most geologists hadn't even started to think about other possible supercontinents before Pangea. One of the first people to do so was a gifted Canadian geoscientist who spent his sabbatical leave at Cambridge in 1965, J. Tuzo Wilson. He proposed that supercontinents might be cyclic, i.e., that an earlier proto-Atlantic ocean may have existed prior to the present Atlantic and that it was closed up and subducted away when Pangea came together, and so on. This accordion-like expansion and contraction of the ocean basins has come to be called the "Wilson Cycle." Wilson also gave the name "transform" to the strike-slip fault boundaries in plate tectonics. And he was the first person to recognize that the San Andreas fault (SAF) is a transform fault. It transforms the oblique strike-slip motion between the North American and Pacific plates in the Gulf of California and under the Imperial Valley of southern California to the Gorda and Juan de Fuca spreading centers offshore of Cape Mendocino in northern California. Here the plate boundaries are a divergent margin between the Pacific and Juan de Fuca plates and a convergent margin between the Juan de Fuca and North American plates (Wilson, 1965).

PANGEA, PART 2: THE ROLE THAT ITS BREAKUP PLAYED IN CREATING THE YOUNGEST AND WESTERNMOST CRUST OF NORTH AMERICA

When Pangea started to break up at about 200 Ma, one of the first ocean basins to form was the central Atlantic as the east coast of North America, from Florida to Nova Scotia drifted away from the northwest coast of Africa (Fig. 3). Remember that at this time the land of the western tier of states and provinces did not yet exist, and the blank strip in Figure 2 was occupied by ocean water, with waves lapping on precambrian crust somewhere in Arizona and Nevada. In fact, it was Earth's job during the next 200 My to create new continental crust or bring in pre-existing crust to fill this gap. At present almost all continental, or sialic, crust is created by subduction. The subduction zone can be just offshore and the magma generated when the down-going slab reaches a depth of about 100 km will come up through continental crust, as in the Andes. Or the subduction zone can be so far offshore that an island arc is created, like Tonga or Fiji. Once an island arc is created, the motion of the oceanic plate it rests on can transport it to a subduction zone at an active margin and suture it

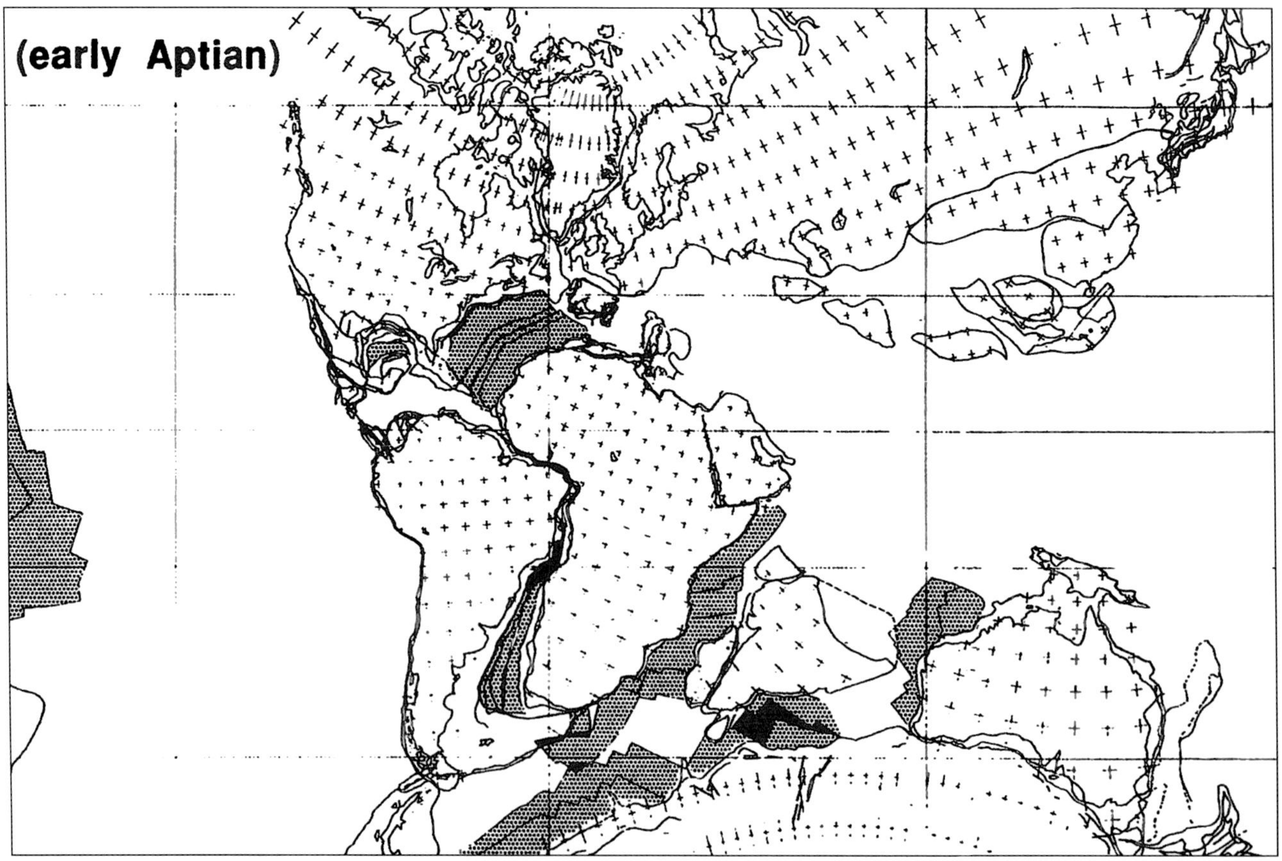

Figure 3. Reconstruction of the northern part of Pangea at about 120 million years ago, or about 80 million years into its breakup. Note that the only parts of the Atlantic that have begun to open is the section of North America between Florida and Nova Scotia, as it pulls away from northwest Africa, and the South Atlantic. The North Atlantic will take another 60 million years to open up, as Greenland pulls away from northwestern Europe. (From Bally et al., 1989)

to the edge of the continent, as an allochthonous or exotic terrane. You will see in the next section that this west coast "gap" was created during the break up of Rodinia, and the oceanic crust occupying this gap probably started forming at about 750 Ma. So the seafloor off of Arizona, Nevada, and the other westernmost tier of states could've been 550 My old by 200 Ma. In oceanic crustal terms, old means cold, and cold means dense. Geotectonic experts are still arguing about what forces move the plates, but the current consensus is that, along with convection currents in the mantle, gravity plays the major role. Plates move away, i.e., gravity pulls them away, from spreading centers because the crust and mantle are heated by the rising magma and the expanded crust of the ridges and rises stands thousands of meters higher than the older, colder crust that floors the abyssal plains. And somewhere far from where it was born, the oceanic crust cools to the point where its density is greater than that of the upper mantle and it sinks, is subducted, or both.

No one knows precisely when the Pacific seafloor first began to subduct beneath the current west coast of North America. The recent discovery of tectonic fragments of Cambrian age island arcs and backarc basins in the western Cordillera shows that the newly created Pacific seafloor was being subducted by the beginning of the Paleozoic. But, fairly large-scale magmatism didn't start in California until the middle of the Mesozoic. The oldest granitic plutons in both the Sierra Nevada and Peninsular Range batholiths are late Triassic, or about 220 Ma (Moore, 2000; Walawender, 2000). Even the precambrian rocks of the San Gabriel Mountains in the central Transverse Ranges were first intruded in Phanerozoic times by the Lowe Granodiorite at about 220 Ma (Ehlig, 1981).

The opening of the central Atlantic at about 200 Ma was caused in part by the westward motion of the North American plate. Such westward motion should result in the North American plate overriding the Pacific seafloor and either initiating subduction or increasing the convergence rate of an already subducting plate. We know that subduction of the Pacific oceanic crust produced enormous volumes of volcanic and plutonic rocks along the western margin of North America (Fig. 4). This chain of batholiths formed in the Mesozoic, primarily in the Cretaceous, and extends 3500 miles from Baja California to Alaska (Moore, 2000). But, even this great span is just a small part of what is currently known as the Pacific Ring of Fire. Because subduction off California was replaced by the San Andreas Fault transform boundary, most of California is no longer part of this still active, subduction-caused chain of active magmatic arcs that stretches from the tip of South America, up through central America and Mexico, thru Oregon and Washington, along the south coast of Alaska, out along the Aleutian Islands, and down the Kamchatka Peninsula and Japan to Taiwan (Dickinson, 2009).

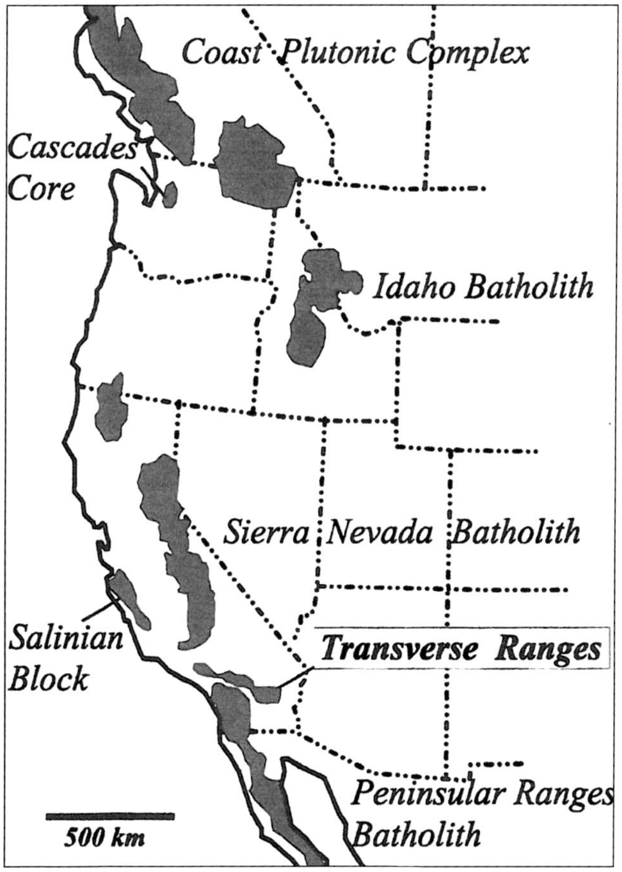

Figure 4. Major Mesozoic Convergent Margin Plutonic Belts and Batholiths. Plutons are mostly of Cretaceous age. The Transverse Ranges are a special case because they comprise Proterozoic crust, like that of western Arizona, that is intruded by Mesozoic plutons. (From Needy et al., 2009)

The rate of subduction of the Pacific seafloor under North America clearly depends on the relative motions between the western North American plate and whatever plate underlay the northeastern Pacific Ocean. In 1970, Tanya Atwater published her famous and revolutionary article that showed that you could use the linear magnetic anomalies recorded off the western margin of North America to show that one of the Pacific plates, the Farallon plate, had been subducting under North America for tens of millions of years, but when the plate on the other side of its spreading center, the Pacific plate, arrived at the trench off of Los Angeles at about 30 Ma, subduction along this suture stopped and the relative motion changed to strike slip—and the San Andreas fault system was born! We all know the tectonic consequences as the right lateral, strike-slip fault system migrated from the trench, across the continental shelf or borderland to its present position. But what about all those millions of years of subduction? If you could calculate the rate and direction of subduction, would this give additional insight into the geologic history of the west? With the help of computers and more knowledge of the linear magnetic anomalies, Engebretson et al. (1985) did exactly that. Figure 5 shows a map of the likely plate geometry and velocities of the northern Pacific Basin at 110 Ma (Engebretson et al., 1985). Note that at this time the Farallon Plate was subducting under the western margin of both North and South America and the Pacific plate was still in the southern hemisphere. Since only remnants of the Farallon Plate, like the Juan de Fuca, Cocos, and Nazca Plates, still exist, you can see how much of it was subducted in the last 110 My. Because subduction has consumed most of the world's seafloor and its bar code older than about 150 Ma, Engebretson et al. could only model the plate tectonics of the Pacific Basin since 175 Ma.

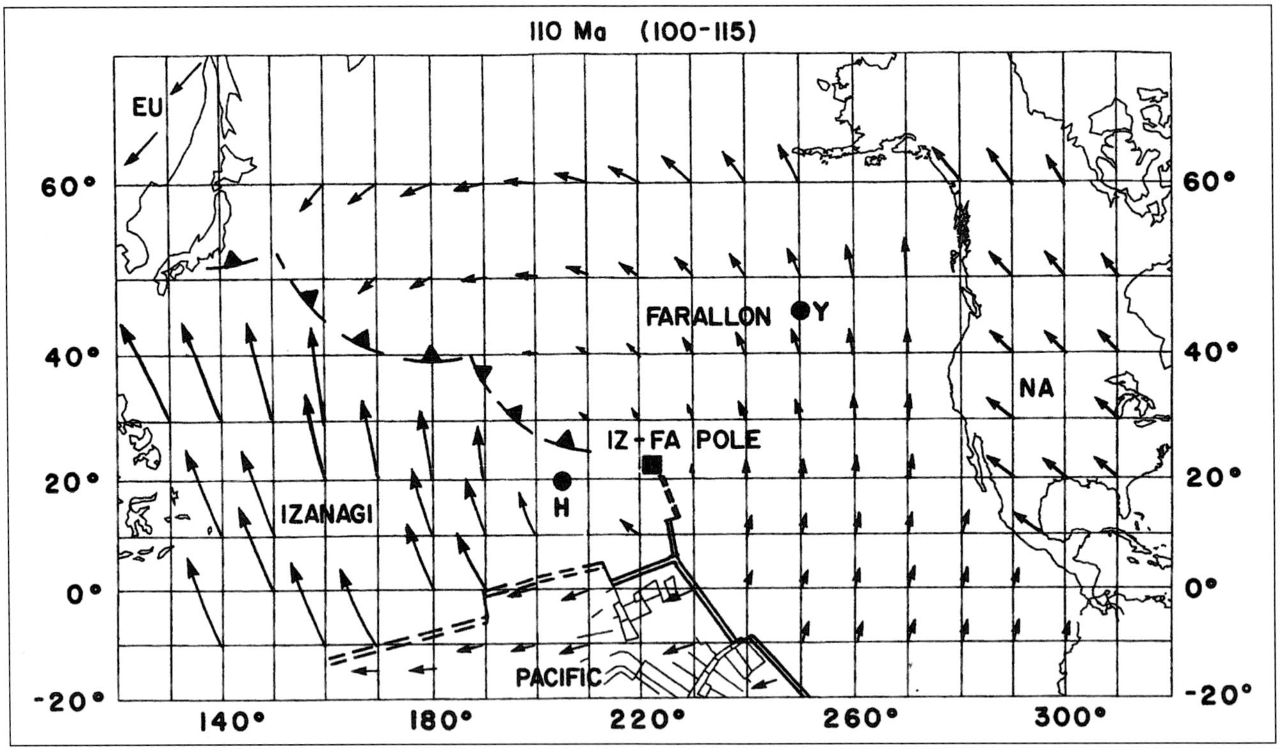

Figure 5. Map of the Oceanic Plates in the North Pacific at 110 Million Years Ago. Arrows show the absolute velocity at points on the plates, and their length is the distance moved in 10 million years. These velocities assume that the hotspots in the Atlantic have remained fixed relative to those in the Pacific, and are calculated using the relative motions between plates as shown by the oceanic linear magnetic anomalies. Dots are the locations of the Hawaiian (H) and Yellowstone (Y) hotspots at that time. Double lines are spreading centers (dashed where inferred). Barbs or teeth mark subduction zones—sites of creation of the far-travelled island arcs discussed in the text. The IZ-FA dark square is the location of the Euler pole for the relative motion between the Izanagi and Farallon plates. (From Engebretson et al., 1985)

Fortunately, the bulk of the magma generation in the Peninsular Ranges Batholith (PRB) and the Sierra Nevada Batholith (SNB) occurred between 120 and 80 Ma, so they could compare their computed subduction results to the timing and volume of subduction-related magmatism in the west coast of North America. Subduction is often oblique, i.e., the velocity vector of the down-going plate has a direction in between the strike and the dip of the subduction zone. Figure 6 shows the two orthogonal components of the subduction rate off our coast since 175 Ma. Figure 6a shows the normal velocity, or the component of the total subduction rate that is perpendicular to the presumed trench axis in kilometers of plate consumed per million years. In ordinary geologic terms, "normal velocity" is simply "dip-slip rate." In Figure 6b, "tangential velocity" means the same as "strike-slip rate" and is simply the component of the subduction rate parallel to the strike of the trench. Assuming that the conditions their model was based on are met and their calculations were correct, they realized that the information in these two figures should tell us a great deal about the tectonic and magmatic effects of the subduction off our coast for much of the last 200 My. Some of the more important points are: 1) during most, if not all, of this time the subducted plate was the Farallon Plate (a third plate, the Kula plate, may have "substituted" for the Farallon plate between 85 Ma and 60 Ma), 2) the normal, or down-dip subduction rate, varied widely between an almost complete stand-still for 10 to 15 million years at about 140 Ma and a very rapid rate of 125 to 150 km/My between 75 and 45 Ma, 3) between 175 and 100 Ma subduction was oblique and had a decreasing component of left lateral strike slip, 4) between 100 and 75 Ma the Farallon Plate was subducting to the NE and had a right lateral, strike slip component of about 35 km/My, 5) beginning at 75 Ma the Farallon Plate was subducted at a very high rate and dove almost directly downdip.

> **RAKE:** Think about oblique slip on a thrust fault where the striations that mark the slip direction lie between the strike and dip of the fault, e.g., have a rake of 45 degrees. For those not familiar with it, *rake* is defined as the angle, as measured in a plane, between the strike of the plane and a line in the plane.

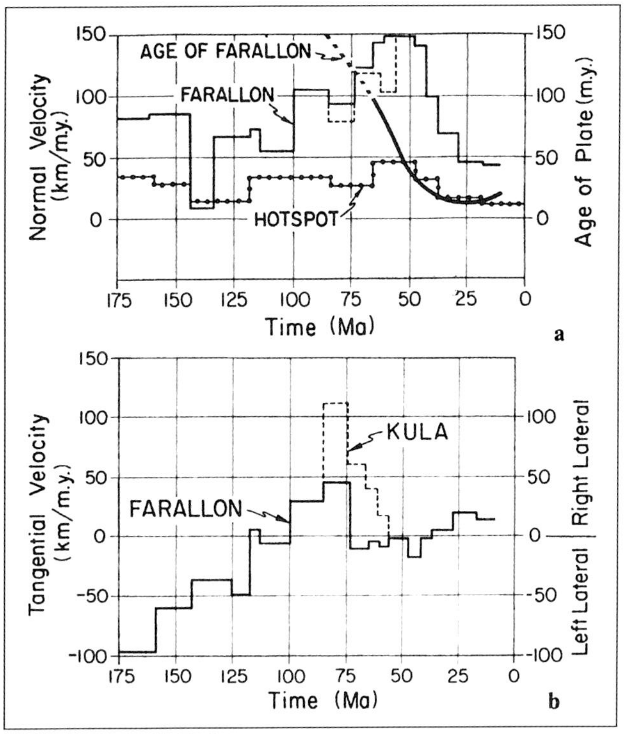

Figure 6. Calculated subduction rate of the Farallon plate for a point near San Francisco from 175 million years ago to the end of subduction at 10 million years ago, when the Triple Junction passed San Francisco (solid line). Data are taken from the computer modeling shown in Figure 5. Figure 6a shows the normal velocity component, i.e., the subduction rate component perpendicular to the presumed trend of the trench of N40W. Geologically, it is simply the down-dip subduction rate. The dashed line shows the normal velocity if the Kula plate were the subducted plate between about 85 and 60 million years ago. The heavy, curved line that begins at about 75 million years ago is the age of the Farallon plate that was entering the trench at the latitude of San Francisco (read age on the right margin of the plot). The dotted line is the motion of North America relative to the hotspots. Figure 6b shows the component of the subduction rate tangential or parallel to the trench axis (or, the strike-slip component of the subduction rate).The right margin shows when the strike-slip component of subduction was either right lateral or left lateral, i.e. the subduction was oblique and directed to the northeast or southeast, respectively. The sharp corners/square wave nature of these plots could be, at least partially, an artifact of the computational procedures and may not be real. (From Engebretson et al., 1985)

Now we can ask ourselves some of the same important questions that they asked themselves. What is the correlation, if any, between the rate and direction of subduction and the magmatism or tectonic activity that it caused on the North American plate above it? An increase in the volume of magma produced, whether it forms volcanoes or plutons, would be expected with increasing subduction rates. Is there an optimum or maximum subduction rate? On the other hand, is there a minimum subduction rate, i.e., a rate so slow that either no magma is produced, or what is produced never reaches the surface?

Engebretson et al. (1985) knew the magmatic chronology of the Sierra Nevada Batholith (SNB), and, since it is so similar to that of the Peninsular Ranges Batholith (PRB), they made some of the same correlations that will be made here. Fortunately, we have a somewhat detailed chronology of the igneous activity in the PRB (Walawender, 2000) and additional information for the SNB (Moore, 2000). Engebretson et al.'s (1985) subduction parameters, as shown in Figure 6, and the igneous chronology of Walawender and Moore are summarized in Table 1. The time span since 175 Ma has been divided into four time intervals based on times of significant changes in the subduction parameters or in the mode of magmatism.

Starting with the first time span, 175 to 120 Ma, note that the average normal or dip-slip rate is relatively low. In fact, as noted above, subduction almost stopped for 10 or 15 My at around 140 Ma. In the Peninsular Ranges, Walawender (2000) shows a magmatic hiatus of 35 My between the end of Jurassic plutonism at 160 Ma and the first, early Cretaceous (125 Ma) plutons of the western belt. In the central Sierra Nevada, Chen and Moore (1982) found a 40 My gap in magmatic activity between about 150 and 120 Ma, save for the injection of the Independence dike swarm at 148 Ma. They are mafic dikes filling fractures due to extension on the east side of the Sierras and probably aren't subduction generated. The correlation between these magmatic lulls and the very low subduction rate centered at about 140 Ma is clear and the magmatic lull is probably a good example of an insufficient subduction rate. But why do these magmatic lulls extend for 10–15 My on either side of the subduction rate minimum? Are the radiometric dates of the plutons incorrect? Or, is there a problem with the computer model? Or is it a combination of the two sources of error?

TABLE 1. Relation between subduction (rate and direction) of the Farallon Plate and magmatism in the Peninsular Ranges Batholith (PRB) and Sierra Nevada Batholith (SNB).

Time Span (Ma)	Average Dip-slip Rate (km/My)	Average Strike-slip Rate (km/My)	Relative Dip-slip Rates	Rake	Magmatic Activity in PRB	Magmatic Activity in SNB
175–120	60	60 (LL)	Low	45°	Very little. Last stage of Jurassic arc. Magmatic lull between 160 and 125 Ma. A few early plutons in western PRB at 125 Ma. and SP volcanics.	Some Jurassic plutons. Volcanic rocks, especially at 160-140 Ma. No plutons between 150 and 120Ma.
120–100	65	10 (LL)	Low	80°	Majority of plutons in western PRB. Much Santiago Peak volcanics.	Many plutons. Volcanics with a pulse at 100 Ma.
100–80	100	40 (RL)	High	70°	Arc jumps to east side. Very large plutons.	Large plutons on east side.
75–30	115	0	Very High	90°	Magmatic activity ended at 90 Ma	Magmatic activity ended at 80 Ma.

See text for definition of rake. LL = left lateral. RL = right lateral. My = million years. Ma = million years ago.

Engebretson et al. (1985) acknowledge that the sharp corners in their curves may not be real. We will see that in the next time period, a fairly constant, down-dip subduction rate of about 60 km/My can produce a good supply of granitic magma. So maybe their square wave is really a broader U-shaped curve that drops below some minimum value, 40 km/My for example, at about 150 Ma, bottoms out at about 140 Ma, and then increases back up to the minimum rate by about 125 Ma.

The time interval of 120 to 100 Ma is important in the magmatic history of these two batholiths, and quite possibly in all the batholiths of North America. This is the time of intrusion of almost all the plutons in the western side of the PRB and most of the smaller plutons in the SNB. The time interval of 100 to about 80 Ma is separated from the prior time interval because in the PRB and SNB the mode of magmatism changed abruptly. In the PRB, there may have been a magmatic lull between 100 and 95 Ma, and at about 95 Ma the axis of magmatism jumped several tens of kilometers east, the average composition changed from tonalite to granodiorite, and the size of the magma chambers increased enormously. This eastern belt of the PRB, which contains the famous La Posta pluton with a surface area of about 1400 square kilometers, extends from Riverside, California, all the way through the state of Baja California (Walawender, 2000). Even more amazing, Walawender (2000) estimates that the volume of this belt, intruded in the short interval between about 95 and 90 Ma is approximately the same as that of the entire western belt that took 20 to 30 My to form! In the SNB the axis of magmatism had already been migrating eastward, and beginning at about 100 Ma unusually large plutons were also emplaced in the eastern crest of the Sierras. This jump in pluton magma chamber size seen in both batholiths corresponds almost exactly to the twofold jump the down-dip in subduction rate seen in Figure 6a at 100 Ma.

The magmatism in the PRB and SNB stopped abruptly at about 90 and 80 My, respectively, even before the down-dip subduction rate jumped to 125 km/My at 75 Ma, and then jumped again to 150 km/My at about 65 My. The traditional explanation for the end of magmatic activity at about 80 Ma in the western batholiths of the US is that such a very high convergence rate caused the North American plate to override the Farallon plate, greatly reducing the subduction angle. Since magma is produced when the subducted plate reaches a depth of about 100 km, the axis of volcanic and plutonic behavior shifted eastward. As the space between the down-going plate and the bottom of the continental lithosphere decreased, the coupling or drag between the two plates increased. The crust was compressed and the Utah, Colorado, and Wyoming area experienced much basement-cored thrust faulting, i.e., the Laramide Orogeny. And, as Engebretson et al. (1985) point out, the rapidly decreasing age of the subducting Farallon plate, as the Farallon-Pacific spreading center approached the trench, was probably also a factor in the production of this orogeny.

Engebretson et al. (1985) offer no explanation for the jumps in subduction rate at about 100 Ma and 75 Ma, but a look at the very important map of Cande et al. (1989) that shows the magnetic lineations/barcode of the world's ocean basins can give us a possible cause. As mentioned earlier, the opening of the "central" Atlantic was due to the rifting of the east coast of North America, from Florida up to Nova Scotia, away from the northwest coast of Africa, and it probably increased the westward motion of the North American plate sometime after 200 Ma. That was the first part of the Atlantic Ocean in the northern hemisphere to open (Fig. 3). The mid-North Atlantic began to open at about 120 Ma, as the section of the east coast of NA from Nova Scotia up to the northern tip of Newfoundland rifted away from the section of the European coast from Spain up to the British Isles. Interestingly, the spreading center bent to the west of the SW coast of Greenland and started to rift it away from Newfoundland. But then at about 60 Ma, the "mid"-Atlantic ridge decided to become more 'mid' and jumped east to begin rifting the east coast of Greenland away from Norway. The creation of the

North Atlantic was complicated and took three rifting events separated by tens of millions of years. Given the fairly close correspondence in time between the sudden increases in Farallon plate subduction rate at 100 Ma and 75–65 Ma and the initiation of the last two rifting events in the North Atlantic, they could be at least a partial cause of those rate jumps.

Local magmatism is not the only way that subduction can add material to a continental margin. The Franciscan Formation in the San Francisco area and its equivalent in the Coast Ranges is a great example of how seafloor sediments and parts of the igneous oceanic crust can be scraped off the subducting plate and form an accretionary wedge. As more and more material is shoved under the wedge, the top of the wedge rises above sea level. A possibly more glamorous way of adding crust to the west coast happened when the subducted oceanic plates were carrying island arcs, detached fragments of continental crust, or oceanic plateaus. When these rock masses were too large or too buoyant to be subducted, they docked and were sutured to the western margin. If these rock masses are sufficiently large they are called exotic or allochthonous terranes. During the Mesozoic, a large part of the Pacific Northwest and Alaska were built by the successive addition of such terranes. Many of the exotic island arcs have sedimentary rocks with tropical fossils, and some have fossils of animals found only in the western Pacific. They are truly far-travelled terranes (remember Figure 5). The topic of allochthonous terranes became very popular in the 1980s. In recent years geologists have questioned how far these terranes have really travelled and consider that some of the island arcs formed not far off the coast. For these they have introduced the term "parautohochthonous" (Colpron et al., 2007).

Figure 7 shows very nicely the strips of allochthonous terranes that were thought to comprise southern California (Howell et al.,1985). You can see that the continental borderland, the Peninsular Ranges, the Transverse Ranges, and even a large strip of the land east of the SAF and extending into SW Arizona were considered exotic terranes. Note that the western and eastern parts of the PRB are shown as being separate terranes. The author and one of

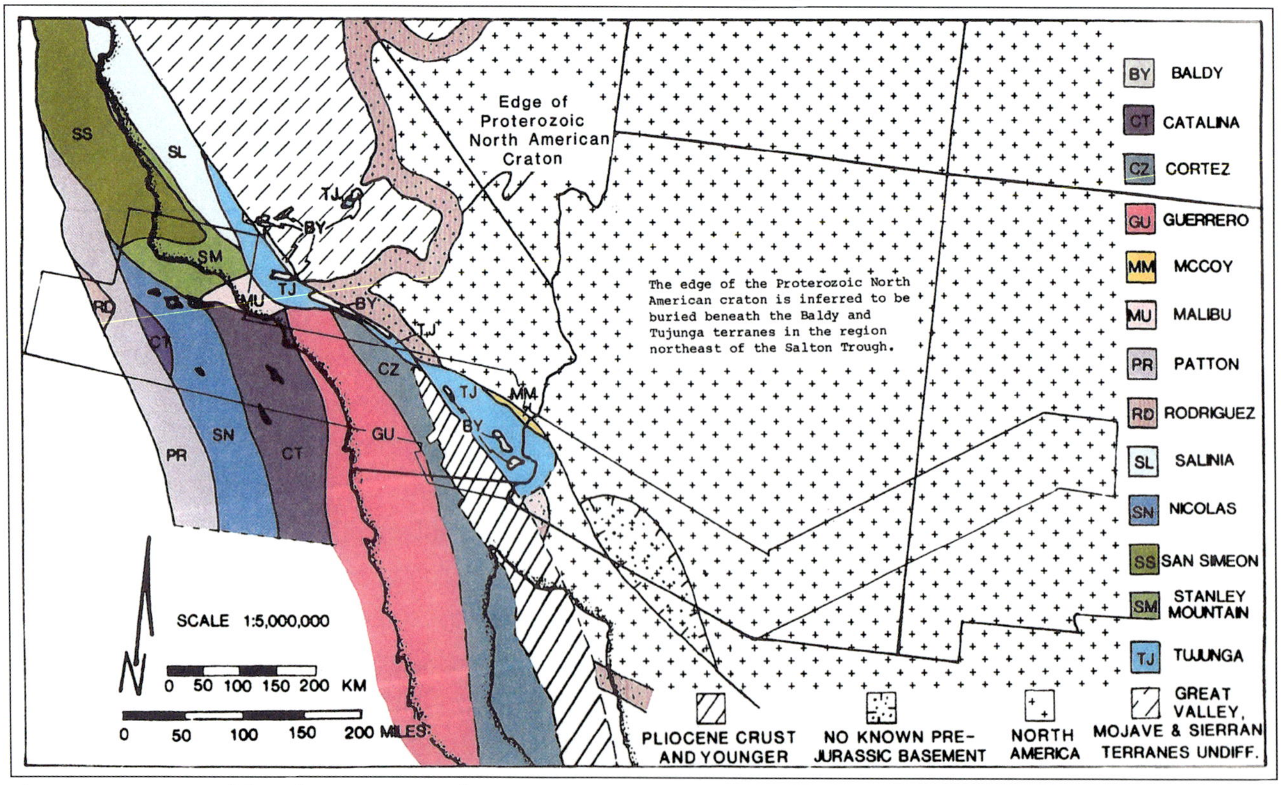

Figure 7. Terranes of Corridor C—3. Map showing possible allochthonous terranes of southern California and northern Baja California and their continental borderlands. (From Howell et al., 1985)

his former grad students, Brad Erskine, are at least partly responsible for this division of PRB terranes. In the course of his paleomagnetic study of the PRB, for his MS degree at SDSU, Brad made two important discoveries: 1) The primary opaque mineral in the western half was magnetite, whereas the primary opaque in the eastern half was ilmenite and there seemed to be no magnetite. Because magnetite is a ferromagnetic mineral, it is capable of having a remanent magnetism, whereas ilmenite is only paramagnetic, so the only reliable paleomagnetic data came from the western plutons. 2) The paleomagnetic data suggested that the western half had migrated more than 1000 km from the south and had rotated 40 degrees clockwise (Erskine and Marshall, 1980; Erskine and Marshall, 1981). Erskine's work was the first to show the existence of a Mt-Ilm line dividing the two halves of the PRB. Other work, such as isotopic ratios, rare earth element patterns, and the very different kinds of inclusions, also showed that the two sides are different (Marshall et al., 2006). Even though some authors have questioned the paleolatitude of the western half, there's no question about the chemical and magmatic differences between the two sides. (Erskine and I still think that the western half is truly allochthonous, and our arguments await a future paper!) The chemistry of the western side shows no sign of the magma passing through continental crust, and the plutons along the coast intrude mostly andesites, so this side apparently formed over oceanic crust (Walawender, 2000). The lithology of the rocks metamorphosed by the intrusion of the eastern plutons shows that they had penetrated the old precambrian crust. The protoliths were precambrian, Paleozoic, and maybe younger, mudstones, sandstones, and carbonates deposited on the continental shelf formed by the break-up of Rodinia, the next supercontinent to be discussed. A final note about the two batholiths discussed in this article. The SNB and PRB are surprisingly similar in many ways: their plutons have essentially the same age range; their average composition, granodiorite, seems to be the same; they both have very large, 100 to 90 Ma plutons at their crests. But there are also some differences: the state aeromagnetic anomaly map doesn't show a magnetite-ilmenite line in the SNB, and the magma chemistry on both sides of the SNB shows that the magmas came up through and interacted with older continental crust (Moore, 2000).

RODINIA, THE PENULTIMATE SUPERCONTINENT

The paleogeography and even existence of pre-Pangean supercontinents remained obscure until the last few decades. One reason for this is because the oldest oceanic crust is only about 160 Ma. The advent of very sensitive isotope measurements, especially U-Pb isotope dating of just parts of single zircon crystals, has begun to shed light on geologic events in the Archean and Proterozoic Eons. In ideal conditions, the ages of rocks several billion years old can now be obtained with a precision, and hopefully accuracy, of several million years. Paleomagnetic studies have also become increasingly sophisticated, and if the original magnetization direction has not been destroyed by heat or metamorphism, a rock's paleolatitude and paleo-orientation, i.e., where paleo-north was, can usually be determined. Paleomagnetism cannot be used to determine paleolongitude, but if two separate terranes have similar ages and paleolatitudes, a study of their rock types and geologic history can at least say whether or not they may have formerly been part of the same continent or supercontinent. So teams of geochronologists, paleomagnetists, petrologists, and geochemists have now proposed two Proterozoic supercontinents. As would be expected, the number of different models and paleogeographies of the component crustal blocks increases with age!

The more recent supercontinent is called Rodinia and it existed from about 1100 Ma to 900 Ma, i.e., during the late Mesoproterozoic to early Neoproterozoic (Condie, 2011). It was formed by a number of continental crustal blocks colliding with North America or Laurentia. The Grenville orogenic belt shown in Figure 2 was formed during this aggregation, when the northwestern part of Europe (Baltica) and a large part of South America (Amazonia) collided with the current east coast of Laurentia at about 1.1 Ga. This orogeny was one of the most intense in the geologic history of the east coast of North America. Many geologists consider that the mountain ranges created by it could've rivaled the present Himalayas. Note too that it extended down through the south and southwest of the US and into Mexico (Fig. 2). When these continent-sized blocks rifted away from our east coast at about 700 Ma, a proto-Atlantic ocean formed, the Iapetus Ocean.

But continental blocks also collided with what is now the west coast of Laurentia. We already saw in Figure 2 that there is almost no land older than about 1 Ga where the western tier of states is now. But that doesn't mean that there was no land there at 1 Ga! In fact, very old rocks occupied this space during the existence of Rodinia. Figure 8 shows a compilation by Bill Dickinson of four of the possible models or paleogeographies of the continental blocks that were there (Dickinson, 2009). Notice that there are no north arrows on these maps, but paleomagnetic data show that what is now the west coast of North America, i.e., Dickinson's "Cordilleran margin," actually trended east-west and was in equatorial latitudes in the late precambrian and early Paleozoic! Paleomagnetic data have shown that continental blocks have not only been translated in N-S directions over geologic time, but have also been rotated. The current western margin and cordillera of North America was not only the *northern* margin of North America at the beginning of the Phanerozoic. According to the reconstruction in Figure 11 it was even the northern margin and at a latitude of 30 degrees N at about 1300 Ma. Interestingly you can see in this figure the continental blocks that we have been calling the "visiting continents," here labeled "Proto-Australia," are just to the north and waiting to crash into the cordillera one or two hundred million years later! For the readers who find it strange or difficult to imagine our current west coast as a north coast, paleomagnetic data show that the North American continent rotated counter clockwise during the Paleozoic so that this coastline was on its west side and oriented basically N-S by Triassic time! Some important points in Figure 8: 1) two or more continents, or parts of continents, were connected to the Cordilleran margin during Rodinia time, 2) land extended for hundreds or thousands of miles north (current west) of what is now the Cordillera, 3) what is now southern California was occupied by either Antarctica (8A, B), Australia (8C), or even Siberia (8D)! Because the Cordilleran margin had been occupied for several hundred million years by possibly four of today's continents, or parts of today's continents, that are now scattered all over the world, I will henceforward call them the "visiting continents" for simplicity!

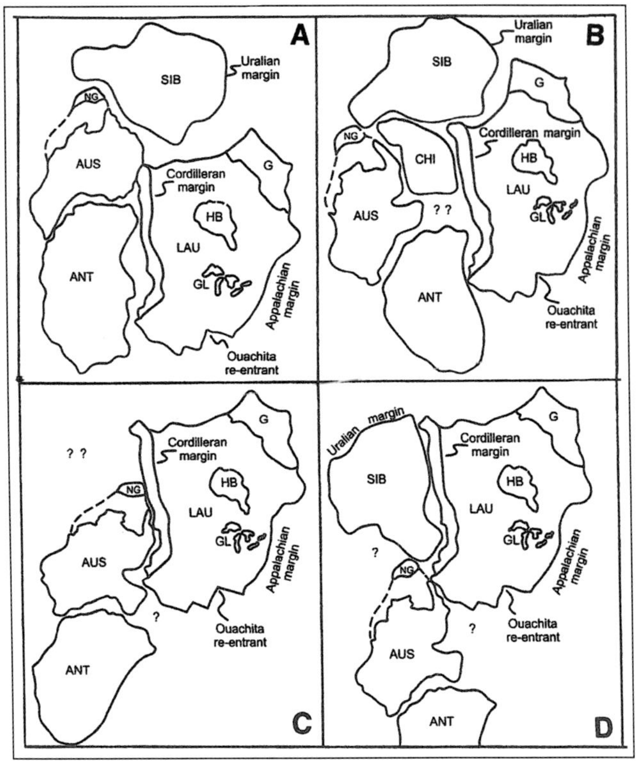

Figure 8. Four Proposed Paleogeographies of Continents That May Have Accreted to the Current Western Margin of Laurentia (LAU). ANT-East Antarctica; AUS-Australia with New Guinea (NG); CHI-South China; SIB-Siberia. Geographic features are: HB-Hudson Bay; GL-the Great Lakes; G-Greenland. Note that a north arrow would point to the left at this time, and the Cordilleran Margin trended E-W and was near the Equator. (Modified from Dickinson, 2009)

So now the questions we can ask are: after the visiting continents rifted away, what shape did the new continental margin and coastline of North America have, and where in the west are they now? Let's take what is happening in the East African Rift Valley as a model for the early stage of continental break-up, and what is known about the geology of the east coast of North America as a model for the end stage, when the fully formed continental margin fronts an ocean basin. But first a note of caution. Just as no two subduction zones are the same, because so many variables are involved, so no two continental rift zones are the same. Kearey et al. (2009) discuss at length the many factors that determine the different modes of continental rifting and whether the rifting will continue to the formation of a new ocean basin, or remains 'incomplete', like the Basin and Ranges.

One of the first signs of rifting is the appearance of basaltic volcanoes and the intrusions of basaltic dikes and sills as normal faults and tension fractures allow basaltic magmas to come up from the mantle. As the crust is stretched and thinned it subsides, forming a chain of shallow, fresh water lakes scattered along a linear topographic depression. When this depression becomes lower than sea level, it is flooded with seawater. The thinning of the continental crust is accomplished by a multitude of normal faults, many of them probably listric. Like the basin and ranges in Nevada and western Utah, there are horsts and grabens, but they are quickly covered by the sediments shed off the highlands on either side of this more or less linear sea. Faulting and intrusion of basaltic magmas probably become more intense as you approach the zone that will become the seafloor spreading center. So the bulk composition of the crust gradually changes from silicic to mafic with time and proximity to the future spreading center. Thus, subsidence is caused not only by the thinning of the continental crust, but also by the increase in its bulk density and the weight of the overlying sediments. This kind of crust is called *mixed crust* and as the percent of basalt increases, perhaps exponentially, away from the coastline, the water depth starts increasing more rapidly. This process results in rifted continental borders that have continental shelves whose slope is only a fraction of a degree, so that the water depth is less than 200 meters out to distances of hundreds of kilometers from the shore. But, at the outer edge of the almost flat shelf, the slope of the seafloor increases to at least several degrees. This part of the continental edge, called the continental slope, continues down to the abyssal seafloor and true oceanic crust. The technical term for the combined continental shelf and slope is the *continental margin*.

The visiting continents are thought to have begun rifting from North America about 800 million years ago. After some millions of years, continental margins should have formed on both the edge (northern, at that time) of North America and on the 'stern' of the retreating continents. They would've been separated by a new, ever-widening ocean, sometimes called the Proto-Pacific Ocean. So, where in today's western North America is this old continental margin? Since the rifting began in precambrian times, we need only to find a map of the western limit of precambrian crust (Figure 9). The heavy line shows the western limits of outcrops of precambrian continental crust, but this crust continues west and forms the basement for thick sections of late precambrian sedimentary rocks. The data from deep oil wells and geophysical surveys suggest that the real western limit is in the vicinity of the dashed line. All the rocks west of the dashed line are considerably younger than 800 Ma—they are the rocks discussed above in Pangea, Part 2. What is now British Columbia, Washington, Oregon, and most of California was occupied by the new Proto-Pacific Ocean, which was floored by newly created ocean crust. The line jogs west into east central California to account for the precambrian rocks in the Death Valley area and those penetrated by magmas in the Sierras,

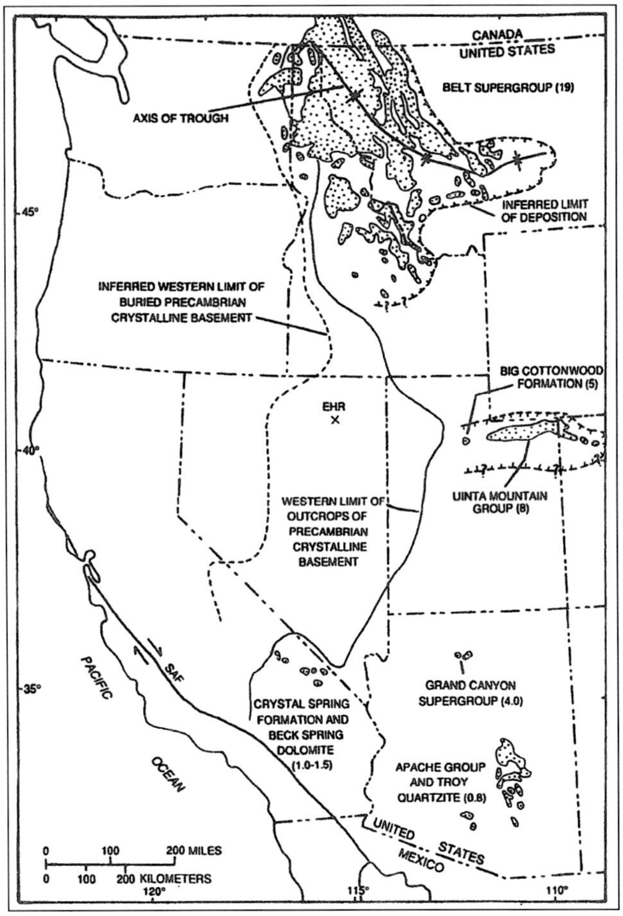

Figure 9. Map of Western North America Showing the Western Limit of Outcrops of Precambrian Basement Rocks—solid line; and the Inferred Western Limit of Precambrian Basement Where It Is Covered by Overlying Younger Rocks—dashed line. Also shown is the distribution and inferred basins of deposition of Middle and early Late Proterozoic sedimentary rocks. Numbers indicate maximum thickness, in kilometers. (From Stewart, 1991. Copied with permission.)

as mentioned above. Where it ran further south is uncertain, but in southern California, the area of this field trip, you can see from the geological map of North America that the San Andreas Fault essentially marks the western limit of precambrian rocks (Fig. 10). The only precambrian rocks to the west of the SAF in this area are the 1.8 to 1.2 Ga gneisses, syenites, and anorthosites of the San Gabriel Mountains. If you restore the 300 km of right lateral slip on the SAF, these rocks would lie next to similar age and composition rocks in the Orocopia and Chocolate mountains and the ranges to the south of the Gila River straddling the United States-Mexico border.

All these early to middle Proterozoic granitic and metamorphic rocks, intruded by Mesozoic plutons, are exposed east of the SAF in southeast California and continue across the desert ranges of southwest Arizona and are buried and hidden by the sedimentary rocks of the Colorado Plateau—except in the bottom of the Grand Canyon.

It comes as a surprise when you first realize that these Precambrian rocks in southeastern California are part of the same group of continental provinces as the almost 2 Ga Vishnu Schist and Zoroaster Granite! Anyone who has rafted down the Grand Canyon gorge knows that the nearly vertical, red walls are composed of hugh thicknesses of carbonate rocks. Remembering that the shelf at this time was located at equatorial latitudes, we shouldn't be surprised to find a thick section of Paleozoic carbonate rocks. But rafters also see a package of tilted Precambrian strata, called the Grand Canyon Supergroup (GCS) that are deposited unconformably on the Vishnu schist and granites and make an angular unconformity with the basal Cambrian Tapeats Sandstone. The oldest part of the GCS, called the Unkar Group, was deposited between about 1200 and 1000 Ma in a crustal downwarp possibly caused by the collision when the visiting continents first arrived. The basin was close to sea level judging by features indicative of shallow marine and subaerial deposition (Hendricks and Stevenson, 2009). The upper part of the GCS, called the Chuar Group, is exposed nowhere else in the western US, and was also deposited in a subaerial and shallow marine environment. But it is between 900 and 700 Ma, so it experienced the breakup of Rodinia. The lower units were most likely deposited in a failed rift or graben and the younger sediments were probably deposited on the newly created continental shelf. According to Ford and Dehler (2003) the Chuar Group is possibly correlative with the Beck Springs Dolomite in Death Valley and the Windermere Group in Montana and Canada (See Figure 9).

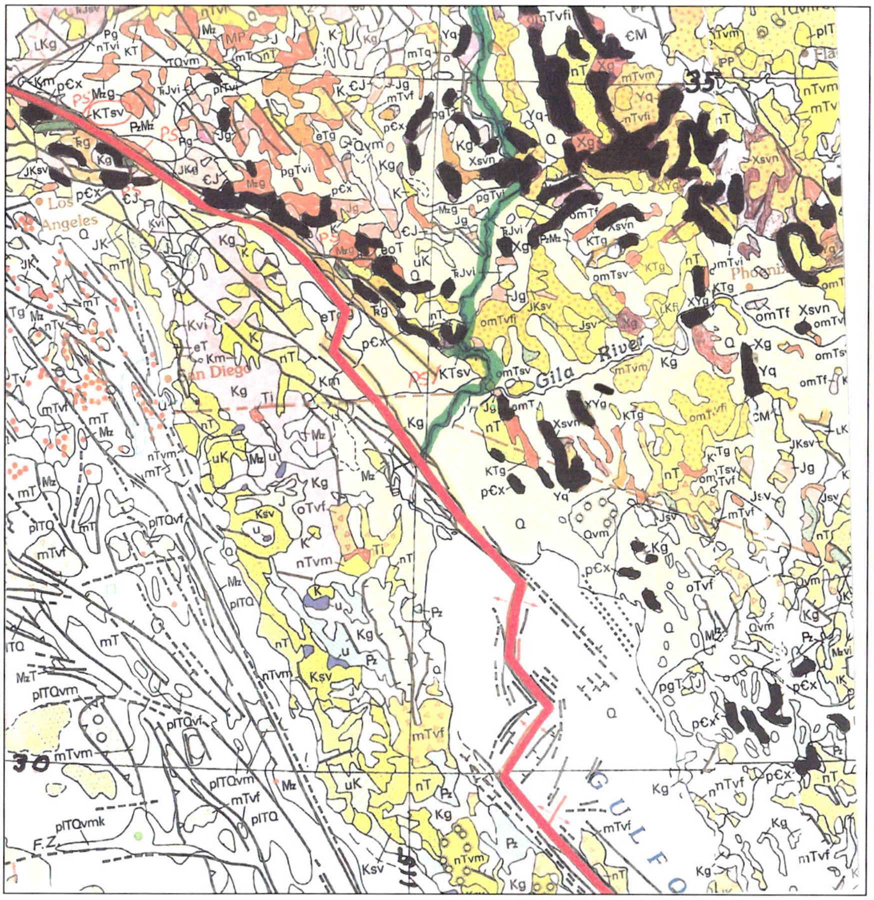

Figure 10. Geological Map of Southern California, Southwestern Arizona, Northern Baja California and Sonora, and the Head of the Gulf of California. Broad red line is the spreading center-transform fault system in the northern Gulf that continues as the San Andreas Fault in southern California. Broad green line is the Colorado River/California-Arizona Border. Blackened areas are Precambrian igneous and metamorphic rocks, mostly of early to middle Proterozoic age. (Geologic Map of North America, 2005, Southern Half, Geological Society of America, Boulder, Colorado)

NUNA OR COLUMBIA, THE OLDEST SUPERCONTINENT

The assembly and breakup of the oldest supercontinent is not as well understood as the later two, but it existed from about 1.9 Ga to about 1.3 Ga, and is called either Nuna or Columbia (Evans and Mitchell, 2011; Condie, 2011). It suffered a long and possibly incomplete breakup, and some parts may have remained together until the formation of Rodinia. One of the most recent paleographic reconstructions is shown in Figure 11 and is for a time late in its life, about 1.3 Ga. Note that present day north is still to the east at this time, and, as mentioned earlier, that our west coast runs E-W and is at about 30 degrees north latitude. Siberia is located next to present day northern Canada, and the Baltic shield and north central Russia are attached to Greenland. The dark strip labeled "Grenville" is the Grenville Province, a part of the current east coast and south of North America that includes the Grenville orogen and the land between it and the Superior Province. The history behind the two names for this supercontinent is that two geologists gave different names to what turned out to be the same assembly of terranes—a single name has yet to be decided upon! Our old friend, the great Canadian geologist, Paul Hoffman used the name "Nuna," because it's Eskimo for "land bordering the northern seas."

He, and other geologists working on the Canadian Shield, realized that the time from about 2.0 to 1.8 Ga was special in the geologic history of North America (Hoffman, 1989). Not only were the many Archean microcontinents in the shield welded or sutured together then, but a series of island arcs and crustal fragments were added to the SW US. Two of these, the Mazatzal and Yavapai orogenic provinces, are shown and labeled in Figures 2 and 11. They were sutured to Wyoming at about 1.78 Ga and make up what was at that time the northwest corner of North America (Fig. 11). Note in this figure a small, unnamed patch of less dark land just north of the Mazatzal

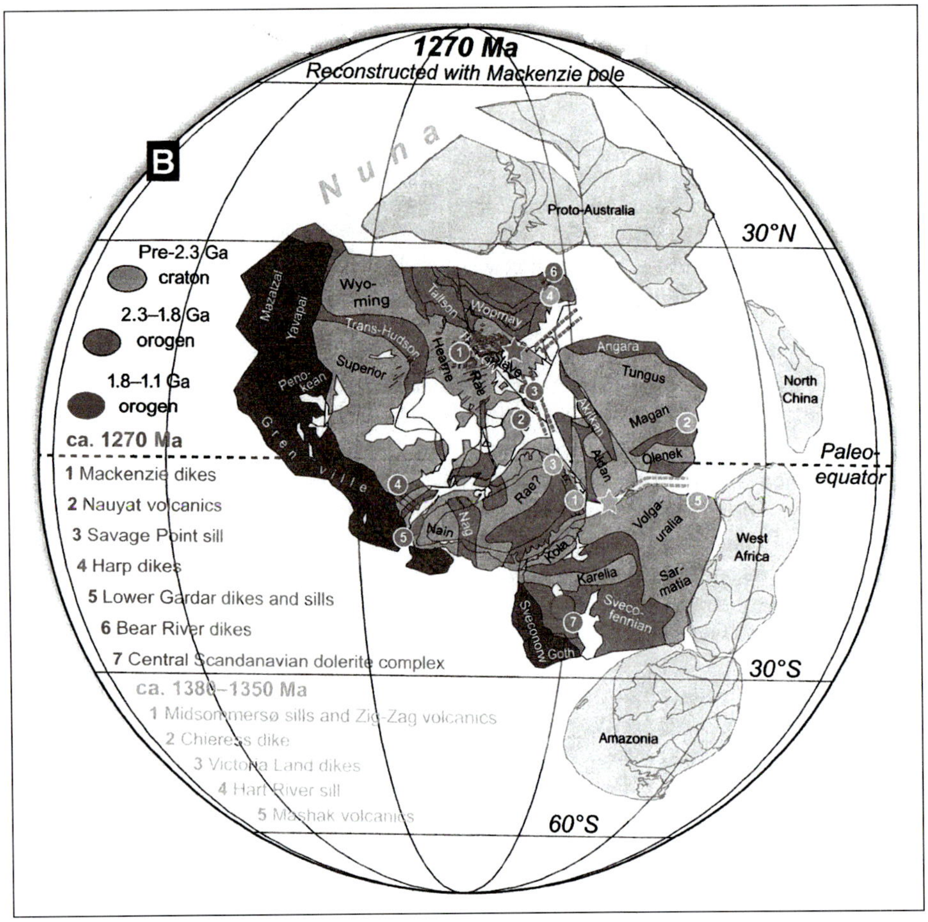

Figure 11. Map of a Possible Paleogeography of Nuna/Columbia Late in its Lifetime, When Breakup Has Already Commenced. Note that at this time (1270 million years ago) the model shows the current western margin of North America as trending E-W and at a paleolatitude of about 30N. (From Evans and Mitchell, 2011)

Figure 12 (below). A Highly Speculative Paleogeographic Reconstruction of the Converging Island Arcs That Possibly Formed the 1.8 Billion Year Old Crust of Much of the Southwest United States. This reconstruction is based on work by geologists at Northern Arizona University and the University of New Mexico. (Courtesy of Ron Blakey, Northern Arizona University, 2008)

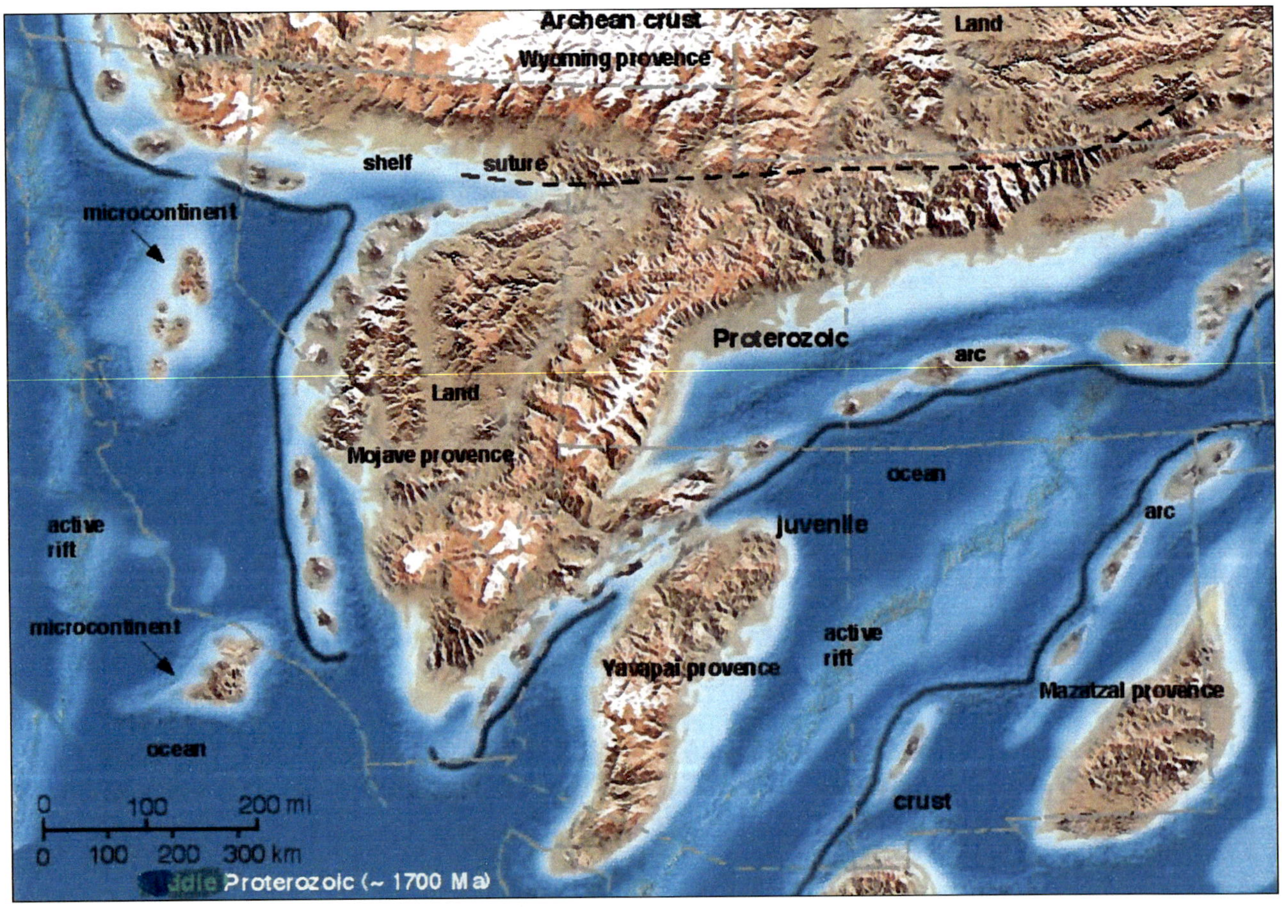

province, which forms the crust of present day SE California and NW Arizona. In Figure 2 the Mojave province isn't even broken out, but it is the SW part of the Yavapai Orogen—the part that lies within the upper left corner of the rectangle. These three provinces formed about 1.8 to 1.7 Ga and have undergone several periods of deformation, probably some of them coincident with their suturing to Wyoming and to one another. Their importance to us is that they form much of the crust of the southwestern US—from SE California through Arizona and New Mexico and extending up into eastern Nevada, Utah, and parts of Colorado. Some geologists think that these provinces were mostly island arcs in a setting like that at present in Indonesia. Figure 12 is a hypothetical view of how the area south of the Archean Wyoming province may have looked just prior to their accretion. Figure 13 shows outcrops of the late Paleoproterozoic granitic and metamorphic rocks of these three provinces/orogenic belts in SE California and northwestern Arizona. The boundary between the Mojave and Yavapai provinces is a 75 km wide zone of mixed affinities (Duebendorfer et al., 2001).

The breakup of Nuna may also have played a role in the deposition of one of the most controversial sections of sedimentary rocks in the Cordillera. Seen at the top of Figure 9 in Idaho and western Montana is the famous Belt Supergroup, which is by far the largest and thickest deposit of Middle and Late Proterozoic sedimentary rocks along the present western edge of the US. According to Stewart (1991), the Belt is about 1.5 Ga, is up to 19 km thick, and rests unconformably on "crystalline basement rocks." If the age is correct, it could have been deposited during the breakup of Nuna. The unusually great thickness strongly suggests that it was deposited either on the outer continental shelf and slope, the miogeocline and eugeocline of Nuna, or that there was another continent attached to the then north coast of Laurentia that was rifting away at this time. In that case, the Belt could have been deposited in a rift basin that was located along the axis of rifting so that the crust it was deposited on gradually lowered until it formed the base of the continental slope. The abyssal floor of today's ocean basins is 15 km deep and the weight of a pile of sediments could isostatically depress it another 4 km. Is the Proto-Australia macrocontinent, lurking just north of Wyoming and Canada at 1270 Ma in Figure 11, the recently rifted culprit in this scenario? If so, it had a love-hate relationship with North America—involving two episodes of aggregation and rifting!

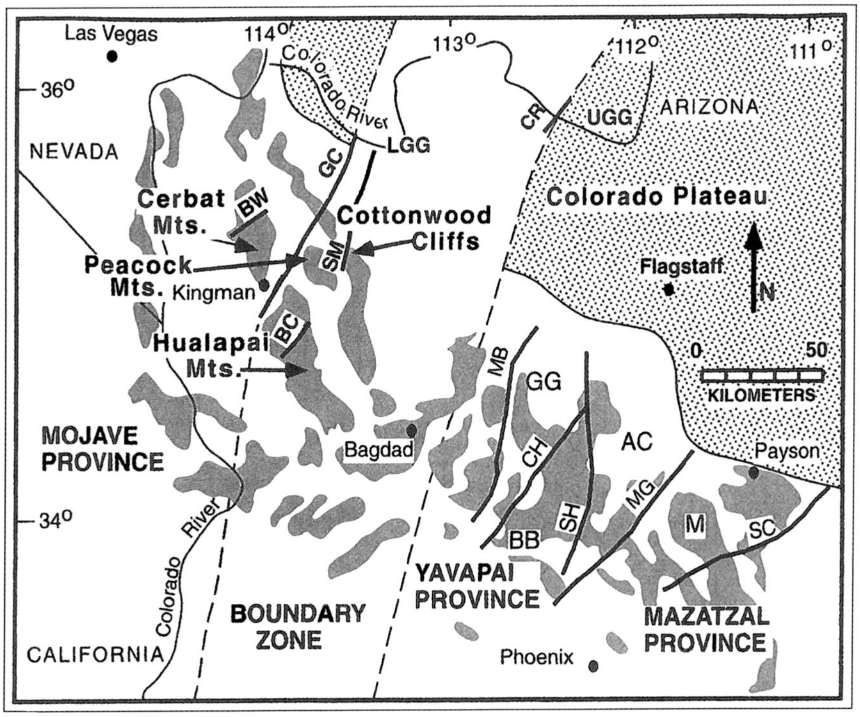

Figure 13. Map of Northwestern Arizona and surrounding states showing the distribution of Proterozoic crustal provinces (the dark patches) and the Mojave-Yavapai boundary zone. LGG-Lower Granite Gorge; UGG-Upper Granite Gorge of the Grand Canyon. Features labeled with double capital letters are tectonic blocks and major shear zones. (From Duebendorfer et al., 2001)

SUMMARY AND CONCLUSIONS

In this article we have seen how much the paleogeography of the earth has changed over the last two billion years. We have also learned that the further we want to go back in time, the more we need even more sophisticated analytical techniques and instruments. We make progress when we make use of the combined data from the fields of geology, geophysics, geochemistry, and even astronomy. The age of the earth, 4.6 Ga, is well-agreed on. But when the first crust formed, whether it was all basalt, all anorthosite, or some other composition is not agreed on. The granitic Acasta Gneiss tells us that there was at least some continental crust by 4.0 Ga. But how it formed is not known. Since zircons are more common in granitic rocks than in mafic rocks, the 4.4 Ga detrital zircons found in west Australia suggest that there may have been some granitic crust formed just 200 million years after the earth itself formed—which would mean that the initially molten earth was cooling very quickly!

We do know that there are about 35 patches of Archean (older than 2.5 Ga) crust scattered around the world's continents (Condie, 2011). Did they form the way most of earth's continental crust forms today, by subduction? The earth's upper mantle was probably some 200 degrees C hotter then than now. If there was subduction, did it operate the same way as it does at present? Basically, the style and even existence of plate tectonics between 4.6 and 2.0 Ga are unknown, and are the topics of much current debate. The margins of those patches of Archean crust show signs of rifting, what we would call today "passive margins". This suggests that there was some agglomeration of these patches into microcontinents or supercratons before 2.0 Ga. But, Nuna/Columbia was apparently the first supercontinent, if we use that word in same sense as for Rodinia and Pangea. Now that we have gone through the details of the three supercontinents, telling their history and impact on the west coast of NA in chronological order may be more meaningful.

The Paleoproterozoic was a time of considerable expansion of Laurentia and one of the biggest additions was the accretion of a series of island arcs, that possibly looked much like Indonesia does today, to the present SW border of North America. Three large island arcs that had formed between about 1.8 and 1.7 Ga were sutured to the Archean Wyoming Province shortly thereafter. They form much of the crust of the present SW United States and their granitic and metamorphic rocks extend from southeastern California to the Rocky Mountains. In SE California they form the crystalline basement/crust from about Death Valley, south through the Mojave Desert and the eastern Transverse Ranges, and down into Sonora, Mexico. In southern California, the San Andreas Fault marks their western limit, except for the San Gabriel Mountains in the central Transverse Ranges.

The breakup of Nuna/Columbia was lengthy and many experts think that some of the continental fragments had not drifted far from Laurentia before they returned to form the next supercontinent, Rodinia, at about 1.2 Ga. The major additions to North America during this event were on the current east and west coasts. The collision of Baltica and Amazonia with the current east coast at about 1.1 Ga added a strip of igneous and metamorphic rocks, called the Grenville Province, that extends from Canada to Florida and lies between the Paleozoic Appalachians and the Archean craton. Which continents were added to the current west side of North America is still debated, but Australia and Antarctica seem the two most likely to have joined the Archean Wyoming Province and the Paleoproterozoic rocks of the former island arcs. This 'western' addition to Laurentia apparently had very different tectonic and magmatic consequences from those on the 'east' coast. The Grenville Province/Orogen comprises highly deformed igneous and metamorphic rocks that apparently are the roots of a Himalayan-type mountain chain. The absence of similar igneous activity in the 'west' is consistent with the "foreign continents" not being far from this coast and therefore having little oceanic crust to subduct. And the lack of tectonic deformation is perhaps due to a less violent collision or even no real suturing. However, the 1200 to 1000 Ma sedimentary rocks that lie unconformably on the 1.8 Ga Vishnu schist and granite in the bottom of the Grand Canyon were perhaps deposited in a part of Laurentia that was down warped by the 'gentle' collision of the foreign continents.

The most important contribution that Rodinia made for us on the west coast of North America was its breakup beginning at about 800 Ma. When the foreign continents rifted away from Laurentia, their departure created a continental margin, complete with a continental shelf and slope. It was on this continental shelf, that trended E-W and was in warm equatorial waters at that time, that a thick layer of late Proterozoic and Paleozoic sediments, including thick sections of carbonates, were deposited. They are probably best exposed in the Grand Canyon and are the rocks that make up the younger section of the Precambrian Grand Canyon Supergroup and the beautiful red-stained limestone layers above them. In the current US southwest, these sedimentary rocks continued to be deposited on the 1.8 Ga basement that formed the continental shelf and slope from late Proterozoic time up into the Mesozoic. Despite several orogenic events, like the Antler and Sonoma in the eastern Cordillera, the traditional view has been that this continental margin was one of quiet deposition during the Paleozoic. The discovery in the western Cordillera of tectonic fragments of island arcs and backarc basins ranging in age from Cambrian to Triassic show that subduction of the newly formed Pacific seafloor probably began as early as Cambrian time. But the tectonic activity and magmatism increased dramatically in the mid-Mesozoic as the last supercontinent began to breakup.

The oldest plutons in the Sierra Nevada and Peninsular Ranges were intruded in late Triassic at 220 Ma. But the major magmatism in California and probably along the whole Cordillera from Mexico to Alaska began in the Cretaceous. Many authors ascribe this increase in magmatic activity to an increase in the westward motion of the North American Plate that was caused by the breakup of Pangea. We saw how the opening of the Atlantic Ocean began with the rifting of the US east coast westward away from NW Africa at 200 Ma. This increase in the westward motion of North America accelerated the eastward subduction rate of the Pacific seafloor and accretion of any island arcs riding on it—causing the Mesozoic magmatic arcs whose rocks occupy much of the western tier of states and provinces in North America.

To better understand the plate tectonics of the Pacific basin and test the subduction rate vs. magmatic activity hypothesis, we used the results of some important computer modeling of the motion of Pacific basin plates by Engebretson et al. (1985). Except for a possible short substitution by the Kula Plate, we found that the plate subducting beneath North America between 175 and 30 Ma was the Farallon Plate. Although they had already compared their computed subduction rates and directions to the magmatic chronology of the SNB, we compared their rates with the plutonic history of both the Sierra Nevada Batholith (Moore, 2000) and the Peninsular Ranges Batholith (Walawender, 2000). We made many of the same observations that they did. The correlation between their calculated subduction rate curves and the timing and volume of magmatic activity was very good! The first time span, from 175 to 120 had an average subduction rate of about 60 km/My. But the subduction rate fell to almost zero between about 135 and 145 Ma, a time that is close to the center of a magmatic lull observed in both batholiths. No plutons were produced from about 155 to 125 Ma. The average subduction rate in the next time span, from 120 to 100 Ma, was slightly higher (65 km/My), but it remained at that rate for the entire time span. And this period was highly productive. Most of the plutons in the western half of the PRB as well as their remaining volcanic cover were intruded/extruded during this time. Many of the SNB plutons were also emplaced at this time. At about 100 Ma there was a dramatic increase in subduction rate and, at the same time, a dramatic increase in the volume of magmatism! The subduction rate during this interval was 100 km/My. In the PRB, the arc jumped several tens of kilometers to the east and unusually large plutons were produced. In the SNB the arc had already been migrating slowly east, but at 100 Ma very large plutons also began to be emplaced along the eastern crest of the Sierras. But this pulse of magmatic activity lasted for only 10 My in the PRB and for 20 My in the SNB. The 5 to 15 Ma time difference between the end of magmatism in these two batholiths and any appreciable change in the subduction rate of the Farallon plate, i.e.,the jump to a 125 km/Ma rate at 75 Ma is interesting. The apparently premature end of plutonic activity in the two batholiths between 80 and 90 Ma could be associated with the substitution of the Kula plate at 85 Ma (Figure 6). Another possibility is that, after producing so much magma between 100 and about 85 Ma, the mantle at 100 km simply became infertile—the water from the down-going slab could no longer produce or flux any granitic magmas from the basaltic mantle wedge.

From 75 to 30 Ma the average subduction rate of the Farallon plate, which was entirely dip-slip, was about 115 km/My, but the 'instantaneous' rate increased from 100 to 150 km/My during about the first third of this interval and then fell to 50 km/My during the last half. The drastic jump in subduction rate at 75 Ma marks for many geologists the beginning of the Laramide Orogeny. This increase in the rate of convergence between the North American and Farallon or Kula plates caused the upper plate to over-ride and flatten the upper part of the down-going plate—a process now called "slab/plate rollback." As the North American plate advanced westward, the part of it over the magma generation zone moved to the east, and so did the magmatic activity at the surface. But, in light of the discussion above, west coast magmatism ended some millions of years before slab rollback and therefore not because of it.

So, in this limited set of data on subduction rates and volume of magma generated, we have seen a good correlation between the two—good and reasonable enough to propose that among the variables that influence magma production, slab subduction rate is probably the most important. We don't have enough examples to give a value for what the minimum constant subduction rate for magma production might be. We can say that increasing the subduction rate from about 60 km/My to 100 km/My increases the magma production rate, which then increases pluton size and changes the crystallization process in the magma chamber. The exact value of the maximum subduction rate for magmatic activity is also somewhat unclear. It may fall in the range between 100 and 125 km/My. A value of 100 km/My can certainly produce magma at a very high rate, but possibly only for a limited time, like about 10 My. Ignoring the effect of the age of the downgoing slab, a subduction rate of about 125 km/My may initiate slab rollback and change the locus of the magmatic activity and possibly reduce the magma generation rate.

In conclusion, the times of supercontinent assembly seem to be both the tectonically most active periods in earth's history and the times of most continental crustal growth. The oceanic crust between the blocks is subducted, causing magmatic arcs on the edges of the continents, as well as bringing in and suturing allochthonous island arcs, oceanic plateaus, and anything else too large to subduct. When the continental crust of the approaching continents finally collides, mountain ranges like the Himalayas are formed and the deformation and faulting penetrate far into the continents. The second most tectonically active time is likely during the break-up of supercontinents. The crust is stretched and thinned, rift valleys are formed, and basaltic magmas can come up to the surface. When rifting results in a new seafloor spreading center, new oceanic crust is produced that on the next continental aggregation will fuel the creation of more continental crust.

Finally, the cause or causes of continental aggregation and split up are still unknown. A popular hypothesis for the cause of supercontinental split up has to do with the effect of gravity on a slight doming of the supercontinent. In this model, the temperature of the upper mantle starts to rise under a very large block of continental crust, both because of the heat generated by the radioactive elements in the continental crust, and because the crust acts as an insulating blanket that inhibits heat escaping from Earth's interior. As the temperature of the continental mantle increases, the mantle expands and raises the crust. The force of gravity then pulls on the domed crust from all sides and it breaks up and the pieces slide away from one another. Once again, gravity could be at the heart of plate tectonics!

You can see that we still have a lot to learn about how the Earth works. We have also learned much since the time of Hutton and Lyell. Hutton realized that the great thicknesses of sedimentary rocks and the grand scenery we see today were produced by the same slow geological processes we observe acting at present; he needed time, lots of time! Lord Kelvin, one of the greatest physicists of his day, thought he had disproved Hutton's theory of uniformitarianism when he calculated the age of the earth at 20 Ma to 40 Ma. Wouldn't both of them be surprised to learn just how much time there really has been, and how much Earth's surface really has changed!

ACKNOWLEDGEMENTS

The initial incentive for writing this overview article came from Jeff Miller when he asked me to give a brief overview talk at the restaurant high up on Mt. San Jacinto at the end of his SCGS 2012 field trip. When my talk was over he said, "By the way, could you write up some of the main points of your talk—just a few pages?" Well, Jeff, that simple request provoked many weeks of work and reflection on my part, and I hope that you and all those who read this will enjoy and learn from what you started. I thank Dave Bloom for his driving on this trip and for his encouragement and good advice while I was struggling with this paper—and thanks to his secretary for her excellent job of typing Table 1. The good times in the field and stimulating and insightful conversations at lunch with both Dave and Bill Elliott have kept my mind and body active in my retirement years!

REFERENCES CITED

Atwater, T., 1970, Implications of plate tectonics for the Cenozoic tectonic evolution of western North America: Geological Society of America Bulletin, v. 81, p. 3513–3536.

Bally, A.W., Scotese, C.R., and Ross, M.L., 1989, North America; Plate-tectonic setting and tectonic elements: *in* Bally, A.W., and Palmer, A.R., eds., The Geology of North America—An overview: Geological Society of America, The Geology of North America, v. A, p. 1–17.

Cande, S.C., LaBrecque, J.L., Larson, R.L., Pitman III, W.C., Golovchenko, X., and Haxby, W.F., 1989, Magnetic lineations of the world's ocean basins: Lamont-Doherty Geological Observatory of Columbia University, Contribution No. 4367, published by the American Association of Petroleum Geologists, Tulsa, OK.

Condie, K.C., 2011, EARTH as an Evolving Planetary System, 2nd Edition: Elsevier, San Diego, 574 p.

Copron, M., Nelson, J.L., Murphy, D.C., 2007, Northern Cordilleran terranes and their interactions through time: Geological Society of America GSA Today, v. 17, p. 4–10.

Dickinson, W.R., 2009, Anatomy and global context of the North American Cordillera: *in* Kay, S.M., Ramos, V.A., and Dickinson, W.R., eds., Backbone of the Americas: Shallow Subduction, Plateau Uplift, and Ridge and Terrane Collision: Geological Society of America Memoir 204, p. 1–29.

Duebendorfer, E.M., Chamberlain, K.R., and Jones, C.S., 2001, Paleoproterozoic tectonic history of the Cerbat Mountains, northwestern Arizona: Implications for crustal assembly in the southwestern United States: Geological Society of America Bulletin, v.113, p. 575–590.

Engebretson, D.C., Cox, A., and Gordon, R.G., 1985, Relative motions between oceanic and continental plates in the Pacific Basin: Geological Society of America Special Paper 206, 59 p.

Erskine, B., and Marshall, M., 1980, A paleomagnetic and rock magnetic investigation of the northern Peninsular Ranges batholiths, Southern California (abs.): EOS, Transactions, American Geophysical Union, v. 61, p. 948.

Erskine, B.G., and Marshall, M., 1981, Magnetite and Ilmenite series sub-melts in the northern Peninsular Ranges batholith: their association with magnetic stability and implications on magmatic source regions, (abs.): EOS, Transactions, American Geophysical Union, v. 62, p.849.

Evans, D.A.D., and Mitchell, R.N., 2011, Assembly and breakup or the core of Paleoproterozoic—Mesoproterozoic supercontinent Nuna: Geology, V. 39, P. 443–446.

Ford, T.D. and Dehler, C.M., 2003, Grand Canyon Supergroup: Nankoweap Formation, Chuar Group, and Sixtymile Formation: *in* Grand Canyon Geology, Second Edition: Beus, S.S. and Morales, M., eds., Oxford University Press, New York, NY.

Hendricks, J.D. and Stevenson, G.M., 2003, Grand Canyon Supergroup: Unkar Group: *in* Grand Canyon Geology, Second Edition: Beus, S.S. and Morales, M., eds., Oxford University Press, New York, NY.

Hoffman, P.F., 1989, Precambrian geology and tectonic history of North America, *in* Bally, A.W., and Palmer, A.R., eds., The Geology of North America—An overview: Geological Society of America, The Geology of North America, v. A, p. 447–512.

Howell, D.G., Gibson, J.D., Fuis, G.S., Knapp, J.H., Haxel, G.B., Keller, B.R., Silver, L.T., and Vedder, J.G., 1985, Centenial Continent/Ocean Transect #5; C-3 Pacific abyssal plain to Rio Grande rift: Geological Society of America.

Irwin, W. P., 1990, Geology and Plate-Tectonic Development, *in* The San Andreas Fault System, California: USGS Professional Paper 1515, U. S. Government Printing Office, Washington, D. C., p. 60–80.

Kearey, P., Klepeis, K.A., and Vine, F.J., 2009, Global Tectonics, Third Edition: Wiley-Blackwell, Hoboken, NJ, 494 p..

Marshall, M., Mace, N., Erskine, B., Jansen, E., 2006, What Regional and Local Paleomagnetic Data Tell Us about the Tectonics of the Coyote Mountains, Southwest Imperial Valley, Southern California: *in* Bloom, D.M., and Ziegler, C.L., eds., Geology and history of Southeastern San Diego County, California: p. 55–62.

Miller, R.B., and Snoke, A.W., 2009, The utility of crustal cross sections in the analysis of orogenic processes in contrasting tectonic settings: *in* Miller, R.B., and Snoke, A.W., eds., Crustal cross sections from the western North American Cordillera and elsewhere: Implications for tectonic and petrologic processes: Geological Society of America Special Paper 456, p. 1–38.

Moore, J. G., 2000, Exploring the Highest Sierra: Stanford University Press, Stanford, California, 442 p.

Needy, S.K., Anderson, J.L., Wooden, J.L., Fleck, R.J., Barth, A.P., Paterson, S.R., Memeti, V., and Pignotta, G.S., 2009, Mesozoic magmatism in an upper- to middle-crustal section through the Cordilleran continental margin arc, eastern Transverse Ranges, California: Geological Society of America Special Paper 456, p. 187–218.

Stewart, J.H., 1991, Latest Proterozoic and Cambrian rocks of the Western United States—An overview: *in* Cooper, J.D., and Stevens, C.H., eds., Paleozoic Paleography of the Western United States-II: Pacific Section SEPM, v.67, p.13-38. (Used with permission.)

Tollo, R.P., Corriveau, L., McLelland, J., and Bartholomew, M.J., 2004, Proterozoic tectonic evolution of the Grenville orogen in North America: An introduction: *in* Tollo, R.P., Corriveau, L., McLelland, J., and Bartholomew, M.J., eds., Proterozoic tectonic evolution of the Grenville orogen in North America: Geological Society of America Memoir 197, p. 1–18.

Walawender, M. J., 2000, The Peninsular Ranges - A Geological Guide to San Diego's Back Country: Kendall-Hunt Publishing Company, Dubuque, Iowa, 123 p.

Wilson, J.T., 1965, A new class of faults and their bearing on continental drift: Nature, v. 207, p. 343–347.

Geologic Setting in the Vicinity of San Gorgonio Pass

Jeffrey A. Miller
South Coast Geological Society
President, 2012

E2ManageTech, Inc.
5001 Airport Plaza Drive, Suite 260
Long Beach, California, 90815

Whatever your background, be you a long-in-the-tooth geologist, an outdoor enthusiast or a big eyed gangly kid, the San Gorgonio Pass Region offers many geo-adventures. To say the geologic story in the vicinity of San Gorgonio Pass is complicated is like saying Palm Springs gets hot — understated! Provided herein is a summary of the geologic setting and history of the San Gorgonio Pass Region.

REGIONAL SETTING

A drive through San Gorgonio Pass along Interstate 10 reveals an area of stark contrasts. The pass itself is at an elevation of 1,591 feet above sea level, located in a roughly east-west trending saddle between the San Bernardino Mountains to the north and the San Jacinto Mountains to the south. Mount San Gorgonio, in the San Bernardino Mountains 12 miles to the north of San Gorgonio Pass, is the highest peak in southern California at 11,503 feet above sea level. San Jacinto Peak, in the San Jacinto Mountains 7 miles to the south of San Gorgonio Pass, is the highest peak in the San Jacinto Mountains at 10,834 feet above sea level. Each peak is approximately 2 vertical miles above the pass! This is due to the convergence of the two mountain blocks resulting

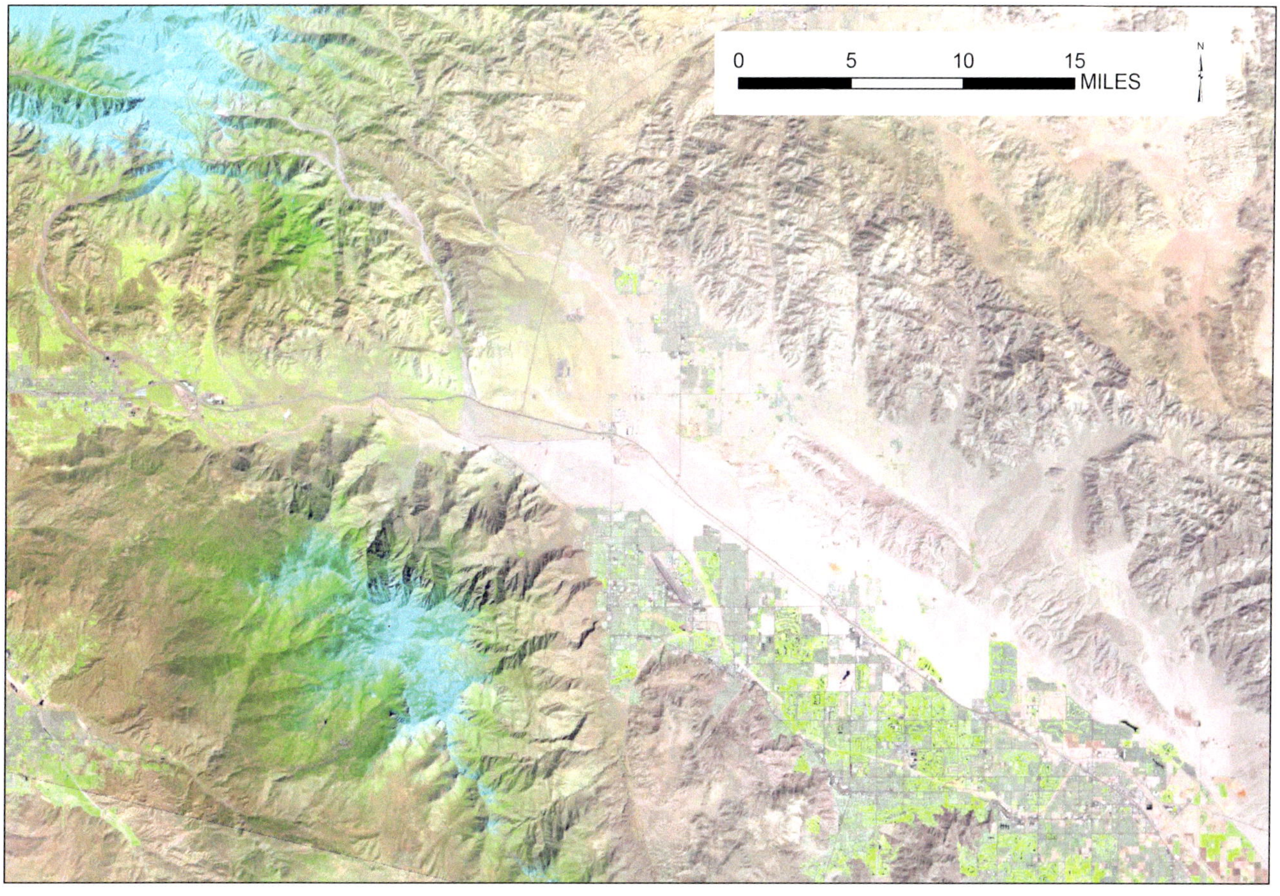

Figure 1. Overview Map. Source: Landsat 8 OLI Level 1 March 2025 from earthexplorer.usgs.gov.

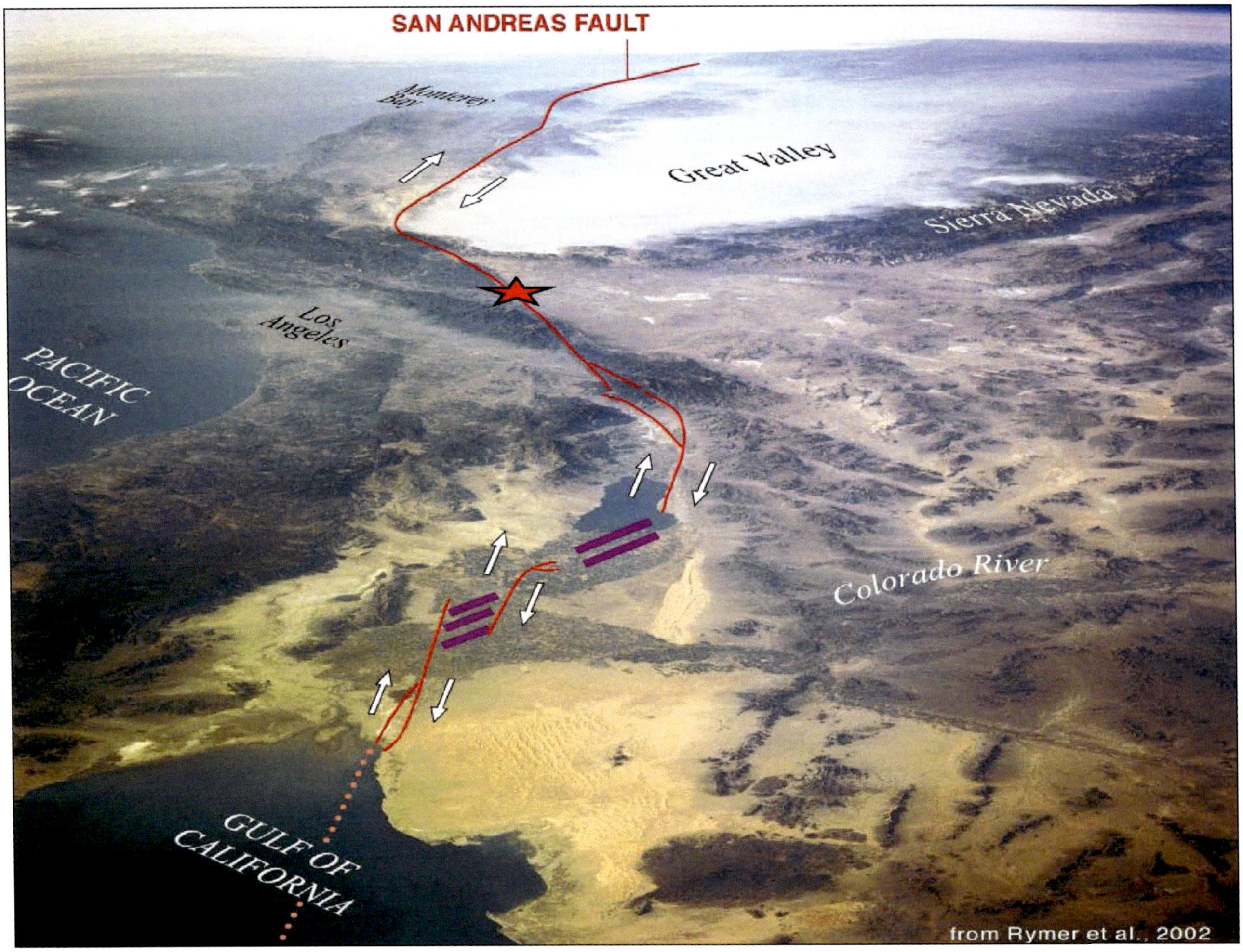

Figure 2. San Andreas Fault, Pacific Plate and North American Plate

from the northwestward movement of the Pacific Tectonic Plate relative to the North American Plate on either side of the through-going San Andreas Fault System. (San Andreas Fault System activity is discussed in greater detail later in this paper.)

> **"*Geomorphology:* a science that deals with the relief features of the earth … and seeks a genetic interpretation of them."**
>
> —*Webster Dictionary*

The San Gorgonio Pass is situated at the convergence of three of California's geomorphic provinces: 1) The Peninsular Ranges to the south and southwest as seen most clearly in the adjacent San Jacinto Mountains, 2) the Transverse Ranges to the north and northwest as seen most clearly in the San Bernardino Mountains, and 3) the Colorado Desert to the east, home to Palm Springs and the desert cities. Each of the three provinces has been developed based on distinctly different trends.

The Peninsular Ranges Province to the south is typified by a series of northwest-southeast trending mountain ranges and valleys extending from the San Gorgonio Pass south into Baja California that have formed roughly sub-parallel to faults associated with the San Andreas Fault System including the San Jacinto Fault Zone and the Elsinore Fault Zone. Though the surface expression and elevation profiles share commonalities with mountain ranges found along the coast in central California, the geology of these mountains share more in common with the plutonic core of the Sierra Nevada Batholith. A view south of the San Gorgonio Pass clearly shows the plutonic rocks at the core of Mount San Jacinto.

The Transverse Ranges Province to the north is typified by an east-west trending group of mountain ranges and valleys extending from the Santa Ynez Mountains in the west to the San Bernardino Mountains of the San Gorgonio Pass Region in the east. The name of the province itself is derived from the unusual east-west trend of these ranges in comparison to the typical north-south structural trend of California as evidenced in the Coast Ranges-Central Valley-Sierra Nevada mountains and the Peninsular Ranges. The Transverse Ranges have developed as a result of the reorganization of the Pacific Plate/North American Plate Boundary from a converging margin to a transform boundary. This boundary is often conceptualized as a finite planar boundary between the two massive plates represented by the San Andreas Fault Zone. However, in the Transverse Ranges (and other regions) this boundary is more appropriately represented by a number of blocks accommodating the northwestern motion of the Pacific Plate relative to the North American Plate. For example, this accommodation is expressed in the western Transverse Ranges by the Santa Ynez Mountains which show evidence of over 100 degrees of clockwise rotation (coincidental with the opening of the Los Angeles Basin) and in the eastern Transverse Ranges by the uplifting of the San Bernardino Mountains in transpression.

The Colorado Desert to the east is a low lying basin, near and below sea level. The Colorado Desert includes the Salton Sea, located approximately 40 miles southeast of the San Gorgonio Pass. The southern terminus of the San Andreas Fault System is located in the vicinity of the Salton Sea. The low elevation therein is the result of the opening of a transform basin, and possibly the creation of a new spreading system, that may ultimately extend the Gulf of California northward as the Pacific Plate continues its northwestward progression.

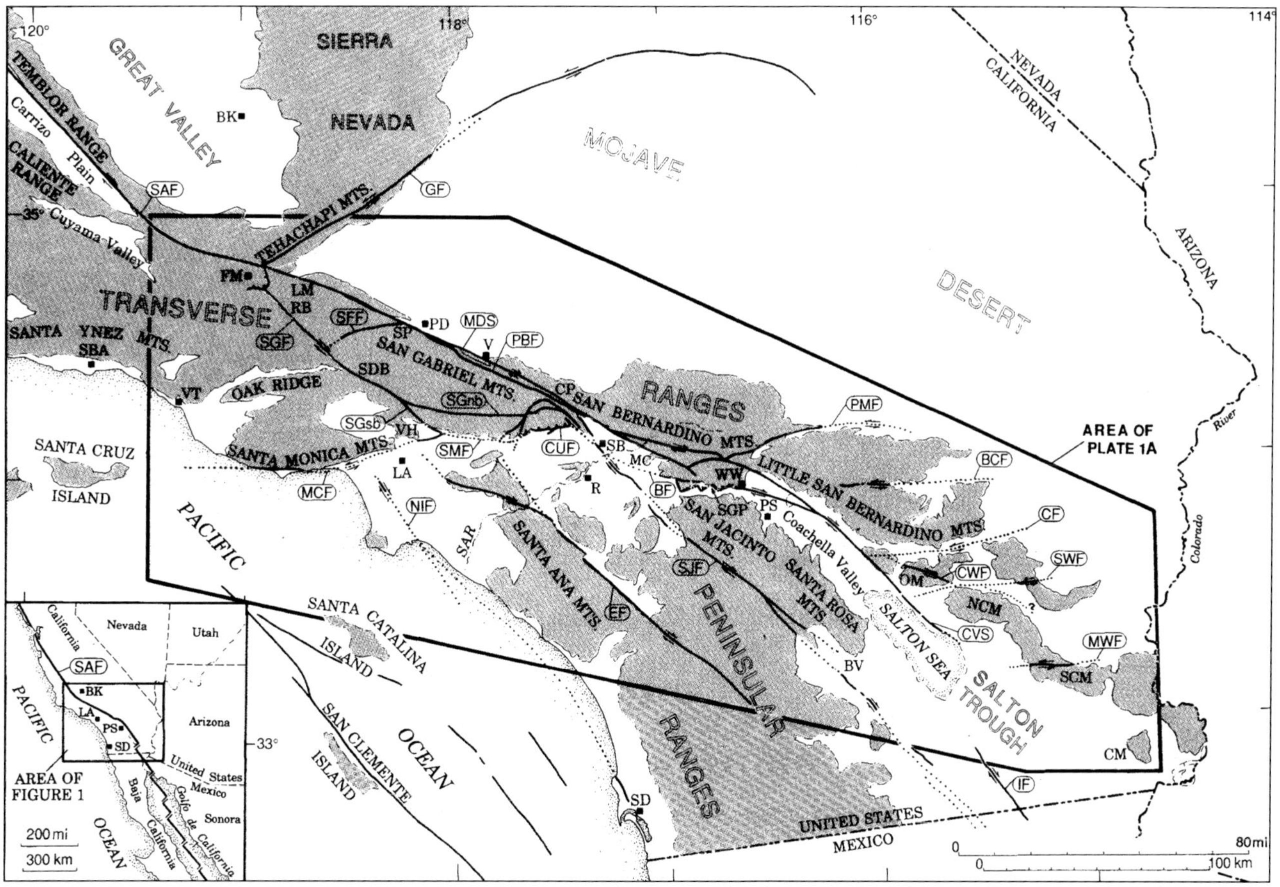

Figure 3. Index Map Showing Geographic Features and Faults

> **"Lithology: the character of a rock formation; also, a rock formation having a particular set of characteristics."** —*Webster Dictionary*

LITHOLOGY

As should be expected during any visit along the San Andreas Fault, the lithology on one side of the fault doesn't correspond very well to the lithology on the other side. This is primarily due to the approximate 300 kilometers (roughly 200 miles) of offset along the fault. In order to gain an understanding of this, imagine having a conversation with a friend in front of you that takes place as the San Andreas Fault moves. In a gross-oversimplification, by the end of your conversation (which would take about 20 million years) your friend who started across from you would now be just west of Bakersfield. Obviously, we do not live long enough to carry out this experiment, but the principle is the same; rocks once adjacent have been displaced hundreds of kilometers, and the rocks now adjacent did not originate next to each other. Because of this, the subsequent section provides information on the lithology of the San Jacinto Mountains Basement, the San Bernardino Mountains Basement and the sedimentary rock sequence found in the San Gorgonio Pass.

SAN JACINTO MOUNTAINS BASEMENT

As mentioned above, the core of the San Jacinto Mountains is similar plutonic material to the Sierra Nevada Batholith. These plutonic bodies formed during the late Mesozoic (approximately 250 to 65 million years ago). However, prior to their emplacement the area was covered by ocean and an extensive sequence of sedimentary rocks had formed. The record of the sedimentary rocks is partial as the uplift of the San Jacinto Mountains has resulted in erosion of approximately 10 to 20 miles of overburden. The rocks that remain have been metamorphosed as a result of burial after deposition and intrusion of the batholith itself. Provided below is a description of the metamorphosed sedimentary rocks (metasedimentary rocks) and the batholithic rocks that remain.

Metasedimentary units of the San Jacinto Mountains consist of a wide array of rocks including schists, gneisses, phyllites, hornfels, metaquartzite and dolomitic and calcareous marbles. Prior to metamorphism, clay, silt, sand, and calcium rich sediments accumulated on the seafloor, were buried and lithified into mudstones, siltstones, sandstones and limestones during the Paleozoic (approximately 250 to 540 million years ago) and late Proterozoic (greater than 540 million years ago). The ages are estimated based on geochemical age dating evidence from the San Jacinto Mountains and fossils found elsewhere in peninsular ranges. Eventually, high temperature and pressure related to the intruding batholith metamorphosed these sedimentary rocks into the metasedimentary rocks identified above. In metamorphosing the rocks, the sedimentary rocks were recrystallized, folded and distended such that original bedding and features were not preserved (Matti and Morton, 2000).

During the Mesozoic, the subduction of the Farallon Plate resulted in the partial melting of overriding rocks and emplacement of plutonic rocks above. These plutonic bodies were emplaced as hot magma but solidified within the crust. The exposed plutonic rocks within the San Jacinto Mountains appear to have been emplaced within a seven million year span approximately 100 million years ago, though there may be rocks not exposed of varying ages (Matti and Morton, 2000).

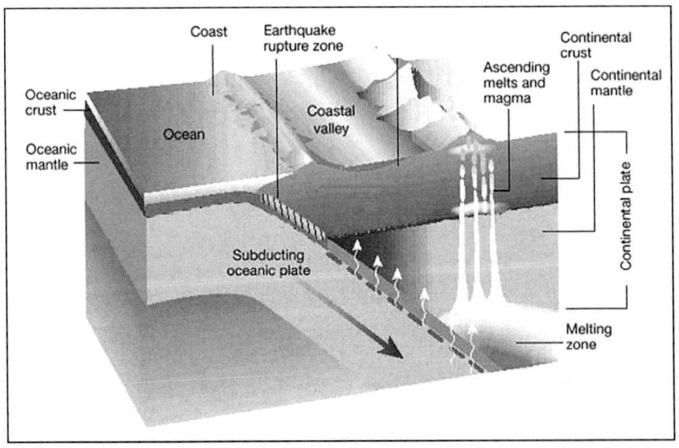

Figure 4. Diagram Showing Subduction Zone Geology.

SAN BERNARDINO MOUNTAINS BASEMENT

In contrast to the San Jacinto Mountains, the San Bernardino Mountains include Proterozoic (greater than 540 million years old) crystalline rocks in addition to metasedimentary and batholithic rocks. The oldest rocks are granitoid and gneissic rocks collectively known as the Baldwin Gneiss. These rocks are primarily found in three different expressions:

1. well foliated to compositionally layered granitic gneisses including augen gneiss characterized by large rounded crystals of potassium feldspar and other quartzofeldspathic minerals probably plutonic in origin,
2. bodies of texturally massive to foliated equigranular to porphyry, probably plutonic in origin, and
3. compositionally layered gneiss that is well-foliated and is probably metasedimentary.

In various locations these rocks are cut by pegmatitic and quartz-rich dikes. The plutonic rocks of this group are at least 1.75 billion years in age based on geochemical dating, and the gneissic rocks they intrude are older, though age cannot be ascertained (see Steno's Law on Cross-Cutting Relationships).

Steno's Principle of Superposition:

"At the time when any given stratum was being formed, all the matter resting upon it was fluid, and, therefore, at the time when the lower stratum was being formed, none of the upper strata existed."

Steno's Principle of Original Horizontality:

"... strata either perpendicular to the horizon or inclined to it were at one time parallel to the horizon."

Steno's Principle of Lateral Continuity:

"Materials forming any stratum were continuous over the surface of the Earth unless some other solid bodies stood in the way."

The Principle of Cross-Cutting Relationships:

"If a body or discontinuity cuts across a stratum, it must have formed after that stratum."

A thick sequence of metasedimentary rock crops out locally in the San Bernardino Mountains. In the eastern part of the range, basal units of these rocks rest unconformably on top of the Baldwin Gneiss. The rocks consist generally of a lower quartzite sequence and an upper carbonate sequence, deposited in shallow marine environments on the continental shelf. The metaquartzite sequence ranges in age from late Proterozoic to early Cambrian and the carbonate sequence ranges in age from early Cambrian through Pennsylvanian (Matti and Morton, 2010).

SAN GORGONIO PASS REGION SEDIMENTARY SEQUENCE

Whereas the basement of the San Jacinto and San Bernardino Mountains consist of plutonic and metamorphic rock of Mesozoic and older age, the sediments and sedimentary rocks exposed in the vicinity of San Gorgonio Pass range from recent alluvium to the upper Miocene (approximately 10 million years). These sedimentary rocks were deposited in marine and non-marine environments, including the current depositional regime of alluvial fans derived from the San Jacinto and San Bernardino mountains on either side of the San Gorgonio Pass. Lava flows and small intrusive bodies are found sporadically in the sedimentary sequence.

The surface of this lowland is covered by various generations of alluvial-fan deposits that mainly are derived from the San Bernardino Mountains. The foothills of the San Bernardino Mountains are underlain by upper Cenozoic nonmarine and marine sedimentary deposits that from oldest to youngest include:

1. Hathaway Formation: sandstone and conglomerate,
2. Imperial Formation: marine mudstone and sandstone,
3. Painted Hill Formation: sandstone and conglomerate,
4. Cabazon Fanglomerate: conglomeratic sandstone and conglomerate,

Formation	Map symbol	Age	Maximum thickness (feet)	Lithology
Alluvium	Qal	Recent		Unconsolidated stream gravels and low stream terrace gravels
----------UNCONFORMITY----------				
Landslide deposits	Qls	Recent		
----------UNCONFORMITY----------				
Undifferentiated terrace gravels	Qt	Quaternary		Terrace gravels and canyon-fill material of doubtful correlation
----------UNCONFORMITY----------				
Burnt Canyon breccia	Qbc	Quaternary	100+	Dissected landslide deposit
----------UNCONFORMITY----------				
Heights fanglomerate	Qh	Quaternary	500	Tan to dark-brown, ill-sorted conglomerate, dominantly dark clasts of gneissic rock
----------UNCONFORMITY----------				
Cabezon fanglomerate	Qc	Quaternary	1500+	Gray to tan ill-sorted conglomerate, rich in clasts of pegmatitic and granitic rocks
----------UNCONFORMITY----------				
Deformed gravels of Whitewater River	Qd	Quaternary(?)		Gray to tan, ill-sorted conglomerate, rich in clasts of pegmatitic and granitic rocks
-------CONTACT NOT EXPOSED-------				
San Timoteo (?) Formation	Pst	Pliocene-Pleistocene	1800+	Tan to gray sandstone, siltstone, and conglomerate, rich in light-colored clasts of granitic rocks; interbedded fresh-water limestone
-------CONTACT NOT EXPOSED-------				
Painted Hill Formation	Pp Ppc Ppb Ppd	Middle Pliocene (?)	3400+	Light-gray sandstone and conglomerate, rich in clasts of granitic and volcanic rocks; Ppc, resistant conglomerate bed; Ppb, interlayered flows of olivine basalt; Ppd, associated dikes
Imperial Formation	Pi	Lower Pliocene (?)	300	Tan to yellow marine sandstone, siltstone, and shale; gypsiferous; fossiliferous
Hathaway Formation				
Upper member	Phu	Lower Pliocene (?)	650	Gray, massive, ill-sorted conglomerate and breccia, rich in clasts of augen and flaser gneiss
Lower member	Phl	Lower Pliocene (?)	1100+	Light-gray sandstone, siltstone, conglomerate, and fresh-water limestone
-------CONTACT NOT EXPOSED-------				
Coachella fanglomerate				
Upper member	Mcu Mcb Mcd Mcm	Upper Miocene (?)	3750+	Massive conglomerate, locally rich in clasts of basalt and dark-green saussuritized gneiss; Mcb, interlayered flows of olivine basalt; Mcd, associated dikes; Mcm, marker horizon containing distinctive clasts
Lower member	Mcl	Upper Miocene (?)	850	Well-indurated massive conglomerate, locally rich in clasts of olivine basalt; basal breccia of gray schist fragments

Figure 5. Sequence and Character of the Sedimentary Rocks of San Gorgonio Pass

5. Heights Fanglomerate: conglomeratic sandstone and conglomerate, and
6. Quaternary Alluvium: alluvial fan deposits.

Each of these sedimentary units has been deposited contemporaneously with the evolution of the San Andreas Fault System. As a result, the units are themselves faulted and folded (Matti and Morton, 2010). A correlation of units based on age is presented in Figure 6.

STRUCTURAL COMPLEXITY

The San Andreas Fault has been active in southern California for approximately 10 million years. At times the San Andreas Fault, as it went through the San Gorgonio Pass, took on the form of a purely strike-slip fault, much like the San Andreas Fault in the lower Colorado Desert to the east. However, as the San Jacinto Block has converged with the Transverse Ranges, over that 10 million year span the trace of the San Andreas Fault System has shifted up to 10 miles north and south as the plate boundary has been reorganized.

Because the system has been reorganized many times there are a number of faults mapped at the surface as part of the San Andreas Fault System. Below is a list of mapped faults and strike-slip displacements:

- Banning Fault 10 ±5 million years ago (unknown displacement),
- Wilson Creek Fault 5–4 million years ago (44 kilometers),
- Mission Creek Fault 4–1 million years ago (89 kilometers),
- San Jacinto Fault 1.5 million years ago to present (25 kilometers),
- San Gorgonio Pass Fault 1.5 million years ago to present (thrust fault; displacement unknown),
- Mill Creek Fault 0.5–0.25 million years ago (8 kilometers),
- San Bernardino Strand 0.1 million years ago to present (3 kilometers),
- New Banning Strand 0.1 million years ago to present (3 kilometers), and
- Garnet Hill Strand 0.1 million years ago to present (unknown).

Figures 8 to 13 show the locations of these various strands and are highlighted when active

	San Timoteo badlands (Frick, 1921; Axelrod, 1950; English, unpubl.)	Western San Gorgonio Pass (this report)	Eastern San Gorgonio Pass (this report)	Indio and Mecca Hills (Dibblee, 1954)
Pleistocene	Bautista fm.	Heights fgl. Cabezon fgl.	Cabezon fgl. Deformed gravels of Whitewater River	Ocotillo cgl.
Upper Pliocene	San Timoteo fm.	San Timoteo(?) fm.		Palm Spring fm. and Canebrake cgl.
Middle Pliocene	Mt. Eden fm.	Painted Hill fm.	Painted Hill fm.	
Lower Pliocene		Imperial fm. Hathaway fm.	Imperial fm.	
Upper Miocene	Jackrabbit ss.		Coachella fgl.	Imperial fm. Mecca f.

Figure 6. Generalized Correlation of Sedimentary Rocks

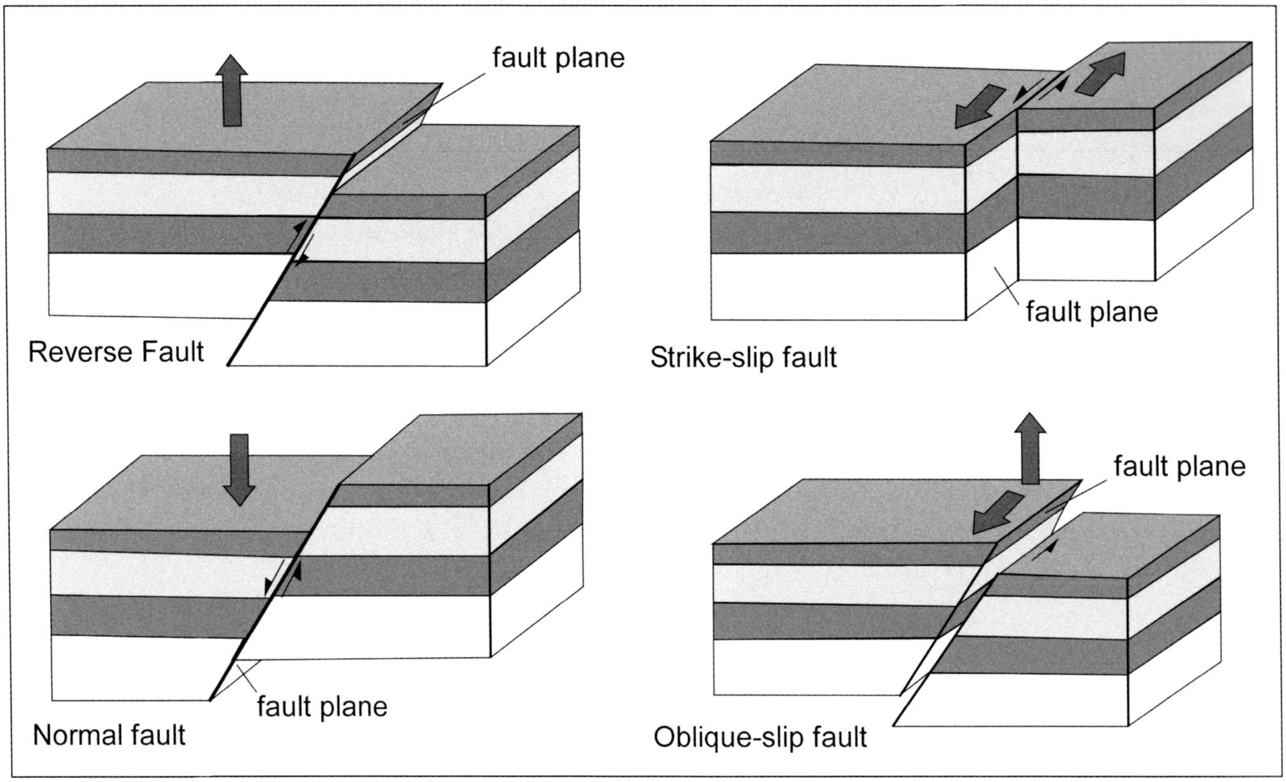

Figure 7. Principal Fault Types

(Matti 1993). As plainly evidenced by these figures, the complexity of the system in the vicinity of San Gorgonio Pass is very complicated. Of critical importance is the variance from the long planar strands of the San Andreas Fault System visible to the east and west of the San Gorgonio Pass as compared to the shorter strands intersecting obliquely in the San Gorgonio Pass.

STRUCTURAL MODEL UNDERLYING THE PASS

The underlying reason for the structural complexity discussed in the previous section is the rigidity of the two mountain blocks bounding the San Gorgonio Pass. Whereas the overall trend of the San Andreas Fault System is roughly northwest-southeast in orientation, the rigid mass of the Transverse Ranges is oriented in an east-west manner. As a result the relative northwest motion of the Pacific Plate has resulted in the San Jacinto Mountains converging with the San Bernardino Mountains. As discussed above, this is partially responsible for the drastic elevation of the peaks in the vicinity of San Gorgonio Pass. As the north-south cross-section shows, the San Jacinto block is actually underthrusting the San Gorgonio Pass and the San Bernardino Mountains (Figure 14).

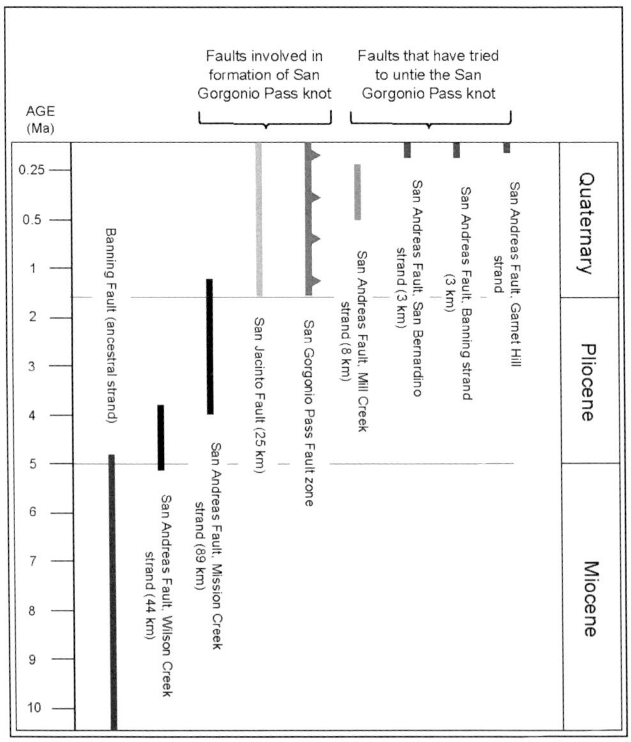

Figure 8. Fault Activity Sequence

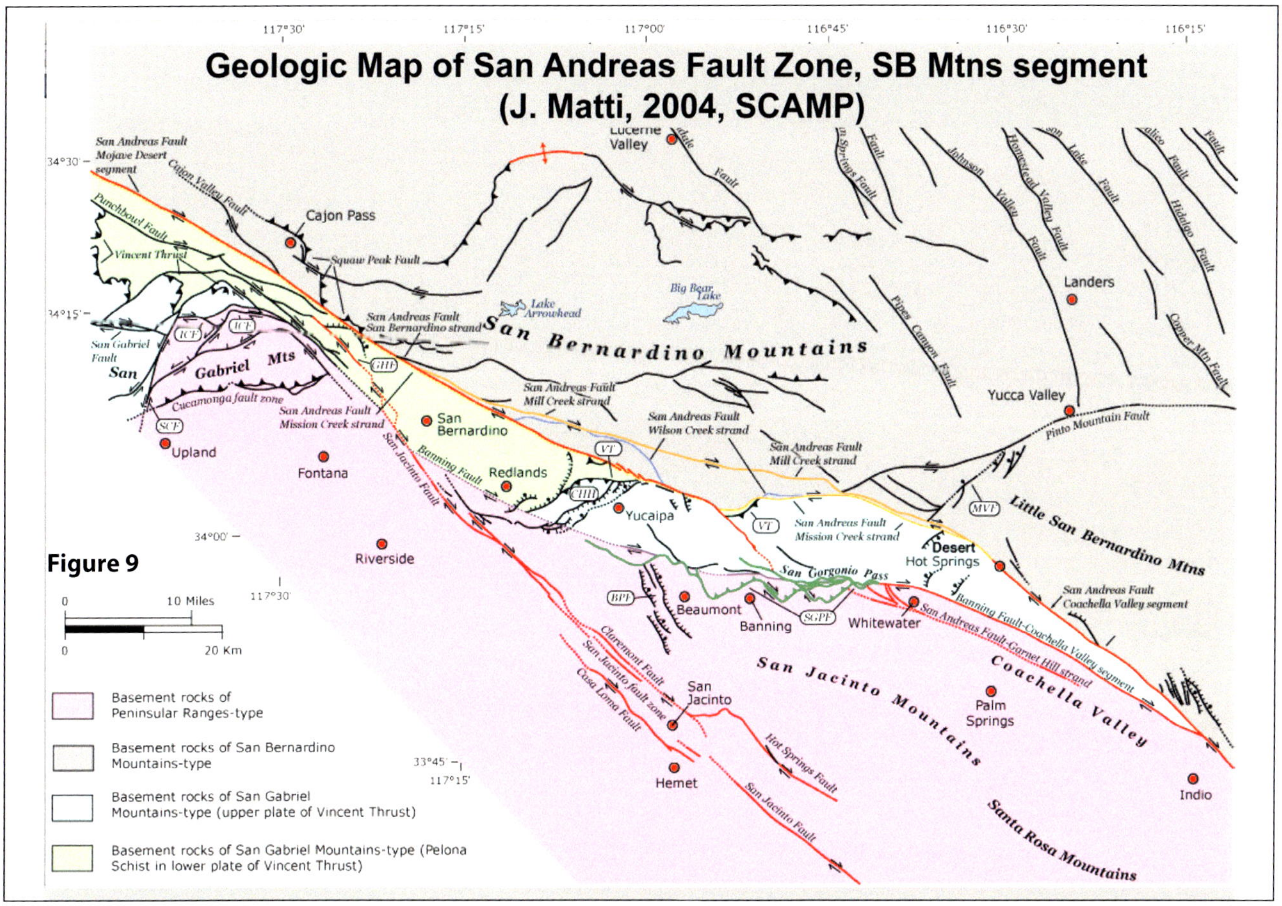

Figure 9

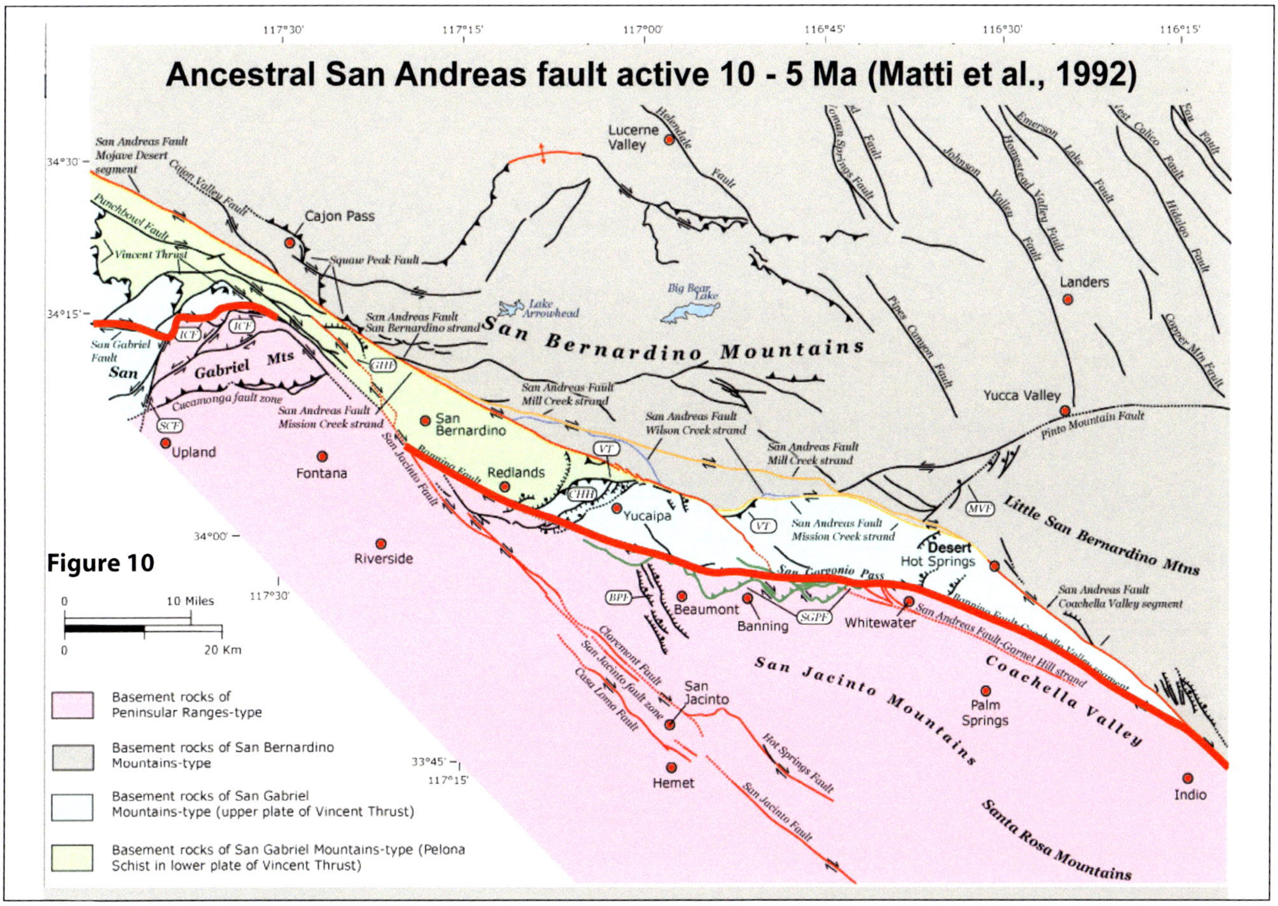

Figure 10

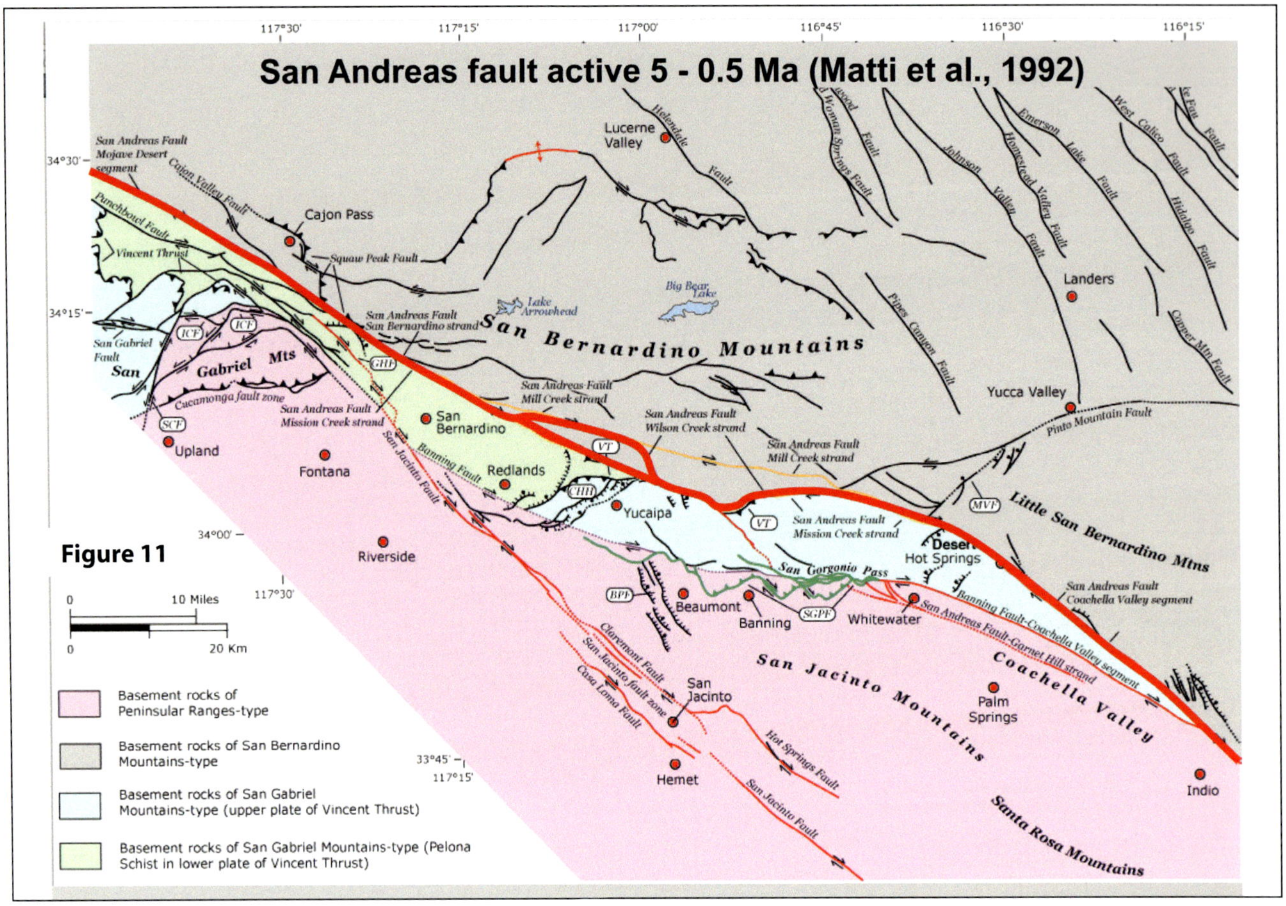

Figure 11

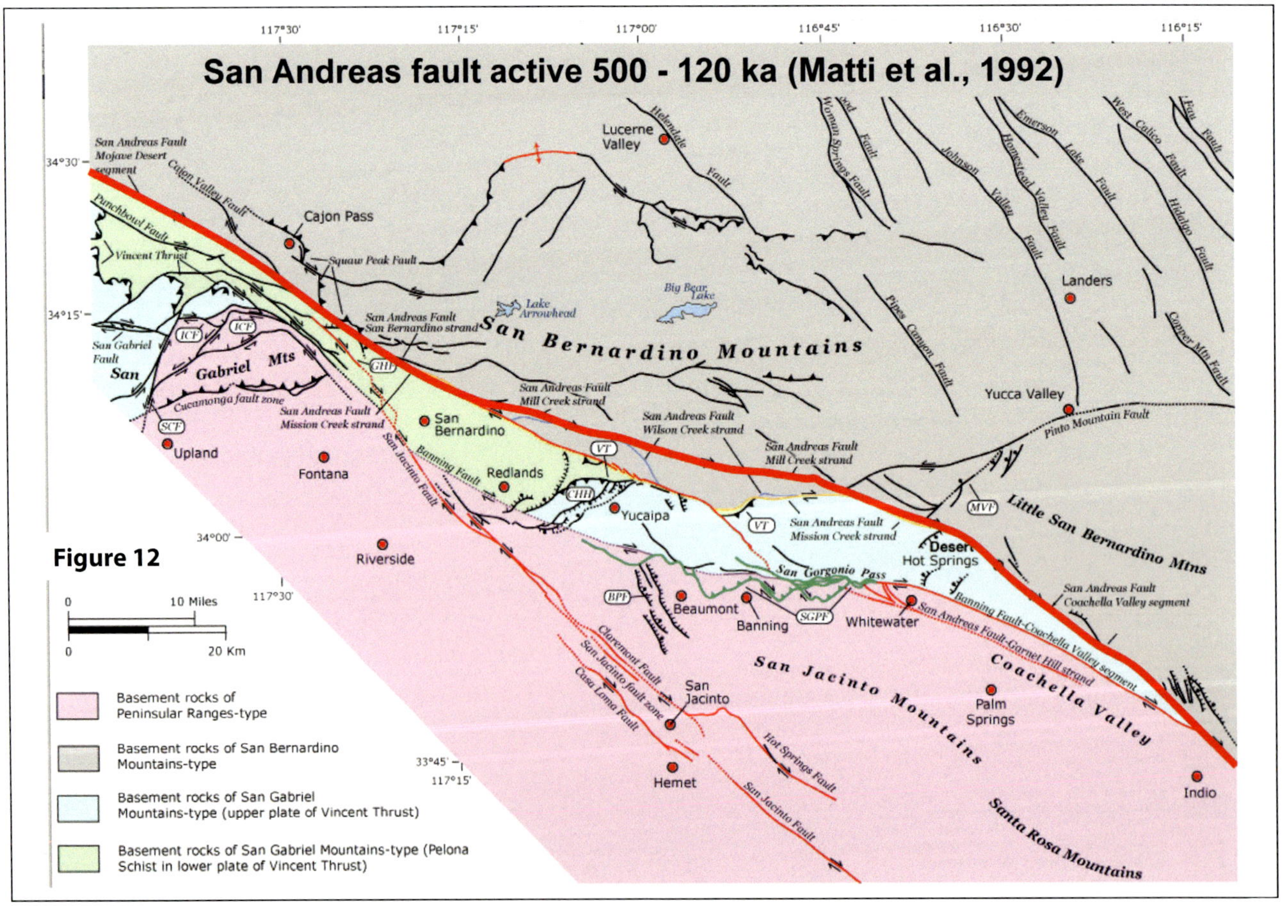

Figure 12

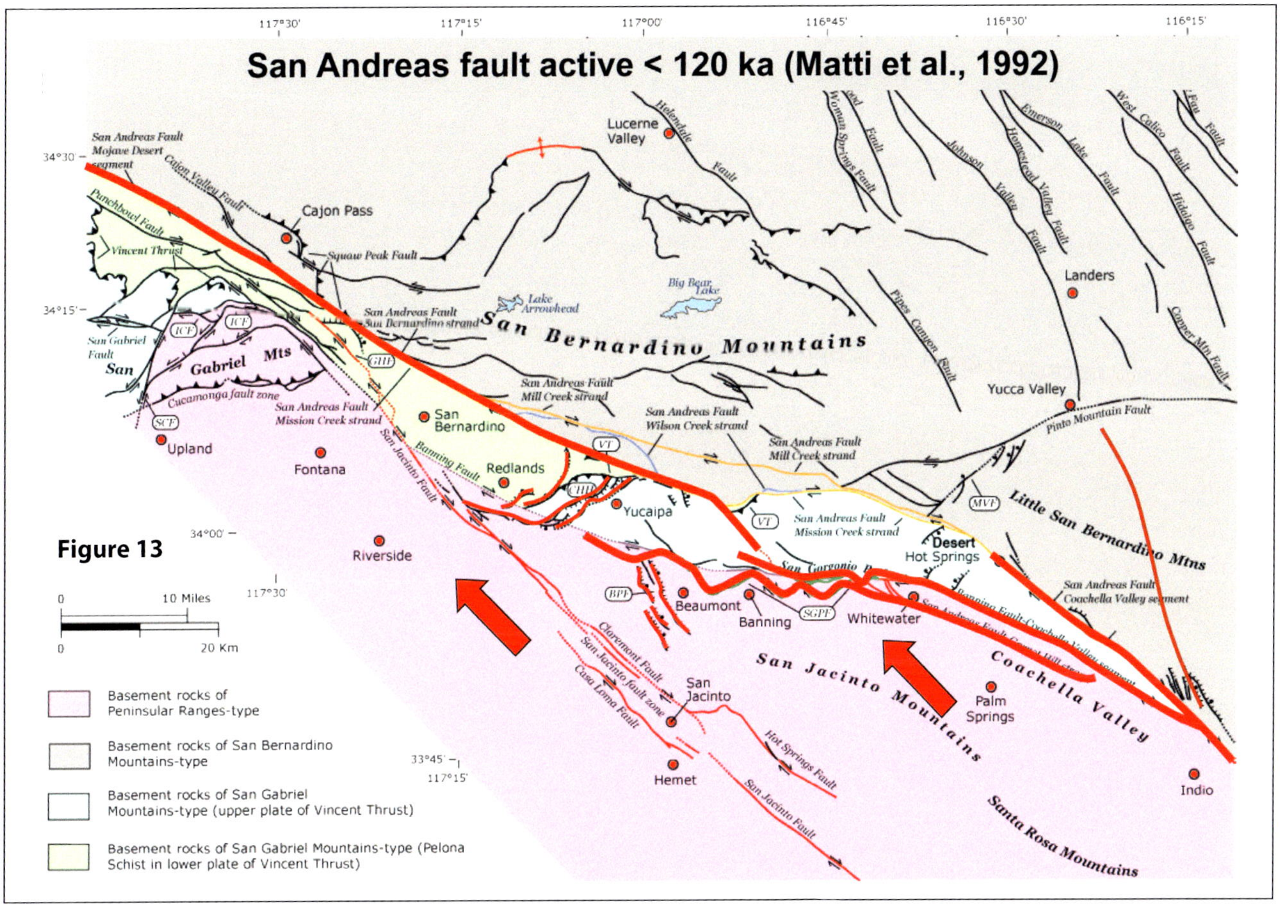

Figure 13

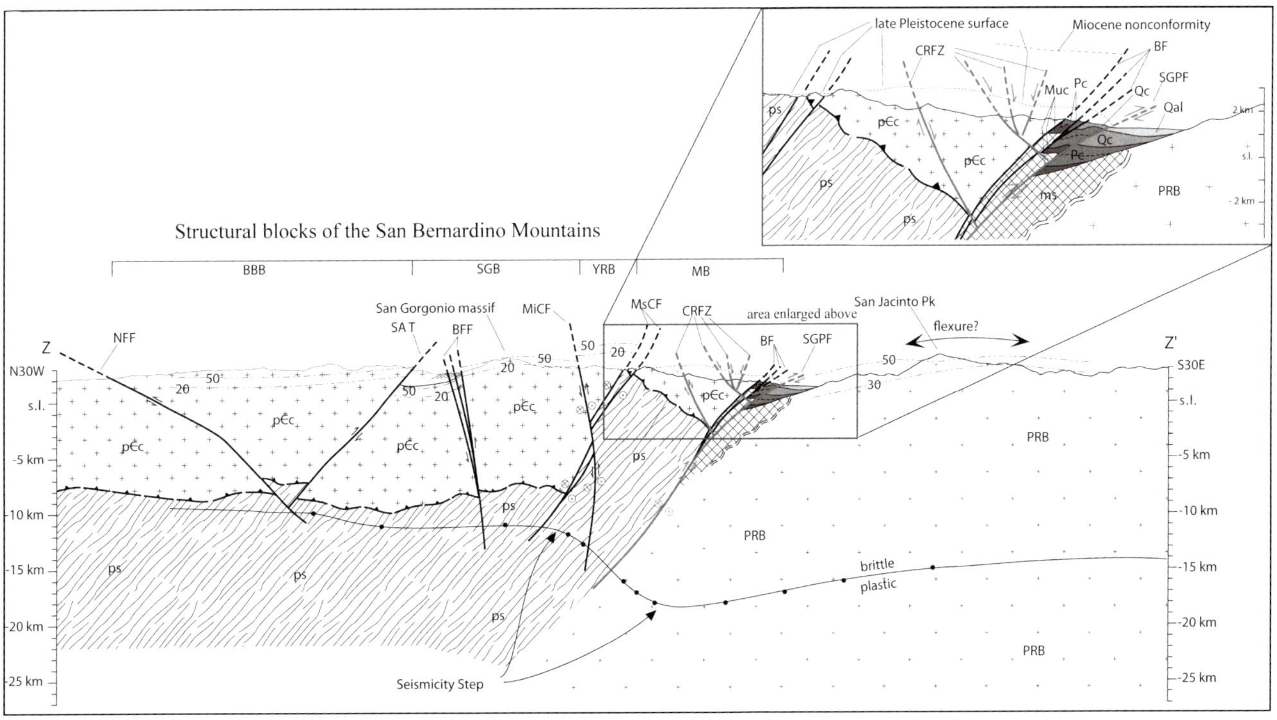

Figure 14. North-South Cross-section through San Gorgonio Pass

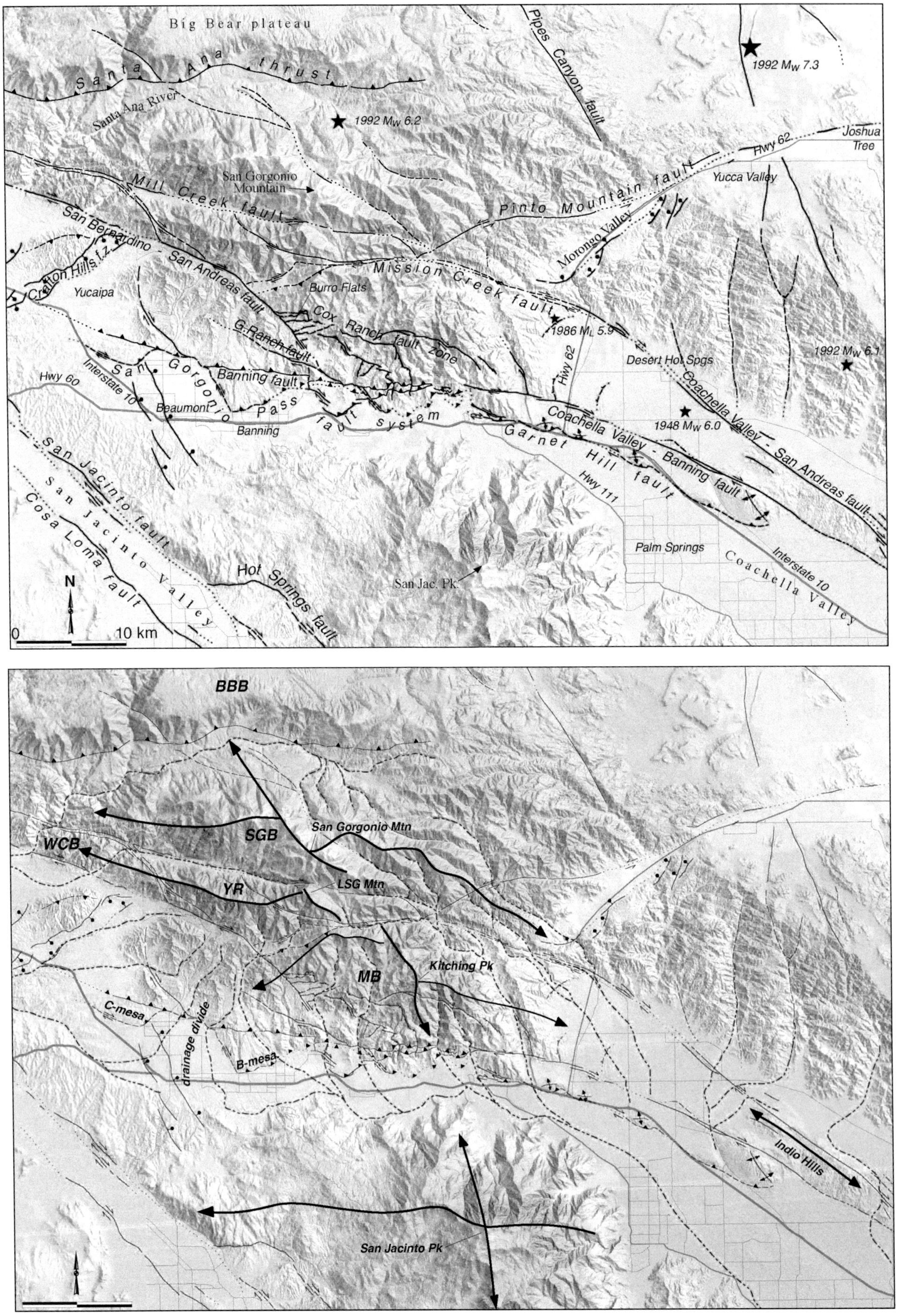

Figure 15. Faults of the San Andreas System in San Gorgonio Pass

EARTHQUAKE RECURRENCE

A detailed map showing recent mapping and identification of active and inactive strands of the system is included as Figure 15. In southern California we live with the knowledge that an earthquake is coming, but not knowing when. The San Andreas Fault is not the only Fault that concerns us, but because of its length it is one of the most feared. Whereas there is no evidence for an earthquake rupturing the entire fault at once, the Fault has recently been broken down into three segments. The northern segment runs from north of San Francisco to Parkfield Area north of Los Angeles. The central section, known as the Mojave section, runs from the Parkfield through Cajon Pass down to San Gorgonio Pass. The southern Section runs from San Gorgonio Pass to the Salton Sea.

Recent modeling has predicted the possibility of a magnitude 7.8 earthquake if rupture were to propagate from the Salton Trough through to Parkfield (Figure 16). The last time we believe this happened was in the 14th Century (Yule, 2012). Subsequently, the central and southern sections have ruptured separately, most recently in 1857 and 1680 respectively. Recurrence rates for these segments are not consistent, however recent evidence based on fault trenching in San Gorgonio Pass indicated that the structural complexity of San Gorgonio Pass acts as a barrier for some events that would otherwise be through going (Figure 14). The main question is "Are we due for three smaller (but still big) events, or one large event?"

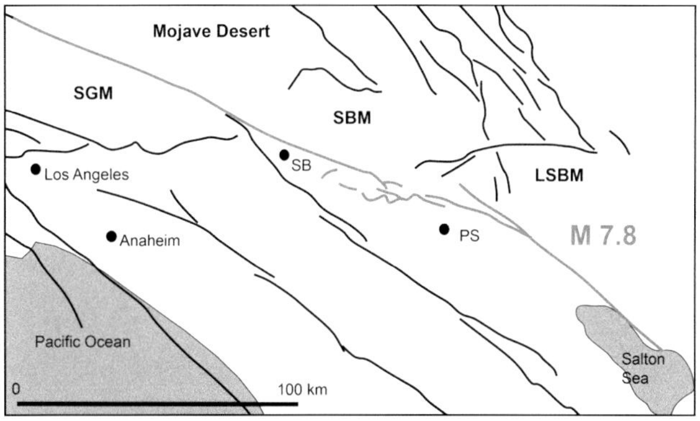

Figure 16. Magnitude 7.8 Rupture Scenario.

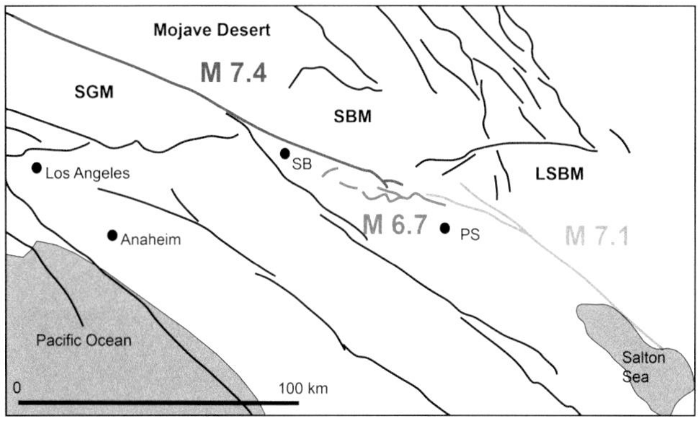

Figure 17. Three smaller quake rupture scenario.

REFERENCES:

Allen, C.R., 1957, San Andreas Fault Zone in San Gorgonio Pass, Southern California, Bulletin of the Geological Society of America. March.

Matti, J.C. and Morton, D.M. 1993. Paleogeographic evolution of the San Andreas fault in southern California: A reconstruction based on a new cross-fault correlation, GSA Memoir 178, Chapter 2, 53 p.

Matti, J.C. and Morton, D.M. 2000. Geologic setting, San Bernardino National Forest. http://www.fs.usda.gov/Internet/FSE_DOCUMENTS/stelprdb5207093.pdf. February 9.

Yule, D. and Sieh, K., 2003, Complexities of the San Andreas Fault, Journal of Geophysical Research, November.

Yule, D. and Spotila, J., 2010, Quaternary Geology of the San Bernardino Mountains and their tectonic margins, 24 p.

SDAG 2009 near the mouth of Deep Canyon.

The Origin Of Morongo Valley, An Extensional Basin in the Eastern Transverse Ranges, Southern California

R. Forrest Hopson Mountain Geosciences, LLC
1200 Riverside Drive, #1269
Reno, Nevada 89503
fhopson@sbcglobal.net

ABSTRACT

Morongo Valley formed as a result of extension between the Morongo Valley and Pinto Mountain faults in response to rotation of the eastern Transverse Ranges and right lateral shear on the San Andreas Fault. In Stage 1, the proto Pinto Mountain Fault curved around the north and west margins of the Pinto and Little San Bernardino mountains where it joined the San Andreas Fault. Stage 2 is marked by right lateral shear on the San Andreas Fault dragging the west end of the Pinto Mountain Fault to the northwest. A kink formed in the Pinto Mountain Fault allowing it to bend clockwise. When rotation of the eastern Transverse Ranges could no longer be accommodated, the Morongo Valley Fault formed to allow the rotation to continue. Consequent extension between the Pinto Mountain and Morongo Valley faults resulted in the down dropping of Morongo Valley in Stage 3. Down dropping on the south was accommodated along the Mission Creek Canyon and Dry Morongo Canyon faults.

INTRODUCTION

Morongo Valley is a narrow wedge-shaped basin in the eastern Transverse Ranges between the Little San Bernardino Mountains and San Bernardino Mountains in southern California (Figure 1). The community of Morongo Valley is in the southeastern part of Morongo Valley about 15 kilometers north of Palm Springs. The community of Yucca Valley is a few kilometers east of Morongo Valley.

Morongo Valley is in a triangular tectonic block informally called the Morongo Valley Block and is adjacent to a left-stepping bend in the San Andreas Fault (Figure 2). This block is bounded by the Pinto Mountain, Morongo Valley, and San Andreas faults. Thus, the presence of fault-bounded Morongo Valley relative to the Pinto Mountain, Morongo Valley, and San Andreas faults implies a kinematic relationship, yet its origin has never been addressed. This paper describes the faults of Morongo Valley, and proposes a kinematic model that relates Morongo Valley to the Pinto Mountain, Morongo Valley, and San Andreas faults.

Morongo Valley is in a tectonically diverse region that is largely related to right lateral shear along the evolving San Andreas Fault (Matti and Morton, 1993). In this region, the San Andreas Fault has accumulated 160 kilometers of right lateral displacement (Matti and Morton, 1993). The San Andreas Fault changes its orientation from N45°W in Coachella Valley to east-west where it is joined by the Pinto Mountain Fault owing to the westward projection of the San Bernardino Mountains by the Pinto Mountain Fault (Matti et al., 1992). North of the San Andreas Fault, transtension along northwest-striking, right-lateral faults in the Mojave Desert has given rise to block-faulted mountains and valleys. South of the Mojave Desert, the eastern Transverse Ranges have experienced 41° of clockwise rotation since middle Miocene time (Carter et al., 1987). The Pinto Mountain Fault is the north boundary of the rotated province

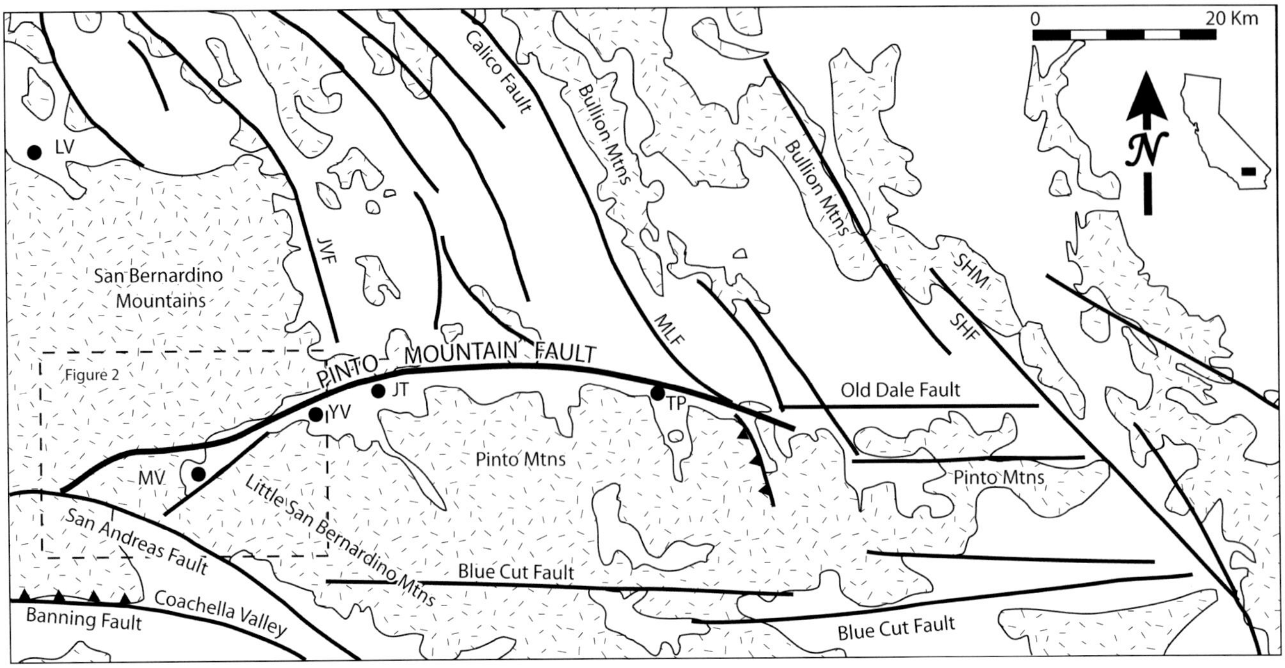

Figure 1. Selected Major Faults and Mountain Ranges near the Morongo Valley Block. SHF denotes Sheep Hole Fault; JVF, Johnson Valley Fault; MLF, Mesquite Lake Fault; SHM, Sheep Hole Mountains; MV, Morongo Valley; LV, Lucerne Valley; YV, Yucca Valley; JT, Joshua Tree; and TP, Twentynine Palms. Barbs indicate upper plate of thrust fault. The dashed rectangle labeled Figure 2 denotes area covered by Figure 2. Modified from Howard (2002), Bortugno and Spittler (1986), and Jennings (1977).

(Richard, 1993; Dickinson, 1996; Hopson, 1996, 1999). In addition, the San Bernardino Mountains are being uplifted by transpression between two frontal thrust faults (Meisling and Weldon, 1989; Matti et al., 1992).

Morongo Valley and adjacent areas are underlain by predominantly crystalline basement rock. Cretaceous granite and granodiorite and Jurassic to Cretaceous plutonic and metamorphic complex of the Little San Bernardino Mountains (LSBC) are the principal basement rock units (Matti et al., 1994; Wooden et al., 2001). The LSBC and granite and granodiorite form the Little San Bernardino Mountains and the lower slopes of the San Bernardino Mountains. A veneer of Quaternary alluvium overlies the basement rocks.

Previous investigators have described the geology of the Morongo Valley area mainly in terms of regional geology and tectonics. Morongo Valley was first recognized as a pull-apart basin by Proctor (1968), who noted the structural nature of the wedge-shaped depression between the Morongo Valley and Pinto Mountain faults. Dibblee (1967), Proctor (1968), and Matti et al. (1992) mapped the geology during reconnaissance. Hopson (1996) and Grimes (1987) mapped the geology in greater detail along the Pinto Mountain and Morongo Valley faults. Dibblee (1968) recognized that Pleistocene alluvial deposits in Morongo Valley are offset 9 kilometers in the left sense by the Pinto Mountain Fault. Dibblee (1982ab) described the Pinto Mountain Fault in Morongo Valley. Matti et al. (1992) recognized that faults in Morongo Valley have normal dip-slip separation, but did not document the kinematic roles of the faults.

FAULT RELATIONSHIPS OF MORONGO VALLEY

Morongo Valley is bounded by the Pinto Mountain and Morongo Valley faults, an inferred fault on the southwest margin, and Dry Canyon Fault (Figure 2, Figure 3). From the northern escarpment of the Little San Bernardino Mountains, the curved Pinto Mountain Fault tracks westward along the

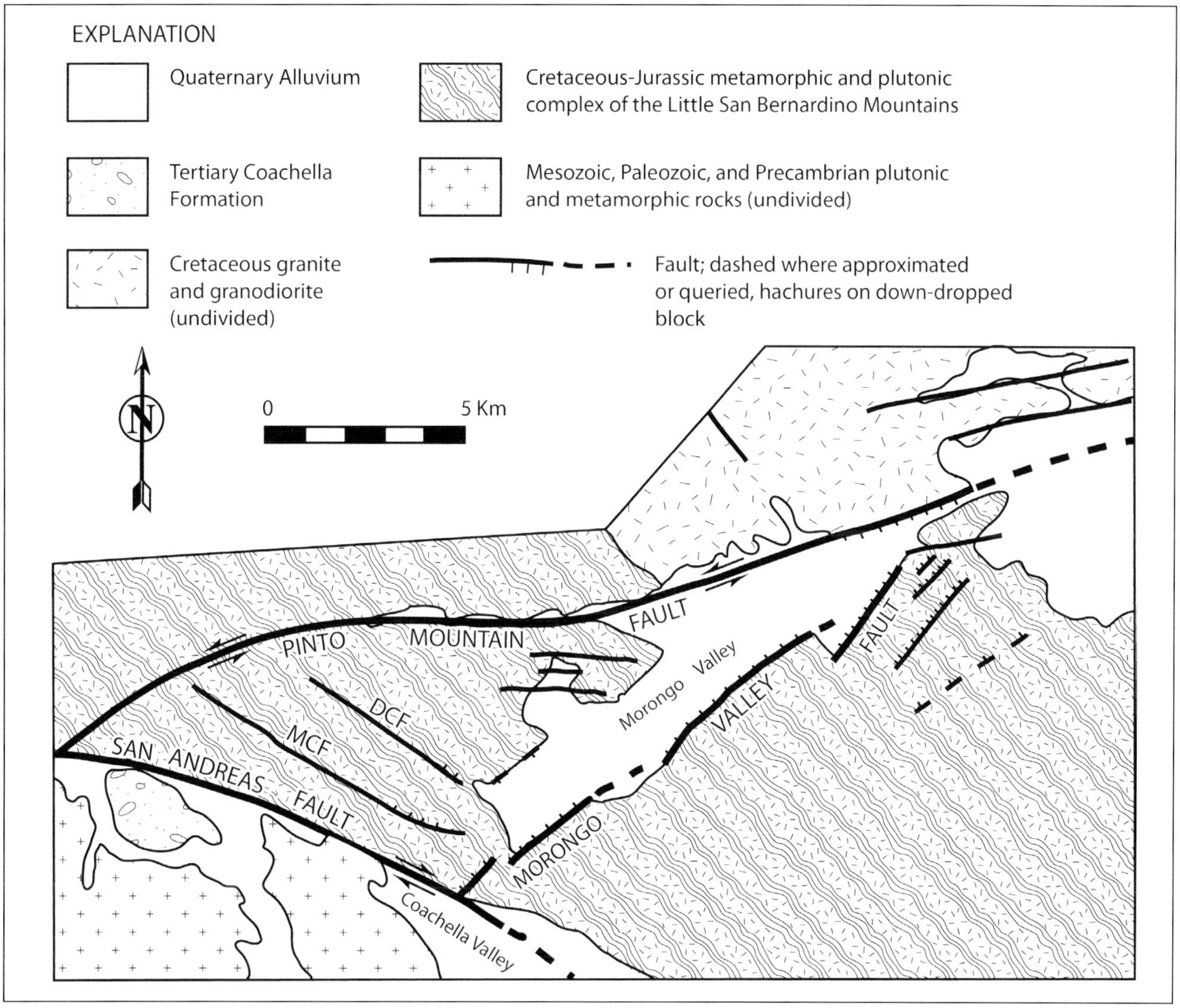

Figure 2. Generalized Geologic Map of the Morongo Valley Block. DCF denotes Dry Canyon Fault and MCF, Mission Canyon Fault. Modified from Dibblee (1968), Grimes (1987), and Hopson (1996).

Figure 3. Morongo Valley Showing the Approximate Locations of the Pinto Mountain and Morongo Valley Faults, Which Bound the Mountain Fronts on the North and South Sides of the Valley. An inferred graben-bounding fault (denoted by dashed white lines) is in the middle distance. PMF and MVF denote Pinto Mountain and Morongo Valley faults, respectively. Photograph by R.F. Hopson.

Figure 4. The Dry Morongo Canyon Fault from the Southeast. The Dry Morongo Canyon Fault cuts Dry Morongo Canyon. Photograph by R.F. Hopson.

Figure 5. Mission Creek Canyon Fault from the Southeast. The Mission Creek Canyon Fault cuts Mission Creek Canyon. Photograph by R.F. Hopson.

northwest side of Morongo Valley and through Big Morongo Canyon where it joins the San Andreas Fault about 20 kilometers to the west (Hopson, 1996). The Morongo Valley Fault splays off the Pinto Mountain Fault about 1 kilometer west of Yucca Valley and strikes southwestward about 15 kilometers (Grimes, 1987; Hopson, 1998). A pair of normal faults, informally called the Mission Creek Canyon and Dry Morongo Canyon faults (Figure 4, Figure 5), splays off the Morongo Valley Fault and strikes northwest between the Morongo Valley and Pinto Mountain faults (Matti et al., 1992).

Both the Pinto Mountain and Morongo Valley faults exhibit normal left separation in Morongo Valley. The Pinto Mountain Fault separates the granite and granodiorite from the Quaternary alluvium that partially fills Morongo Valley, and cuts the LSBC. In the northeastern part of Morongo Valley, a zone of crushed granite lies approximately 1 kilometer north of the projected Pinto Mountain Fault trace. The Morongo Valley Fault cuts the LSBC and separates it from the Quaternary Morongo Valley fill.

Dry Morongo Canyon and Mission Creek Canyon faults track through Dry Morongo and Mission Creek canyons also cut the LSBC. Although the faults are covered by alluvium, their presence is recognized by strong geomorphic lineaments formed by Mission Creek and Dry Morongo canyons. Down-to-the-east normal displacement on these faults appears to increase to the south as evidenced by the southeast-increasing difference between the elevations of the adjacent ridge-crests. Because the Dry Morongo Canyon and Mission Creek Canyon faults subparallel the San Andreas Fault, a right-lateral component would be expected but has not been recognized.

The Pinto Mountain and Morongo Valley faults are made up of three strands each (Figure 2) (Grimes, 1987). Two of the Pinto Mountain Fault strands (upper right hand corner of Figure 2) do not appear to have played a major role in the origin of the Morongo Valley Block because they do not project into Morongo Valley. The main strand appears to have accommodated most of the downdropping of the Morongo Valley Block. The role of the southeastern strand of the Morongo Valley Fault in this downdropping is uncertain, but it may have been minimal (Grimes, 1987). This strand has normal separation but dies out beneath fanglomerate or slope wash (Grimes, 1987).

MODEL FOR MORONGO VALLEY

The origin of Morongo Valley may ultimately be linked to right-lateral shear on the San Andreas Fault and clockwise rotation of the eastern Transverse Ranges. Morongo Valley appears to have developed in at least three stages as summarized in Figure 6 (next page).

In Stage 1 the western Pinto Mountain Fault, which bounds the San Bernardino Mountains and Mojave Desert on the south, formed a curved trace along the north side of the Pinto and Little San Bernardino mountains. The westernmost part of the fault occupied the position of the Morongo Valley Fault and joined the proto San Andreas Fault, which was oriented N45°W (Dickinson, 1996). In Stage 2, right lateral shear on the San Andreas Fault dragged the west end of the San Bernardino Mountains to the northwest, relative to the main part of the range, which was being transported southeastward. Furthermore, the San Bernardino Mountains were being tectonically shoved to the west owing to a combination of transtension in the Mojave Desert and left slip on the Pinto Mountain Fault (Matti et al., 1992). A kink formed in the Pinto Mountain Fault near present day Morongo Valley to allow the westernmost part of the fault to bend clockwise. As the westernmost Pinto Mountain Fault continued to bend, it was rotated out of a position favorable for the eastern Transverse Ranges to rotate. The Morongo Valley Fault therefore formed to accommodate rotation of the eastern Transverse Ranges on the west. Continued right shear on the San Andreas Fault and subsequent bending of the westernmost Pinto Mountain Fault created extensional forces between the Pinto Mountain and Morongo Valley faults. In Stage 3, Morongo Valley was tectonically dropped down between the Pinto Mountain and Morongo Valley faults and the Dry Morongo and Mission Canyon faults owing to the differential movement between the Little San Bernardino and San Bernardino mountains. During the down dropping, the Morongo Valley Block was tilted to the south. Evidence for this tilting is an apparent southward increase in normal displacement on the Dry Morongo Canyon and Mission Creek Canyon faults.

DISCUSSION AND CONCLUSION

The model provides a simple explanation for the origin of Morongo Valley. It proposes that the extension across Morongo Valley is kinematically related to right lateral shear on the San Andreas Fault, clockwise rotation of the eastern Transverse Ranges, and normal displacement on the Dry Morongo and Mission Canyon faults. The driving force for the extension is right lateral shear on the San Andreas Fault.

Although the Pinto Mountain Fault is predominantly left lateral, it is not unreasonable that local normal displacement has occurred because of extensional forces acting across Morongo Valley. As with the Morongo Valley Fault, the Pinto Mountain Fault was a weak zone that allowed the Morongo Valley graben to develop.

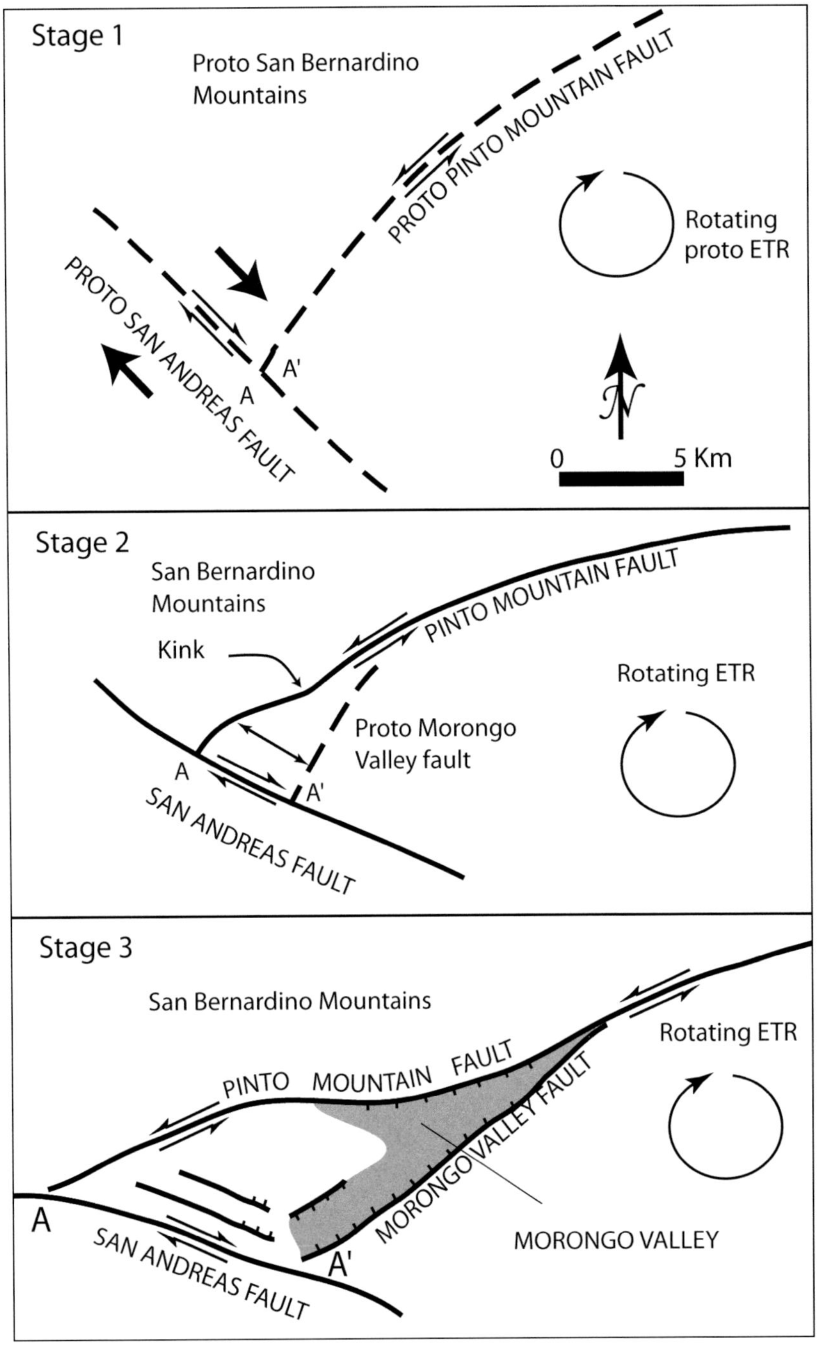

Figure 6. Evolution of Morongo Valley. A-A' represents fixed points across the San Andreas Fault. "A" moves northwest with respect to the Pinto Mountain and Morongo Valley faults at each stage. ***Stage 1***. The western part of the proto Pinto Mountain Fault during early stages of development circa 7 million to 5 million years ago.. The eastern Transverse Ranges are beginning to rotate. Right lateral motion is occurring on the proto San Andreas Fault, which at this time is oriented approximately N45ºW (Dickinson, 1996). ***Stage 2.*** A kink (arrow) forms in the westernmost Pinto Mountain Fault as right lateral shear drags the west ends of the San Bernardino Mountains and Pinto Mountain Fault northwestward. Morongo Valley Fault forms to allow the eastern Transverse Ranges to rotate. Extension is occurring between the Pinto Mountain and Morongo Valley faults. San Andreas Fault, in this vicinity, is deformed into an east-west orientation as the San Bernardino Mountains are directed across it (Matti et al. 1992). ***Stage 3***. Continued extension between the Pinto Mountain and Morongo Valley faults results in the downdropping of Morongo Valley. Normal displacement occurs along the western Mission Canyon and Dry Morongo Canyon faults to accommodate the downdropping of Morongo Valley on the southwest. Gray represents Morongo Valley.

Down dropping on the south side of the valley would not necessarily be confined to the main strand of the Morongo Valley Fault. Normal faults mapped by Matti et al. (1992) and Grimes (1987) just south of the Morongo Valley Fault could have easily accommodated some of the down dropping as well. However, the Morongo Valley Fault can now be considered the active trace of the western Pinto Mountain Fault Zone as suggested by recent earthquakes along the fault that are consistent with left lateral movement (Nicholson, 1996).

In conclusion, Morongo Valley is an extensional basin whose origin is attributed to dragging of the west end of the Pinto Mountain Fault during right lateral shear along the San Andreas Fault.

ACKNOWLEDGEMENTS

I am grateful to Kim Bishop, Arthur G. Sylvester, Clifford Hopson, and Steve Lipshie for helpful reviews. Perry Ehlig (deceased), Jonathon Matti, Joe Wooden, and Steve Lipshie are thanked for discussion and Craig Nicholson for reprint.

REFERENCES

Bortugno, E.J. and Spittler, T.E. (compilers), 1986, Geologic map of the San Bernardino quadrangle: California Division of Mines and Geology Regional Geologic Map Series No. 3A, scale 1:250,000.

Carter, J. N., Luyendyk, B.P., and Terres, R.R., 1987, Neogene clockwise tectonic rotation of the eastern Transverse Ranges, California, suggested by paleomagnetic vectors: Geological Society of America Bulletin, v. 98, p. 199–206.

Dibblee, T.W. Jr., 1967, Geologic map of the Morongo Valley Quadrangle, San Bernardino and Riverside Counties, California: U.S. Geological Survey Miscellaneous Geologic Investigations Map I-517.

Dibblee, T.W. Jr., 1968, Evidence of major lateral displacement on the Pinto Mountain Fault, southern California: Geological Society of America Special Paper 115, p. 322.

Dibblee, T.W. Jr., 1982a, Regional geology of the Transverse Ranges Province of southern California *in* D.L. Fife and J.L. Minch, Geology and Mineral Wealth of the California Transverse Ranges: South Coast Geological Society, p. 7–26.

Dibblee, T.W. Jr., 1982b, Geology of the San Bernardino Mountains, southern California, *in* D.L. Fife and J.L. Minch, Geology and Mineral Wealth of the California Transverse Ranges: South Coast Geological Society, p. 149–169.

Dickinson, W.R., 1996, Kinematics of transrotational tectonism in the California Transverse Ranges and its contribution to cumulative slip along the San Andreas transform fault system: Geological Society of America Special Paper 305, 46 p.

Grimes, G.J., 1987, Geology and convergence of the Pinto Mountain Fault and Morongo Valley Fault, southern California: [M.S. thesis]: California State University, Los Angeles, 102 p.

Hopson, R.F., 1996, Basement rock geology and tectonics of the Pinto Mountain Fault, San Bernardino County, southern California [M.S. thesis]: California State University, Los Angeles, 134 p.

Hopson, R.F., 1998, Quaternary geology and neotectonics of the Pinto Mountain Fault, Mojave Desert, southern California: California Geology, v. 51, p. 3–13.

Hopson, R.F., 1999, A mechanical model for the Pinto Mountain Fault, southern California: preliminary results: Geological Society of America Abstracts with Programs, v. 31, p. 63.

Howard, K.A., 2002, Geologic map of the Sheep Hole Mountains 30′ x 60′ quadrangle, San Bernardino and Riverside Counties, California: U.S. Geological Survey map MF-2344, (1:100,000).

Jennings, C.W. (compiler), 1977, Geologic map of California: California Division of Mines and Geology (1:750,000)

Matti, J.C. and Morton, D.M., 1993, Paleogeographic evolution of the San Andreas Fault in southern California: A reconstruction based on a new cross-fault correlation, *in* Powell, R.E., Weldon, R.J. II, and Matti, J.C., eds., The San Andreas Fault system: Displacement, palinspastic reconstruction, and geologic evolution: Geological Society of America Memoir 178, p. 107–159.

Matti, J.C., Morton, D.M., and Cox, B.F., 1992, The San Andreas Fault system in the vicinity of the central Transverse Ranges Province, southern California: U.S. Geological Survey Open-File Report 92-354, scale 1:250,000.

Matti, J.C., Wooden, J.L., and Powell, R.E., 1994, Late Cretaceous plutonic and metamorphic complex in the Little San Bernardino Mountains, southern California: Geological Society of America Abstracts with Programs, v. 26, p. 70–71.

Meisling, K.E. and Weldon, R.J., 1989, Late Cenozoic tectonics of the northwestern San Bernardino Mountains, southern California: Geological Society of America Bulletin, v. 101, p. 106–128.

Nicholson, C. 1996, Seismic behavior of the southern San Andreas Fault zone in the northern Coachella Valley: Comparison of the 1948 and 1986 earthquake sequences: Bulletin of the Seismological Society of America, v. 86, p. 1331–1349.

Proctor, R.J., 1968, Geology of the Desert Hot Springs-Upper Coachella Valley area, California: California Division of Mines and Geology Special Report 94, 50 p.

Richard, S.M., 1993, Palinspastic reconstruction of southeastern California and southwestern Arizona: Tectonics, v. 12, p. 830–854.

Wooden, J.L., Sumac, Fleck, R.J., Matti, J.C., Powell, R.E., Barth, A.P., 2001, Late Cretaceous intrusive migmatites of the Little San Bernardino Mountains, southern California: Geological Society of America Abstracts with Programs, v. 33, p. A-65.

Offset Holocene alluvium in a San Andreas fault trench near the intersection of Dillon Road and Avenue 44, Indio, California, March 19, 2006. Bill Elliott, kneels on the North American side of the plate boundary.

The Role of the Hot Springs Fault in the Development of the San Jacinto Fault Zone and Uplift of the San Jacinto Mountains

Nate Onderdonk
Dept. of Geological Sciences
California State University, Long Beach
1250 Bellflower Blvd.
Long Beach, CA 90840
nonderdo@csulb.edu

ABSTRACT

The Hot Springs Fault is a strand of the San Jacinto Fault Zone that marks the western edge of the San Jacinto Mountains. Although the Hot Springs Fault is currently not the dominant structure in the area, it has played a major role in the development of the San Jacinto Fault Zone and the uplift of the San Jacinto Mountains during Pleistocene time. Topographic contrasts along the fault, uplifted Pleistocene basin sediments, and elevated fluvial terraces indicate a strong component of dip-slip displacement along the fault in addition to strike-slip offset. The distribution of Pleistocene sediments and structural relationships along the Hot Springs Fault suggest that the fault is a dying strand that accommodated both strike-slip displacement and local topographic development during the early development of the San Jacinto Fault Zone.

INTRODUCTION

The San Jacinto Fault Zone is a major strike-slip structure in southern California and an integral part of the San Andreas Fault System (Figure 1). The fault zone is oriented parallel to the southern San Andreas Fault and may be accommodating the majority of the plate boundary strain in the area. It is the most seismically active fault in southern California and the kinematics, earthquake history, and slip-rate of the fault zone indicate that it is a major seismic hazard for the surrounding communities.

An equally important aspect of the San Jacinto Fault Zone is its influence on regional topography and drainage networks. Displacement along this fault zone has had a great impact on the physiography of the area and interactions between the multiple fault strands have created local topographic highs and basins along the fault zone. Initiation of displacement along the San Jacinto Fault Zone resulted in the uplift of the eastern San Gabriel Mountains and has contributed to basin inversion and uplift of the Timoteo Badlands in the San Bernardino Area (Morton and Matti, 1993). The fault zone also bounds the southwestern edge of the San Jacinto and Santa Rosa Mountains, but exactly how much uplift of these ranges can be attributed to fault strands along this trend is unknown.

The San Jacinto Mountains are delineated on their southwest side by the Hot Springs Fault (Figure 1). The topographic differences that occur across this range-front fault suggest that it has been the dominant structure controlling uplift of the range along the west side. However, little is known about the Hot Springs Fault and its influence on topographic development along the San Jacinto Fault Zone. This paper describes the structural framework and kinematics of the Hot Springs Fault, its relation to the rest of the San Jacinto Fault Zone and evaluates its role in the uplift of the San Jacinto Mountains.

THE SAN JACINTO FAULT ZONE

The Hot Springs Fault is a strand of the San Jacinto Fault Zone. The San Jacinto Fault Zone splays from the San Andreas Fault in the eastern San Gabriel Mountains and strikes southeastward 200 kilometers to the western Imperial Valley. The fault

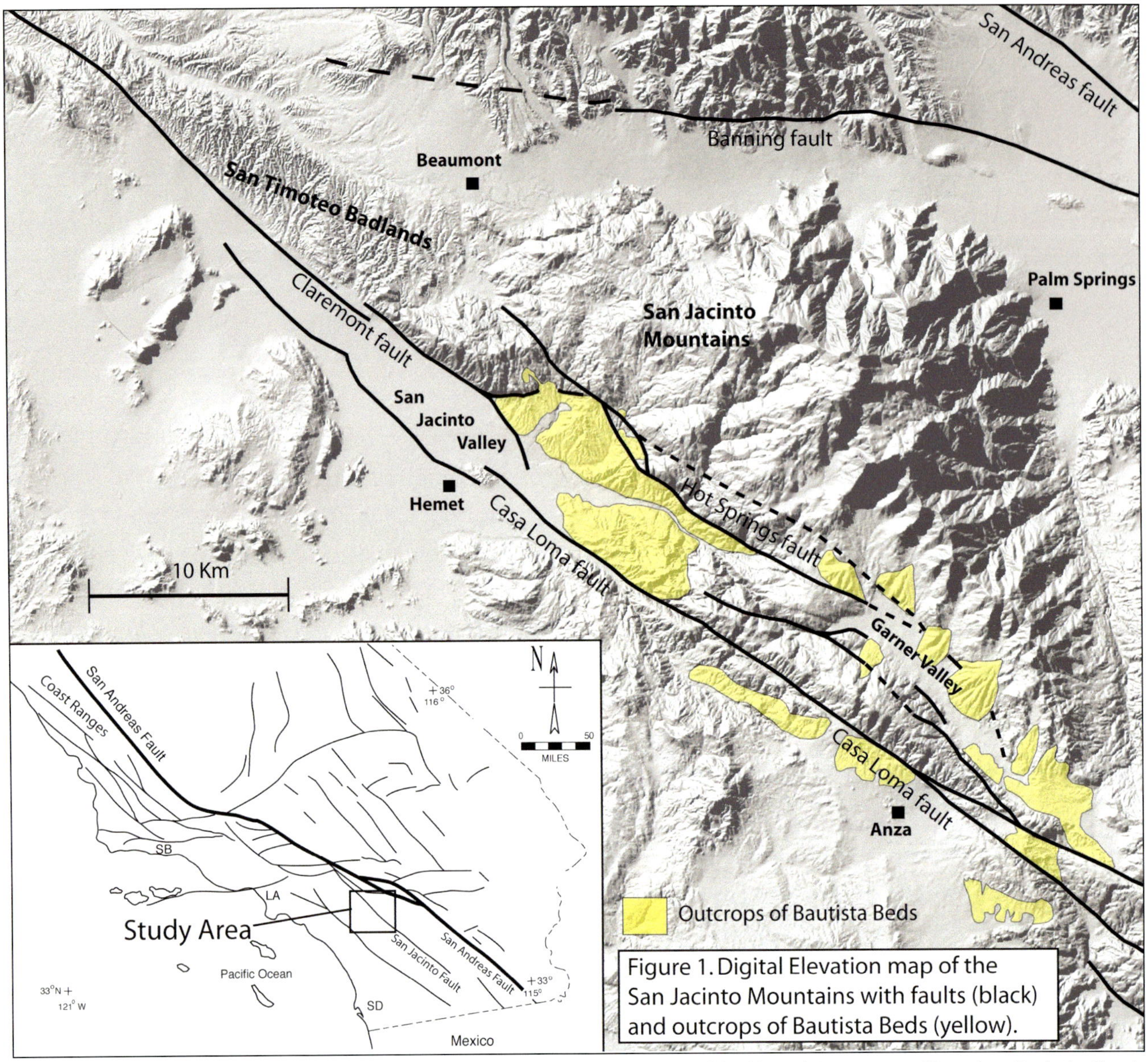

Figure 1. Digital Elevation Map of the San Jacinto Mountains with Faults (black) and Outcrops of Bautista Beds (yellow).

zone is currently the most active component of the San Andreas Fault System in southern California in terms of microseismicity and frequency of larger magnitude earthquakes (Given, 1981; Sanders and Kanamori, 1984; Sharp, 1981; Thatcher et al., 1975; Wesnousky, 1986). At least nine earthquakes of M>6 have occurred on faults in this zone since 1890 (Wesnousky, 1986), and seismic recording stations detect a high level of microseismicity within the zone.

The total offset has been estimated to be between 24 and 30 kilometers of right-lateral displacement across the Claremont and Casa Loma faults in the central and northern part of the fault zone (Sharp, 1967; Morton and Matti, 1993), and about 21 kilometers across the Coyote Creek and Clark faults in the southern part of the fault zone (Sharp, 1967; Dorsey, 2002). Estimates of slip rate along the San Jacinto Fault Zone range from 8 mm/year to over 20 mm/year (e.g., Sharp, 1967; Rockwell et al., 1990; Morton and Matti, 1993; Dorsey, 2002). Recent studies have revealed slip rates at the higher end of that range and suggest that Holocene slip rates on the San Jacinto Fault may be faster than along the southernmost San Andreas Fault (e.g., McGill et al., 2007; Fletcher et al., 2006; Kendrick et al., 2002; Morton and Matti, 1993). Geodetic studies indicate that the San Jacinto Fault is accommodating a larger amount of the plate boundary displacement than the southernmost San Andreas Fault (e.g., Becker et al., 2005). This suggests that the San Jacinto Fault Zone is the dominant plate boundary structure in southern California.

The fault zone is a series of fault segments that splay and overlap. This arrangement is most likely a reflection of the young age of the fault zone and the limited time it has had to develop into a single continuous structure. The age of the San Jacinto Fault Zone is inferred to be between 1 million and 2.4 million years old, with the younger age favored in more recent studies (e.g., Sharp, 1981; Morton and Matti, 1993; Kendrick et al., 2002). The variations in age are the result of determinations of slip amounts and slip-rates on different fault segments within the zone. The splaying and overlapping fault segments make it difficult to infer total offset, age, or slip-rate for the fault zone before data is available on all the segments. The lack of studies along the Hot Springs Fault are especially problematic due to the fact that this fault overlaps with two other major strands of the fault zone, the Claremont Fault and the Casa Loma Fault. These two faults, along with the Hot Springs Fault and the Thomas Mountain Fault make up the fault zone along the western edge of the San Jacinto Mountains and are the only known Quaternary structures that may have affected the uplift of the mountains.

The Claremont Fault, which is the northernmost extent of the San Jacinto Fault Zone, extends southeastward from the eastern San Gabriel Mountains to the city of Hemet in the San Jacinto Valley. In the Hemet Area, the Claremont Fault has created a straight mountain front that bounds the northeastern edge of the San Jacinto Valley (Figure 1). The steep and straight nature of this mountain front suggests continued active uplift along the southern Claremont Fault. The San Jacinto Valley is approximately 2 kilometers deep and C^{14} dates from wood samples collected from drill holes indicate that the valley has been subsiding at a rate of 3 to 5 mm/year for the last 40,000 years (Morton, 1995). The San Jacinto Valley has been identified as a strike-slip pull-apart basin resulting from a right-step between the Claremont Fault and the Casa Loma Fault (Sharp, 1967; Crowell, 1974). The Casa Loma Fault lies at the southwestern side of the San Jacinto Valley and extends about 55 kilometers to the southeast into the Anza Plateau. Both of the faults are active with abundant microseismicity and strong geomorphic expressions.

THE HOT SPRINGS FAULT ZONE

The Hot Springs Fault, originally named by Fraser (1931), steps eastward off of the southeastern extent of the Claremont Fault, 3 kilometers north of the city of Hemet. It is expressed as a zone of faults that can be traced for about 50 kilometers southeast along the mountain front to Garner Valley where the main trace is lost beneath alluvium. Due to the lateral discontinuity of a single fault segment, and in order to include the large number of secondary faults and faults within stopovers along the structural trend, the term "fault zone" is used herein to describe the larger structural zone. A dominant southeast-striking fault can be identified along most of the length of the zone, however, and the name "Hot Springs Fault" is applied to this feature.

Geology along the Fault Zone. Along much of its length, the Hot Springs Fault marks the contact between Pleistocene sedimentary strata and older crystalline rock of the San Jacinto Mountains. The crystalline rocks include tonalite to granodiorite plutonic rocks of the Peninsular Range Batholith and older metasedimentary rocks that consist of various grades of micaceous schist, gneiss, migmatite, and arkosic quartzite. The Pleistocene sedimentary strata along the Hot Springs Fault Zone are known as the Bautista Beds (Frick, 1921). The Bautista Beds are a nonmarine formation with a total thickness of approximately 2,500 meters. The unit is exposed along the San Jacinto Fault Zone from Hemet southeast to the northern Anza-Borrego Valley and ranges from coarse fanglomerate to fine lacustrine shale (Figure 1). These strata were derived from the rising San Jacinto Mountains and were deposited as alluvial fans, stream deposits and lake deposits (Dibblee, 1981).

In the Soboba Hot Springs Area (Figure 2) and along the sides of San Jacinto Canyon (Figure 3) the Bautista Beds consist of poorly indurated fine sand and shale with sparse lenses of pebble conglomerate. The visible thickness is about 350 meters and the unit is exposed for approximately 35 square kilometers in the foothills northeast of Hemet and above San Jacinto Canyon. The fine-grained nature of the Bautista Beds next to the steep mountain front indicates that they were not deposited in the rugged environment that currently exists. At the time when the Bautista Beds were being laid down, stream gradients must have been considerably lower than those of the present-day streams that contain pebble and boulder sized clasts. Where not disturbed by faults, the beds have consistent attitudes with low dips of about 5° to 25°. To the south and southeast of the Soboba Springs area, the Bautista Beds become coarser and grade laterally into thick conglomerate beds that are well exposed in Bautista Creek and along Highway 74 in San Jacinto Canyon. The strata in this area have undergone a considerable amount of deformation since their deposition. Beds are shattered throughout by small faults and folded up to steep angles at the base of the larger bounding faults. Gentle anticlines and synclines exist within the beds and trend approximately east-west along the fault zone. Small remnants of Bautista strata are also found at higher elevations within the crystalline rock at two locations along the fault zone. Their current positions indicate that the strata have been vertically separated by displacement along the Hot Springs Fault Zone.

Structural Framework. The northwest end of the Hot Springs Fault Zone forms a restraining step-over with the Claremont Fault (Figures 1 and 2). The main Hot Springs Fault does not splay off the Claremont Fault as depicted on previous maps (Jennings, 1977; Sharp, 1967), but rather continues to the northwest and can be traced as far as the San Jacinto Potrero (Figure 2; Onderdonk, 1998). The uplifted terrain within the step-over is cut by several northwest-striking faults and is bounded on the south side by a major east-striking reverse fault herein referred to as the Soboba Fault. At its west end, the Soboba Fault connects to the Claremont Fault and is marked by a prominent scarp approximately 9 meters high. Outcrops show that the fault dips nearly vertical and striations on the fault surface indicate an approximately equal amount of north-side-up and right-lateral displacement during the last movement. The fault marks the contact between the uplifted crystalline rock on the north side and fine-grained sandstone beds of the Bautista Formation on the south side that have been drag-folded up against the fault at angles of 85° to 60°. The dips of these beds abruptly shallow to angles of about 35° to 20° away from the fault. The Soboba Fault extends 2 kilometers to the east and is delineated by scarps developed in both the crystalline rock and overlying Holocene colluvium. At its east end, the fault curves northward and most likely merges with one of the northwest striking faults within the step-over. There is no direct connection between the Soboba Fault and the Hot Springs Fault, although a set of smaller reverse faults that branch off of the Hot Springs Fault are aligned with the strike of the Soboba Fault.

Along most of its 50 kilometer length, the Hot Springs Fault Zone delineates a steep linear mountain front that forms the west side of the San Jacinto Mountains. The northwestern section of the fault zone in the Indian Canyon Area (Figure 3) is composed of multiple anastomosing active fault strands with the easternmost marking the edge of the mountains. The faults are marked by deflected and beheaded streams, scarps, springs, vegetation lineaments, and outcrops showing breccia zones, gouge, and fault planes. Fault strands observed in outcrop dip 70° or steeper to the east. Several isolated pods of Bautista Beds are cut and uplifted by fault movement. In addition to being cut by

Figure 2. Simplified Geologic Map of the Soboba Springs Area at the North End of the Hot Springs Fault.

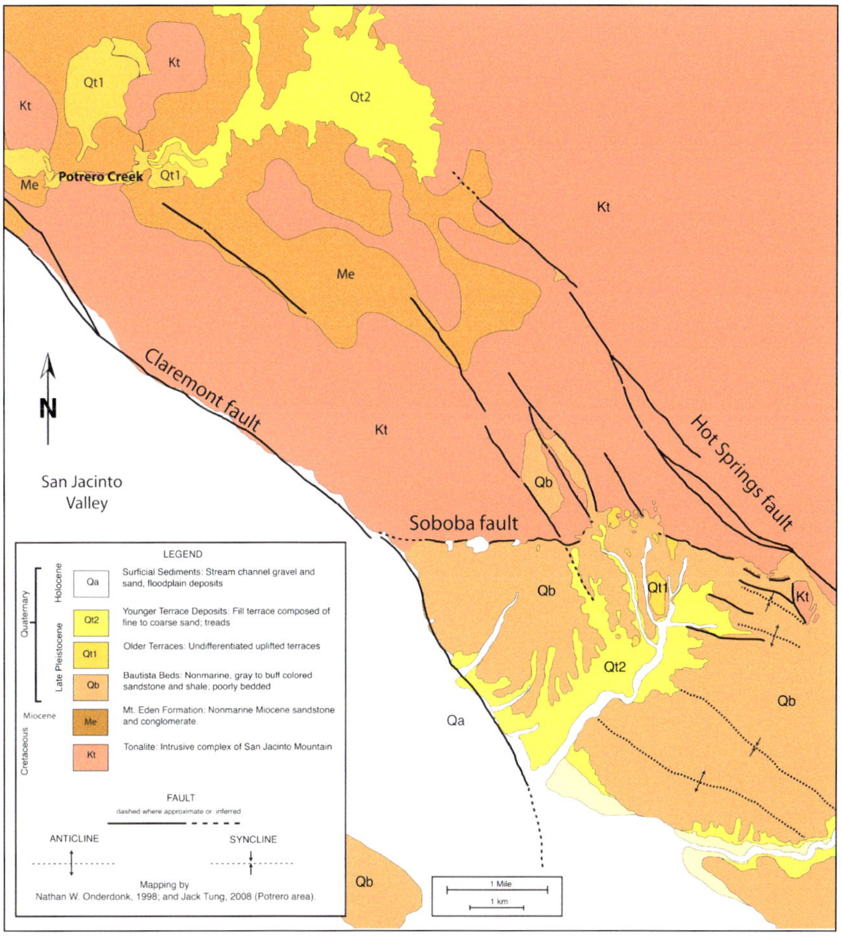

Figure 3. Simplified Geologic Map of the San Jacinto Canyon Area, Central Hot Springs Fault Zone.

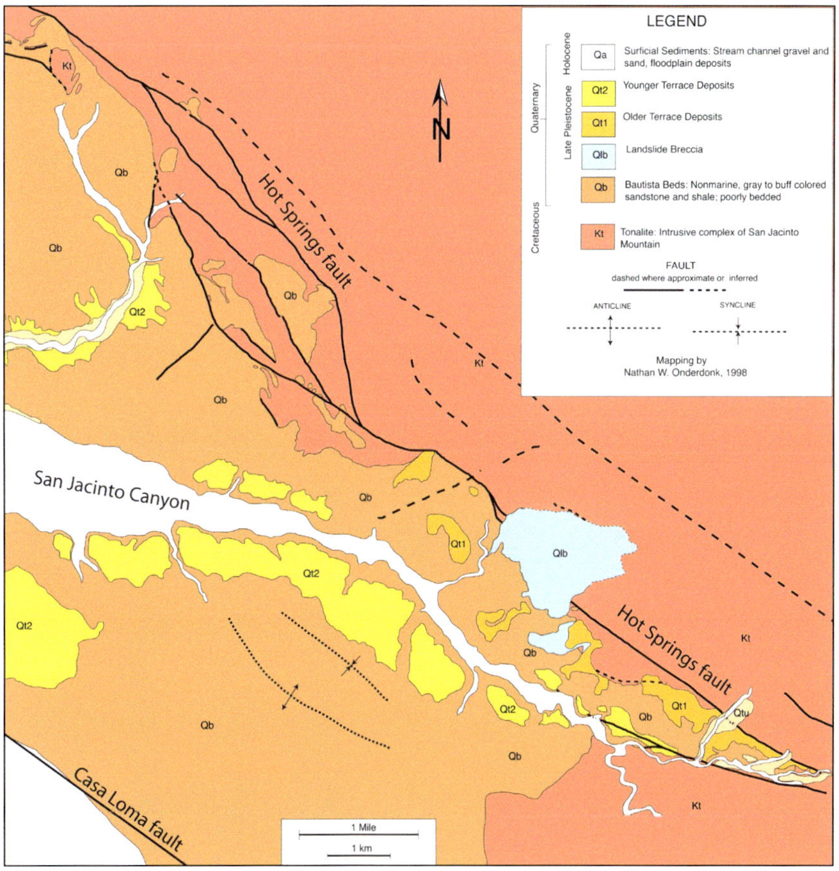

numerous small faults, the beds are folded into a large anticline and syncline. The axes of these folds trend parallel to the fault zone.

Through San Jacinto Canyon (Figure 3), a single Hot Springs Fault can be traced along the base of the mountain. The fault is represented mainly by the straight mountain front, but is also delineated by breccia zones, springs, vegetation lineaments, and outcrops of shear zones and fault surfaces. The fault planes dip steeply to the south where exposed and striations on the fault surfaces have rakes of 40° to 55°, indicating a significant amount of dip-slip displacement in addition to strike-slip. The main fault bounds the Bautista Beds along much of its length although the contact is obscured by landslides in many places. The Bautista Beds coarsen up the canyon to the southwest and dip away from the fault except where folds exist. Overlying the Bautista Beds are several terrace deposits of Quaternary Age. The older terraces (grouped as "Qt_1") are composed of boulders and coarse gravel deposited on a beveled surface in the Bautista Beds and have been deformed by fault movement. These deposits are only found on the northeast side of the canyon and represent older alluvial fans that formed along the mountain front. A set of younger terraces (grouped as "Qt_2" on Figure 3) is present at lower elevations about 65 to 100 meters above the present-day canyon bottom. These terraces still retain their original surface and are used as orchard sites in the lower canyon. They are unpaired across the canyon, indicating a period of constant uplift. No attempt has been made to numerically date either the younger Qt_2 or the older Qt_1 terraces and their designation as Quaternary is based on soil development and the fact that they overlie the Bautista Beds. Other (older?) strands of the fault zone are present to the east in the uplifted terrane above San Jacinto Canyon and are manifested as prominent linear slope breaks. These structures are

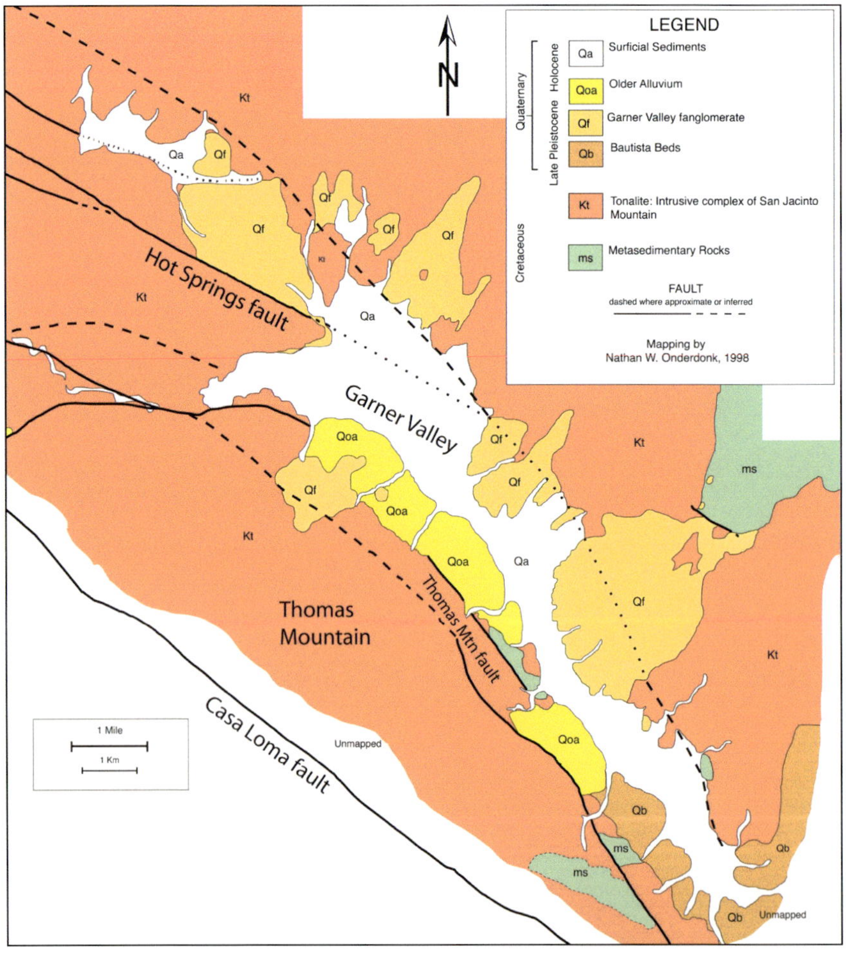

Figure 4. Simplified Geologic Map of the Garner Valley Area, Southern Hot Springs Fault Zone.

identified based on topographic lineaments and are difficult to see on the ground due to the homogeneity of the rocks they break. At least one of these structures extends into the Garner Valley Area where a linear alignment of truncated drainage divide ridges is present along the northeast side of the valley.

Between Highway 74 and Garner Valley the main strand of the Hot Springs Fault Zone created a linear slope that marks the north side of Baldy Mountain (Figure 4). Along the eastern half, the fault has thrust tonalite over the fanglomerate of Garner Valley. This relationship is exposed in an outcrop along Highway 74 above the Herkey Creek Campground where the fault dips 60° to the south. Southeast of Highway 74, the fault disappears under the recent alluvium of Garner Valley and detailed mapping of the east side of Garner Valley along strike with the Hot Springs Fault revealed no conclusive outcrop evidence of its continuation. A small fault is present, however, at the top of Morris Ranch Road where the fault juxtaposes fanglomerate and granitic rock against metasedimentary rocks. This fault is along strike with the Hot Springs Fault and is exposed as a thin clay gouge zone dipping 75° to the south in a quarry cut. In addition to this fault, a linear mountain front of crystalline rock that has been overlapped by Quaternary fan deposits is on alignment with the fault, and is most likely the result of early displacement along the trend of the Hot Springs Fault.

The southern end of the Hot Springs Fault Zone is linked to the rest of the San Jacinto Fault Zone via the Thomas Mountain Fault, which is located one kilometer south of the Hot Springs Fault Zone across Garner Valley (Figure 4). The Thomas Mountain Fault truncates Pleistocene sediments in the southern part of Garner Valley and juxtaposes Quaternary alluvium against tonalite at its northern end, but appears to be overlain by older Quaternary alluvium deposited at the base of Thomas Mountain. At its southeast end, the Thomas Mountain Fault merges with the Buck Ridge Fault in the area of Burnt Valley. At its north end, near Lake Hemet, the fault splits into two strands that merge to the northwest with the Casa Loma Fault. The steep canyon that drains Lake Hemet and Garner Valley was eroded into the gouge of the northern fault splay. Excellent exposures in the spillway of the Lake Hemet Dam show that the Thomas Mountain Fault dips south at approximately 50°. The fault most likely merges at depth with the Casa Loma fault located along the opposite side of Thomas Mountain.

Strike-Slip Displacement and Slip-Rate. Displacement along the Hot Springs Fault is assumed to be dominantly strike-slip due to the fault zone structure and its relation to the rest of the San Jacinto Fault Zone. Hill (1981) estimated a total right-lateral displacement of about 5 kilometers across the Hot Springs Fault Zone based on offset intrusive rock contacts along the eastern extent of the zone. The age of the northern San Jacinto Fault Zone is believed to be between 1 million and 1.5 million years old based on mapping and stratigraphic studies (Morton and Matti, 1993; Dorsey, 2001). If the age of the Hot Springs Fault Zone is assumed to be the same, the slip-rate averaged over this time period would be between 5 mm/year and 3.3 mm/year. However, geomorphic expression and seismic activity suggests that although the Hot Springs Fault is still active, it is not as active as the parallel Casa Loma Fault to the southwest. If the Hot Springs Fault is dying out, the pre-Holocene slip-rate was most likely higher than the estimates given above. Deflected and offset streams in numerous locations should provide a means to estimate Holocene displacement and slip-rates along the fault. However, many of these offset streams are developed in the granitic bedrock and do not offset younger deposits that can be easily dated. In the few locations where deflected streams are cut into Quaternary terrace deposits, the deposits have not been numerically dated and slip rate estimates cannot be made at this time.

Dip-Slip Displacement and Topographic Development along the Hot Springs Fault Zone. In addition to strike-slip displacement along the fault zone, a considerable amount of dip-slip displacement has also occurred. Dip-slip displacement is expressed in kinematic indicators along the fault with striations on exposed fault surfaces consistently inclined to a rake angle around 45°. Steep mountain fronts, elevated plateau-like topography, deep and narrowly incised canyons, and vertically-displaced Bautista Beds and younger terrace deposits also attest to vertical movement along the Hot Springs Fault Zone. Accordant peaks and an elevated plateau that may represent an old fluvial beveled surface are present along the uplifted side of the Hot Spring Fault. Matching this uplifted surface to the bedrock surface on the down-dropped side results in about 430 ± 60 meters of vertical separation. A lower surface on which Bautista Beds have been deposited is vertically displaced approximately 140 ± 20 meters across the fault in the Soboba Springs and Indian Canyon areas. The Bautista Beds in the Anza area, about 23 kilometers south, contain the Bishop Ash dated at 760 thousand years old (Sharp, 1981). Using this as a maximum age of cessation of deposition, a minimum uplift rate of about 0.2 mm/year is obtained.

Quaternary uplift of the San Jacinto Mountains has also resulted in the preservation of a sequence of uplifted fluvial terraces on the west side of the range. Most of these terraces are preserved on the western side of the Hot Springs Fault and therefore represent vertical separation along fault strands farther west, such as the Claremont and Casa Loma faults. Two levels of terraces are preserved in the Soboba Reservation Area and at least three levels of unpaired terraces are present along the sides of San Jacinto Canyon. None of these terraces have been dated or systematically described. These deposits should provide markers with which to estimate Quaternary uplift rates, but until numeric ages are obtained they can only serve as evidence of Quaternary (and most likely Holocene) uplift of the San Jacinto Mountains along the San Jacinto Fault Zone. Although the older (higher) terraces are cut by the Hot Springs Fault, none of the deposits are preserved on the uplifted side of the fault to serve as a marker for vertical separation across this structure.

Uplift along the Hot Springs Fault has also occurred at the northern end of the fault due to a restraining step-over arrangement with the southern end of the Claremont Fault. Uplift within the step-over is recorded by elevated topography bounded by the faults, tilted Miocene-Pliocene strata on the north side of the elevated topography within the step-over, and by a sequence of terraces that have developed along the major drainage (Potrero Creek) passing though the area into the San Jacinto Valley. Tung (2008) identified four levels of terrace deposits along Potrero Creek (Figure 2) and obtained numeric ages on two of the levels using radiocarbon dating on detrital charcoal. He calculated an uplift rate of about 1 mm/year within the step-over area based on the relative heights of the two dated terraces. This rate most likely underestimates the uplift within the step-over because Potrero Creek is not graded to the floor of the San Jacinto Valley due to the presence of waterfalls developed in resistant crystalline rocks downstream from the terraces. However, the second highest terrace (Qt_2 of Tung, 2008) can be followed out along the drainage at a slope of about 25 meters per kilometer to the point where the terrace would intersect the Claremont Fault. At this point the Qt_2 terrace would be approximately 67 meters above the active floor of the valley. Using the radiocarbon age of 26,000 ± 850 years for Qt_2 (Tung, 2008) the uplift rate across the Claremont Fault in the step-over is about 2.6 mm/ year.

Recent Faulting. Along most of its length, the main Hot Springs Fault marks the contact between the Mesozoic plutonic rocks of the San Jacinto Mountains, and the Pleistocene Bautista Beds. The

Bishop Ash has been recognized within the Bautista Beds at two locations (Sharp, 1981; Dorsey, 2002) indicating that the upper boundary of the unit is younger than 760 thousand years. The Hot Springs Fault and other faults in the zone bound and deform these strata so that the maximum age of the movement that offset the Bautista Beds is Late Pleistocene. Younger stream terrace deposits that lie unconformably on the Bautista Beds are also displaced by faults at many locations. These offset terraces have not been numerically dated but are presumed to be pre-Holocene based on the degree of soil development and their heights above active stream channels. A large number of deflected and beheaded stream channels, knickpoints in stream channels, and scarps indicate Late Quaternary offset on the Hot Springs Fault although most of these features are developed in resistant tonalitic bedrock such that the offsets may be pre-Holocene. The only feature that requires Holocene surface rupture within the Hot Springs Fault Zone is a 2 meter scarp in colluvium along the Soboba Fault that branches off the Claremont Fault at the southern edge of the step-over. The fault does not link to the main Hot Springs Fault and Holocene offset of the colluvium is most likely related to activity splaying off the south end of the Claremont Fault. A small amount of seismicity (primarily at the southern end of the Hot Springs Fault) attests to continued activity at depth along the fault zone, but further work is need to numerically document the last surface rupture.

DISCUSSION

Age of the Hot Springs Fault and Its Relation to the Rest of the San Jacinto Fault Zone. The Hot Springs Fault is interpreted to be less active than the nearby strands of the San Jacinto Fault Zone based on a comparison of the number and relative age of tectono-geomorphic features, microseismicity, and locations of major historic earthquakes. Both the nearby Claremont Fault and Casa Loma Fault are better expressed geomorphically and exhibit younger geologic offsets than are evident along the Hot Springs Fault. Dated offsets (Sharp, 1981) and trenching work near Anza (Rockwell et al., 1990) have provided evidence of Holocene surface rupture on the Casa Loma Fault and abundant microseismicity reveals continued movement at depth. Several large (M>6) earthquakes between 1890 and 1923 are believed to have occurred along the Casa Loma and/or Claremont faults in the area (Thatcher et al., 1975). No paleoseismic data exists for the Claremont Fault in the San Jacinto Valley Area, but offset Holocene deposits and seismic activity indicate that this fault is active. Trenching studies along the fault to the north have revealed evidence of Holocene surface rupture (Wesnowsky et al., 1991; Fumal personal communication, 2007) and the fault may exhibit slip rates as high as 20 mm/year (Kendrick, 2002).

Although Holocene displacement along the Hot Springs Fault is not precluded by the relationships along the fault, it is deemed unlikely and most of the movement along the fault appears to have occurred during Pleistocene time. The age of inception for the Hot Springs Fault Zone cannot be conclusively determined. However, geologic relationships suggest that some faults within the zone may have been active during early deposition of the Bautista Beds. Along most of the length of the Hot Springs Fault Zone, the faults bound the eastern edge of the Bautista Beds. The absence of these sedimentary rocks along the eastern side of the fault zone is either the result of erosion off the uplifted side, or early movement along faults in the zone that created the topographic gradients and led to the deposition of the Bautista Beds. The lack of other structures between the eastern edge of the Bautista Beds and the higher areas of the San Jacinto Mountains supports the idea that early movement along faults in the zone elevated the mountains and resulted in the deposition of the Bautista Beds. Continued oblique displacement resulted in further uplift of the San Jacinto Mountains on the east side of the fault and offset the Bautista Beds such that much of the eastern edge of the basins have been lifted and eroded. Geologic relationships in Garner Valley support this hypothesis. A linear mountain front developed in the crystalline rocks of the southern San Jacinto Mountains is aligned with the Hot Springs Fault to the northwest. This mountain front has been covered with Pleistocene gravels deposited as fans at the mouths of the canyons that have been identified as Bautista Beds (Axelrod, 1966; Dibblee, 1981; Dorsey, 2005). These deposits appear to thicken as they cross the linear escarpment. This partially buried mountain front may be interpreted as a degraded scarp from earlier motion along an older segment of the Hot Springs Fault Zone that ceased during Late Pleistocene time during the deposition of the Bautista Fans.

Movement along the Hot Springs Fault Zone is interpreted here to have occurred mainly during the earlier history of the San Jacinto Fault Zone and decreased through Late Pleistocene time as the parallel Casa Loma Fault began to accommodate most of the displacement. Slip along the San Jacinto Fault Zone was transferred to the northern Hot Springs Fault across the restraining step-over with the Claremont Fault that resulted in uplift within this step-over. At the southern end of the Hot Springs Fault, slip most likely was transferred across Garner Valley to the Thomas Mountain and Buck Ridge faults. As more slip was diverted from the Hot Springs Fault to the Casa Loma Fault, dip-slip displacement and uplift was also accommodated to a greater degree by the Casa Loma and Claremont faults. This westward migration of uplift is reflected in the topography along the Casa Loma Fault and recorded by the terraces in San Jacinto Canyon that lie between the Casa Loma Fault and the Hot Springs Fault.

Uplift of the San Jacinto Mountains along the Hot Springs Fault. Uplift of the San Jacinto Mountains by displacement on the Hot Springs Fault occurred during the Pleistocene. Oblique slip along faults in the zone created the steep mountain fronts and elevated plateaus that are present along the eastern side of the fault zone and account for the topographic differences that define the western edge of the mountains. The onset of deposition of the Bautista Beds probably marks the beginning of movement along the Hot Springs Fault Zone. The Bautista Beds are present in a linear alignment along the western side of the San Jacinto and Santa Rosa Mountains and were probably deposited in strike-slip basins associated with the early development of the San Jacinto Fault Zone (e.g., Sharp, 1967). In addition to any strike-slip motion on the early fault zone, the data presented here suggest that there was a significant component of west-side-down displacement on early faults in the zone that elevated the mountains and triggered deposition of the Bautista Beds along the mountain front. Dorsey and Roering (2005) presented paleocurrent data from Bautista Beds in the Anza Area that show a dominantly westward transport direction. The morphology of the Bautista-age fans in Garner Valley and along the Hot Springs Fault also indicate a westward transport direction and show that these sediments were shed from the rising mountains across the Hot Springs Fault Zone. Dorsey and Roering (2005) also noted that the transport directions do not reflect the current topography along the Casa Loma and Thomas Mountain faults and indicates that the uplift of Thomas Mountain along these faults occurred after deposition of the Bautista Beds. These relationships in turn support the hypothesis that the Hot Springs Fault is older than these other structures and the corresponding topography.

The Hot Springs Fault Zone was one of several fault segments that lifted the San Jacinto and Santa Rosa mountains. To the southeast, the western edge of the Santa Rosa Mountains is marked by the southern half of the Buck Ridge Fault and the Santa Rosa Fault. Like the Hot Springs Fault, both of these faults exhibit large topographic contrasts, relatively low amounts of seismicity, and less geomorphic evidence of Holocene activity than other faults within the San Jacinto Fault Zone. Vertical displacement along the Santa Rosa Fault is about 400 meters at its south end based on offset of mylonite in the Santa Rosa Mountains and this correlates with a 400 to 500 meter high escarpment of the mountains (Dorsey, 2002). Exactly how much uplift occurred along the Hot Springs Fault Zone is not known because no markers have been identified on the uplifted side. Geomorphic evidence points to at least 430 meters of uplift and differences in helium cooling ages across the Hot Spring Fault suggest 800 meters of vertical offset on the fault (Wolf et al., 1997).

The Bautista Beds along the northern half of the Hot Springs Fault Zone are composed mainly of fine-grained sandstones and lacustrine shales. The current location of these fine-grained deposits next to steep topography along the fault and the absence of coarser deposits indicate that displacement along the Hot Springs Fault continued until at least 760 thousand years ago (age of the Bishop Ash) and has juxtaposed low-energy environment deposits against high-relief mountain fronts. The offset terraces that overlie the Bautista Beds show continued uplift. However, the locus of uplift has since migrated westward to the Claremont and Casa Loma faults as displacement along the Hot Springs Fault Zone decreased. The stream terraces present along the east side of the Claremont Fault and in San Jacinto Canyon east of the Casa Loma Fault provide evidence of Latest Pleistocene and Holocene uplift along these faults.

The timing and mode of uplift of the San Jacinto Mountains remains to be resolved. Geologic and geomorphic relationships along the Hot Springs Fault indicate that motion along this structure has contributed to the uplift, possibly more than 800 meters, but cannot account for most of the topographic development of the range. Despite the fact that the eastern side of the range is extremely steep, the morphology of the east side is not indicative of active faulting and no active structures have been mapped along the mountain front. This suggests that most of the uplift of the San Jacinto Range must have occurred prior to Pleistocene time. George and Dokka (1994) suggested that the San Jacinto Block experienced rapid exhumation and cooling during the Late Cretaceous based on apatite fission-track ages. However, Wolf and others (1997) have argued against this interpretation based on helium cooling ages and prefer a younger Middle Tertiary age of uplift. Much of the topography of southern California is related to the opening of the Gulf of California around 5 million years ago (Atwater, 1970), and local extension within the Salton Trough occurred from the Late Miocene to the Pleistocene (Axen and Fletcher, 1998). Uplift of the Santa Rosa Mountains may have occurred along the West Salton Detachment Fault that is exposed along the southern east side of the range and was most active between 5 million and 2 million years ago (Axen and Fletcher, 1998). This extensional episode may have led to uplift of the San Jacinto Mountains in the footwall of the extensional structure, although the location and amount of required uplift is problematic and a specific structure along the east side of the San Jacinto Mountains has not been identified.

CONCLUSION

The Hot Springs Fault Zone marks the western edge of the San Jacinto Mountains and was involved in the Pleistocene initiation and development of the San Jacinto Fault Zone. The zone exhibits right-lateral oblique slip kinematics that partially lifted the San Jacinto Mountains and controlled the deposition of the Bautista Beds during Pleistocene time. Continued movement on faults within the zone have truncated the Bautista Beds and vertically displaced them in some locations. The Hot Springs Fault shows little evidence of Holocene displacement and much of the strain accommodated by the fault during Pleistocene time has probably jumped westward to the Casa Loma Fault.

REFERENCES

Atwater, T., (1970). Implications of plate tectonics for the Cenozoic tectonic evolution of western North America. *Geol. Soc. Am. Bull.*, v. 84, p. 1375–1391.

Axelrod, D., (1966). The Pleistocene Soboba Flora of Southern California. *California University Publications in the Geological Sciences*. v.60, 79 pages.

Axen, G.J., and Fletcher, J.M., (1998). Late Miocene- Pleistocene extensional faulting, northern Gulf of California, Mexico and Salton Trough California. *Inter. Geol. Review*, v. 40, p. 217–244.

Becker, T.W., Hardebeck, J.L., and G. Anderson, (2005). Constraints on fault slip rates of the southern California Plate Boundary from GPS velocity and stress inversions. *Geophys. J. Int.*, 160, 634–650.

Biasi, G.P., Weldon, R.J., Fumal, T.E., and Seitz, G.G., (2002). Paleoseismic event dating and conditional probability of large earthquakes on the southern San Andreas Fault, California. *Bull. Seism. Soc. Am.*, 92, 2761–2781.

Crowell, J.C., (1974). Origin of Late Cenozoic basins in southern California: *in* Ernst, W.R., Ed., Tectonics and sedimentation: Society of Economic Paleontologists and Mineralogists Special Publication 22, 190–204.

Dibblee, T.W., Jr., (1981). "Geology of the San Jacinto Mountains and Adjacent Areas": *in* Geology of the San Jacinto Mountains, South Coast Geological Society Annual Field Trip Guidebook No. 9.

Dorsey, R. J., (2001). Two-stage evolution of the San Jacinto Fault Zone: Crustal response to Pleistocene oblique collision along the San Andrea fault. *Eos (Transactions, American Geophysical Union)*, v. 82, p.F933.

Dorsey, R. J., (2002). Stratigraphic record of Pleistocene initiation and slip on the Coyote Creek Fault, Lower Coyote Creek, southern California. Geol. Soc. Am. Special Paper #365.

Dorsey, R. J., and Roering, J.J., (2005). Quaternary landscape evolution in the San Jacinto Fault Zone, Peninsular Ranges of southern California: Transient response to strike-slip fault initiation. *Geomorphology*, v. 73, p. 16–32.

Fletcher, K.E., Johnson, G., Kendrick, K., Hudnut, K., and Sharp, W., (2006). Long-term slip rate on the southern San Andreas Fault determined by Th-230/U dating of pedogenic carbonate. American Geophysical Union, Fall Meeting, Abstract #T13B-0506.

Fraser, D.M., (1931). Geology of the San Jacinto Quadrangle south of San Gorgonio Pass, California. *California Journal of Mines and Geology* v.27, n.4, p.494–540, map scale 1:125,000.

Frick, C., (1921). Extinct vertebrate faunas of the badlands of Bautista Creek and San Timoteo Canyon, southern California. University of California Publications, *Dept. of Geol. Sci. Bull.*, v. 12, n. 5, p. 277–424.

Given, D., (1981). "Seismicity of the San Jacinto Fault Zone" in Geology of the San Jacinto Mountains, South Coast Geological Society Annual Field Trip Guidebook No. 9.

Hill, R. L., (1981). "Geology of Garner Valley and vicinity" in Geology of the San Jacinto Mountains, South Coast Geological Society Annual Field Trip Guidebook No. 9.

Jennings, C., (1977). Geologic Map of California. State of California Division of Mines and Geology.

Kahle, J.E., (1987). The San Jacinto Fault Zone (The Claremont, Casa Loma, and Related Faults) in the Lakeview and El Casco Quadrangles, Riverside County, California: California Division of Mines and Geology Fault Evaluation Report FER-179.

Kendrick, K.J., Morton, D.M., Wells, S.G., and R.W. Simpson, (2002). Spatial and temporal deformation along the northern San Jacinto Fault, southern California: Implications for slip rates. *Bull. Seism. Soc. Am.*, 92, 2782–2802.

McGill, S., Kendrick, K., Weldon, R., and Owen, L., (2007). Pleistocene and Holocene slip rate of the San Andreas Fault at Badger Canyon, San Bernardino, California. *SCEC Annual Meeting Proceedings and Abstracts*, V. 27, p. 144.

Morton, D.M., Matti, J.C., (1993). Extension and contraction within an evolving divergent strike-slip complex: The San Andreas and San Jacinto fault zones at their convergence in southern California. *in* Powell, R.E., Weldon, R.J., II, and Matti, J.C., eds., The San Andreas Fault System: Displacement, Palinspastic Reconstuction, and Geologic Evolution: Boulder, Colorado, Geological Society of America Memoir 178.

Morton, D.M., (1995). "Subsidence and ground fissures in the San Jacinto Basin. USGS Open-file report 94-532, p.29–31.

Onderdonk, N.W., (1998). The tectonic structure of the Hot Springs Fault Zone, Riverside County, California. MA Thesis, University of California, Santa Barbara, 75p.

Rockwell, T.K., Loughman, C.C., and P.M. Merrifield, (1990). Late Quaternary rate of slip along the San Jacinto Fault Zone near Anza, Southern California. *J. Geophys. Res.*, 95, no. 6, 8593–8605.

Sanders, C., Kanamori, H., (1984). A Seismotectonic Analysis of the Anza Seismic Gap, San Jacinto Fault Zone, Southern California. *J. Geophys. Res*, vol.89, no. B7, p.5873–5890.

Savage, J.C., and W.H. Prescott, (1976). Strain accumulation on the San Jacinto Fault near Riverside, California. *Bull. Seism. Soc. Am.*, 66, 1749–1754.

Sharer, K.M., Weldon, R.J., Fumal, T.E., and Biasi, G.P., (2007). Paleoearthquakes on the Southern San Andreas Fault,

Wrightwood, California, 3000 to 1500 B.C.: A new method for evaluating paleoseismic evidence and earthquake horizons. *Bull. Seism. Soc. Am.*, 97, 1054°1093.

Sharp, R.V., (1967). San Jacinto Fault Zone in the Peninsular Ranges of Southern California: *Geol. Soc. Am. Bull.*, 78, 705–730.

Sharp, R.V., (1981). Variable Rates of Late Quaternary Strike Slip on the San Jacinto Fault Zone, Southern California. *J. Geophys. Res.*, 86, B3,1754–1762.

Thatcher, W., Hileman, J.A., and T.C. Hanks, (1975). Seismic slip distribution along the San Jacinto Fault Zone, southern California, and its implication. *Geol. Soc. Am. Bull.*, 86, 1140–1146.

Tung, J., (2008). Examining the style and uplift along the Claremont Strand of the San Jacinto Fault Zone. MS Thesis, California State University, Los Angeles, 75p.

Wesnousky, S.G., Prentice, C.S., and K. Sieh (1991). An offset Holocene stream channel and the rate of slip along the northern reach of the San Jacinto Fault Zone, San Bernardino County, California. *Geol. Soc. Am. Bull.*, 103, 700–709.

Wolf, R. A., Farley, K. A., Silver, L. T., (1997). Assessment of (U-Th)/He thermochronometry: The low-temperatures history of the San Jacinto Mountains, California. *Geology*, v. 25, p. 65–68.

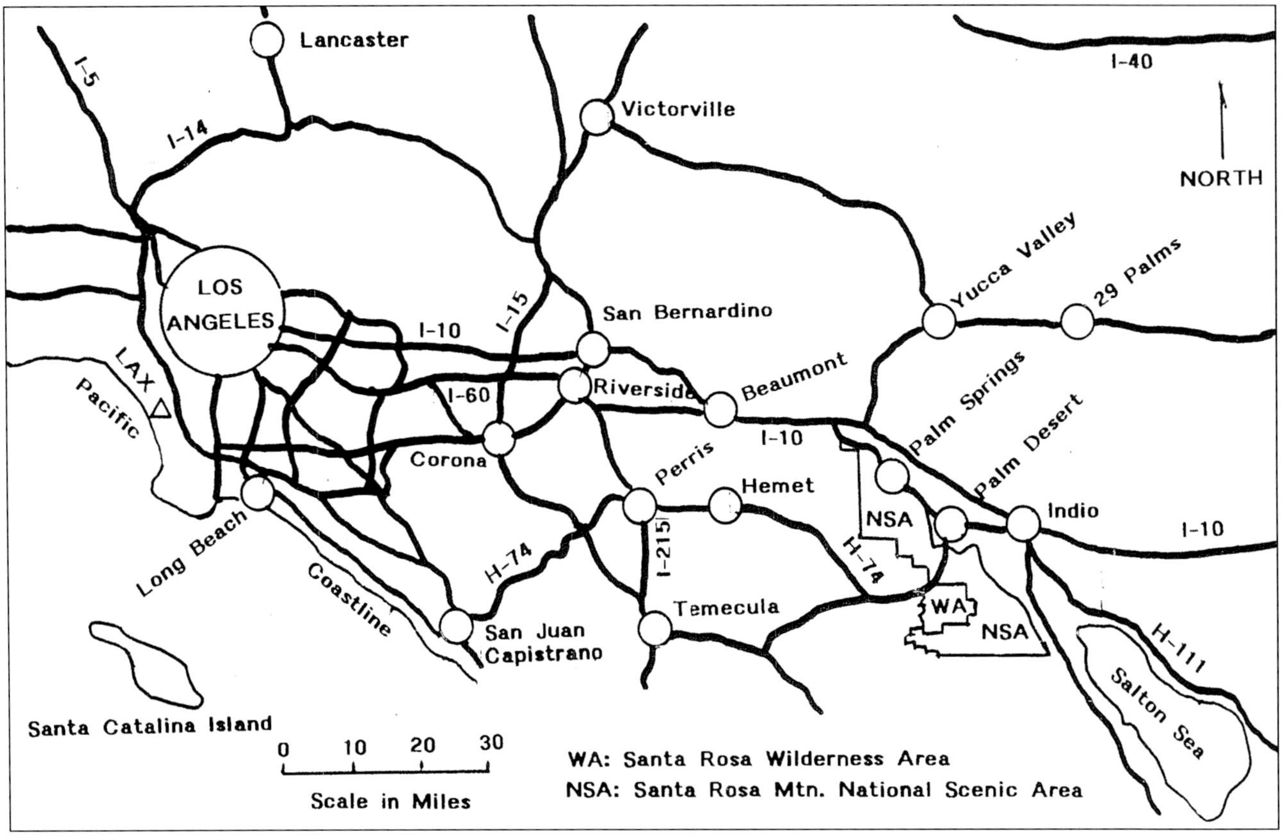

Figure 1. Site location map.

An Introduction to the Geology, Hydrogeology, Economic Potential and Geologic Features of the Santa Rosa Wilderness Area and Santa Rosa Mountain National Scenic Area, Riverside County, California

Harry M. Quinn *Originally self-published in 1995*

I. INTRODUCTION

The Santa Rosa Wilderness Area and the Santa Rosa Mountain National Scenic Area* are located about a hundred and twenty miles east-southeast of Los Angeles (Figure 1). Access to these areas is limited, however, Highway 74 does abut the Wilderness Area and cuts through the Scenic Area. Both areas are basically only accessible by hiking or on horseback.

The Santa Rosa Wilderness Area consists of approximately 20,000 acres of land located in the northeast portion of the Santa Rosa Mountains. It is bounded on the north, east and south by lands belonging to the Santa Rosa Mountain National Scenic Area and on the west by private and Forest Service lands (Figure 2). The Santa Rosa Mountain National Scenic Area contains about 200,000 acres of land that extends along the mountain front from Anza Borrego Desert State Park on the southeast to near the Snow Creek area on the northwest. The Santa Rosa Mountain National Scenic Area surrounds the Santa Rosa Wilderness Area on three sides, abuts the east and part of the north edges of the San Jacinto Wilderness Area and shares a common boundary with the Anza-Borrego Desert State Park on the south (Figure 2). The northeast boundary of the Scenic Area is about coincidental with the southwestern boundary of the Colorado Desert Geomorphic Province. The San Jacinto Wilderness Area, Santa Rosa Wilderness Area and the Santa Rosa Mountain National Scenic area are all located along the extreme northeast boundary of the Peninsular Range Geomorphic Province. This well defined geomorphic and geologic province includes the southwestern portion of California, both onshore and offshore, and the Baja California Peninsula. The geologic composition of the Peninsular Range Province consists principally of pre-late Cretaceous metasedimentary and metavolcanic rocks that have been intruded by late Cretaceous age plutonic rocks, mainly of granitic composition. These late Cretaceous plutonic rocks make up what is commonly referred to as the Southern California Batholith. Only scattered, usually thin deposits of sedimentary rocks are present within the wilderness and scenic areas.

* The United States Congress established the Santa Rosa Wilderness Area in 1984 with the passage of the California Wilderness Act (Public Law 98-425), managed jointly by the Bureau of Land Management (BLM) and U.S. Forest Service. According to BLM, the Santa Rosa Wilderness Area currently contains 61,600 acres of BLM and Forest Service lands.

The Secretary of the Interior designated approximately 194,000 acres of the Santa Rosa and San Jacinto Mountains as the Santa Rosa Mountains National Scenic Area on March 31, 1990.

Since the original publication date of this paper, in October 2000 the U.S. Congress provided greater protection to the area's natural qualities with the establishment of the Santa Rosa and San Jacinto Mountains National Monument (Public Law 106-351). BLM and Forest Service share management of the monument's 280,071 acres. The National Monument designation replaces the National Scenic Area designation. The Wilderness Area, however, remains.

GENERAL GEOLOGIC SETTING

The Santa Rosa Wilderness Area and Santa Rosa Mountain National Scenic Area combined, contain approximately 220,000 acres of land that is situated in the northeast portion of the Peninsular Range Geomorphic Province. . It is characterized by elongate ranges and valleys whose general northwest-southeast trend is terminated abruptly on the north by the east-west trending Transverse Range Province. The northeast side of the Peninsular Range Province is coincident with the Colorado Desert Province. The portion of this province that lies above sea level is approximately 900 miles long, 140 miles in maximum width and 55 miles in average width. An additional large portion is located below the Los Angeles Basin and submerged beneath the Pacific Ocean. Offshore outcrops of the province lithologies are represented by the southern Channel Islands, such as Santa Catalina and San Clemente Islands.

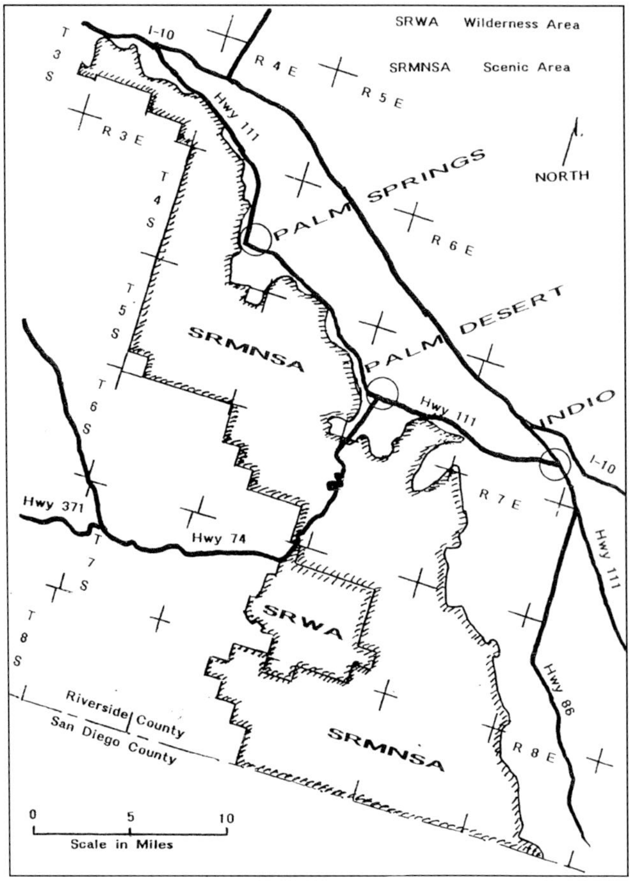

Figure 2. Site access map.

The higher parts of the Peninsular Range Province include the Santa Ana, San Jacinto, Santa Rosa, Agua Tibia, and Laguna Mountains in California and the Sierra Juarez, Sierra Pedro Martir and other peaks that form the "backbone" of the Baja Peninsula. San Jacinto Peak, near the northern end of the province is the highest point, rising to an elevation of 10,850 feet above mean sea level. Santa Rosa Mountain's Toro Peak rises to an elevation of 8,716 feet. These and other high mountains occupy the northeastern and eastern parts of the province and are bounded on the east by striking scarps, 6,000 to more than 9,000 feet high. These scarps abut, and form the boundary with, the Colorado Desert Province in California and they tend to be displaced *en echelon* along subparallel traces of major fault zones. The scarps are locally cut by some tortuous, very steep-walled canyons, one of which (Deep Canyon) traverses both the Santa Rosa Wilderness Area and the Santa Rosa Mountain National Scenic Area.

The general topography of the Peninsular Range Province becomes less rugged toward the coast. It is characterized by combinations of subdued upland surfaces, prominent ridges and peaks, longitudinal valleys which are in part fault controlled, numerous basins and broad, mature valleys, and some tortuous, very steep-walled canyons. Farther to the west is an irregular coastal plain, ranging from a few hundred feet to as much as 30 miles in width, on which marine and fluviatile terraces are displayed. This coastal plain is underlain chiefly by sedimentary and volcanic rocks of late Mesozoic and Cenozoic age, and its surface is interrupted here and there by ridges and other projections of older, more resistant rocks, such as the Palos Verdes Hills.

Lithologically the Peninsular Range Province consists chiefly of pre-late Cretaceous metasedimentary and metavolcanic rocks that have been altered by the intrusion of late Cretaceous age plutonic igneous rocks. The metasedimentary rocks are mainly phyllites, schists, slates, quartzites and marbles, but locally consist of amphibolites and calc-silica hornfels. The metavolcanics are chiefly meta-basalts, meta-rhyolites, and meta-andesites. The intruded plutonic igneous rocks are mainly granitic, such as tonalities, monzogranites, and granodiorites, with some scattered gabbroic rocks. Both the metasedimentary and plutonic rocks are locally intruded by quartz and/or feldspar dikes, veins, and veinlets.

Continued uplifting of the region has limited the deposition of sedimentary rocks within the wilderness and scenic areas. Because of the continued uplift, erosion has prevented the deposition of any thick sedimentary rock sequences. Only Pleistocene and younger rocks are found within the two areas.

The generalized geology of the Santa Rosa Wilderness Area and the Santa Rosa Mountain National Scenic Area is shown on Figure 3. More detailed geology is presented in the text and additional detail can be found in some of the material referenced in this paper. The Bureau of Land Management is presently preparing a more detailed Geologic Map of the Santa Rosa Mountain National Scenic Area.

Pre-Late Cretaceous Metasedimentary Rocks. The oldest rocks in the Santa Rosa and San Jacinto Mountains may be some highly metamorphosed rocks, amphibolite schists and gneisses, of questionable Precambrian age. These rocks may be in part equivalent to the San Gorgonio Gneiss found over in the Transverse Range Province. Rocks of this group may be limited to the southern portions of the scenic area and may not be present in the wilderness area.

A group of metasedimentary rocks, consisting mainly of quartzite, mica-schist, recrystallized limestones, slates and granitic gneisses of questionable Paleozoic age, are the most common non-plutonic rocks to be found within the boundaries of the Wilderness and Scenic Areas. These exhibit a lower degree of metamorphism than the ones mentioned above.

The quartzites appear to have been formed from earlier deposited sandstones and argillaceous

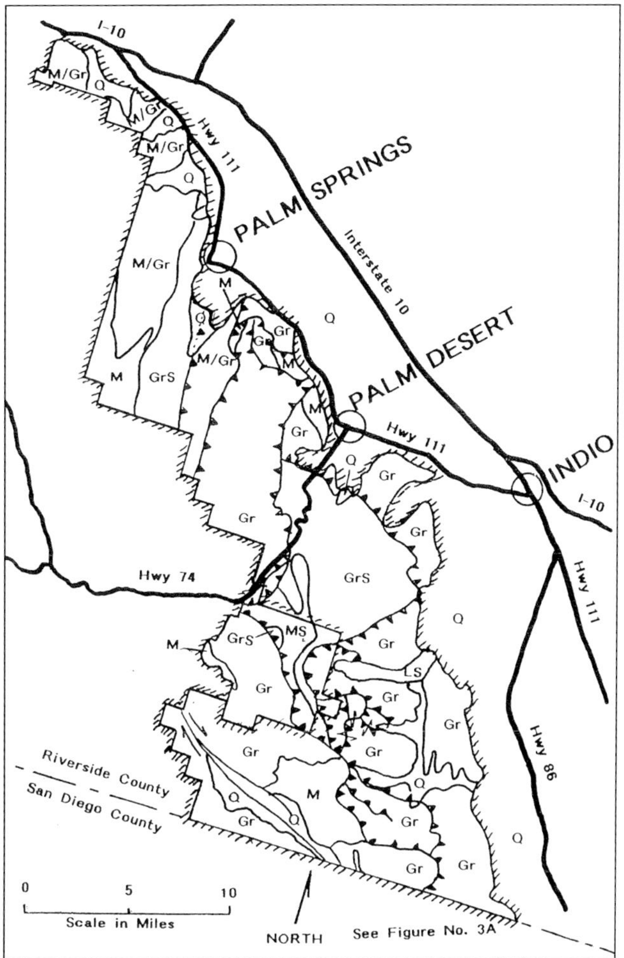

LEGEND

Approximate Boundaries of the Wilderness and Scenic Areas

Strike-Slip-Fault: Portion of the San Jacinto Fault Zone, Right Lateral

Thrust Fault: Tick Marks on the Up-Thrown Side

LS Martinez Mountain Rock Slide: Second largest such feature in North America

Q Quaternary Sediments, may include some Pliocene/Pleistocene rocks as mapped

Gr Granitic Intrusive Rocks: Late Cretaceous in Age

GrS Sheared Granitic Rocks: Locally part of the Santa Rosa Mountain Shear (Mylonite) Zone

M Metamorphic Rocks: Pre-Late Cretaceous in Age

M/Gr Metamorphic and Granitic Rocks Undifferentiated

MS Sheared Metamorphic Rocks: Locally part of the Santa Rosa Mountain Shear (Mylonite) Zone

Figure 3. Generalized geologic map of the Santa Rosa Wilderness and Santa Rosa Mountain National Scenic Areas, Riverside County, California.

sandstones. Some of the quartzites appear to be rather massive. Where the argillaceous material was more abundant, the rocks appear as alternating units of quartzite and quartzitic, mica-schist. The mica-schists locally grade into slates, probably where the argillaceous material had formed shale sequences.

The mica schists consist of muscovite, biotite, feldspar, and quartz. They commonly alternate with quartzites, slates, and gneisses. They tend to be platy and tan to dark brown on the weathered surfaces. Locally the schists contain some scattered garnets.

The recrystallized limestones (marbles) tend to be dolomitic, at least they are in the Pinyon Flats area. The thickness of these limestone beds is quite variable, ranging from a few feet to over one hundred feet. They are commonly coarse-grained, but appear to range from fine- to coarse-grained. They tend to be white to tan and to contain scattered small, black graphite blobs.

The interbeds of granitic gneiss have been locally referred to as sills. However, it is highly possible that some of these gneisses are just metasedimentary rocks that were formed from argillaceous, arkosic sandstones. These granitic gneisses locally grade into biotite, felspathic, mica-schists, which may indicate a pre-metamorphic depositional change rather than an intrusive sill.

In the northern portion of the Scenic Area a series of dikes and sills can be found intruding into the metasedimentary rocks (Procter, 1968, pp. 17–18). These sills and dikes commonly contain xenoliths, or fragments, of the country rock.

There are some minor amphibole-rich hornblende rock bodies that result from altered gabbroic or basaltic rocks. Where sheared, these rocks locally contain slip-fiber asbestos. These bodies tend to be thin and discontinuous.

The total thickness of this metasedimentary sequence is not known, but it appears to be in excess of 8,000 feet (Proctor, 1968, p. 18). The age of these rocks is also unknown; however, similar type altered carbonate rocks to the south have yielded Ordovician conodonts (Miller, 1983, pp. 410–412) and some possible crinoid columnal fragments have been reported from altered carbonate (limestone) rocks near Cathedral City (Miller, 1944, pp. 21–25). Based on the few fossil remains and some lithologic correlations to similar rocks in the San Bernardino Mountains, these metasedimentary rocks are probably Paleozoic in age. The metamorphic rocks appear to have been derived from a highly varied geologic terrain.

Cretaceous Plutonic Igneous Rocks. The plutonic igneous rocks present in the Wilderness and Scenic Areas are part of the Southern California Batholith. This igneous material has been radiometrically dated at 90–110 million years, or about mid-Cretaceous, indicating an approximate age for the plutons' emplacement (Bushee, et al, 1963, pp. 803–806). The plutonic type rocks in the wilderness and scenic areas consist of granodiorites, leucogranites, quartz monzonites, tonalities, and monzogranites.

Other less common igneous rocks include quartz, feldspar, aplite, and pegmatite dikes, veins and veinlets. These cross-cut the igneous plutonic and to a lesser extent some of the metasedimentary rocks. They were derived from a silica rich magma and were intruded after the main batholithic rocks were in place.

Post-Cretaceous Sedimentary Rocks. Only sedimentary rocks of Pleistocene and younger age have been mapped within the Wilderness and Scenic Areas. The beds are poorly consolidated, sometimes cross-bedded, and locally held together by clay. They consist primarily of alluvial deposited arkosic sands, gravels, silts, and clays with common to scattered boulders. Based on scattered land vertebrate remains, the oldest sedimentary beds are considered to be early Pleistocene in age. These lithologic units are discontinuous and were probably deposited in that manner, but the similarity of the sediment type and contained faunas indicates that they were deposited under similar conditions and at about the same time. Sharp (1967) states that "these deposits accumulated under fluviatile and lacustrine conditions in a series of coalescing intermountain basins which formerly covered much of the granitic and metamorphic terrain."

Younger terrace deposits sit unconformably upon the Bautista beds and bedrock areas. These beds have been in turn dissected by later erosion, indicating the presence of a post-terrace uplift and erosion episode in the Santa Rosa Mountain region. Colluvium has been developed on many of the terrace deposits, as well as some of the bedrock areas.

The most recent sediments are found In the stream and canyon bottoms. These are recent alluvial deposits, consisting of clay to boulder size material that is poorly consolidated. This material is presently being eroded and re-deposited by intermittent and occasional torrential stream activity moving down them.

GEOLOGIC AND STRUCTURAL HISTORY

Geologic History. During the late Precambrian(?) to Late Paleozoic(?) the area now occupied by the Santa Rosa and San Jacinto Mountains National Monument was a marine basin. This marine basin was receiving sediments from a highland area, probably to the northeast. These sediments had to be deposited as a depositional sequence well in excess of 8,000 feet thick, based on the present thickness of the metasedimentary rocks. The sediments were then subjected to regional metamorphism. Based on the metamorphic character, an almandine-amphibolite facies, this material needed to be subsided into the crust some 4 to 10 miles. However, the heat from the later intruding granitic magmas may have helped altered these sediments to the almandine-amphibolite facies without requiring them to have been pushed that deep into the crust.

Igneous plutonic magmas, mainly of a granitic composition, were generated during the Cretaceous. These magmas were intruded into the metamorphosed sediments prior to their being cooled and solidified. Based on radiometric dates, the magmas solidified around 90 to 110 million years ago, or during the mid-Cretaceous. After these plutonic rocks had cooled, silica rich dike material, mainly quartz, aplite, and pegmatite dikes, were intruded throughout the entire region.

After the intrusive activity had ceased, deep seated shearing began to take place. This shearing at depth, most likely caused by deep seated faulting, cataclastically deformed the metasediments and plutonic rocks. This shearing is believed to have begun around 70 million years ago and to have ceased by about 62 million years ago (Erskine and Wenk, 1985, pp. 275).

Uplifting of the Peninsular Ranges probably did not begin until after the shearing ceased. This would indicate that the tectonic uplifting of the Santa Rosa/San Jacinto Mountains began later than 62 million years ago. By late Lower or early Middle Miocene time, the Santa Rosa Mountain area was shedding alluvial sediments into the adjacent Salton Sink area to the east. While some Eocene-age rocks are known from the east side of the Salton Trough, no detrital material older than Miocene is known from the western side. This depositional pattern would indicate that the uplifting of the northeastern portion of the Peninsular Ranges did not begin until late Oligocene or early Miocene time, or about 60 to 55 million years ago.

The oldest sedimentary rocks mapped within the Scenic Area are fluvial deposits assigned to the Pleistocene Bautista beds. These occur along uplifted terraces near the base of the mountain front, commonly adjacent to the mouth of the canyons. Most of the sedimentary rocks present within the Wilderness and Scenic Areas are of recent origin, probably deposited during the last 10,000 years. This lack of sedimentary deposits indicates that uplift and erosion have continued through to the present.

Structural History. Postdating the igneous and metamorphic activity was some deep seated shearing that has been uplifted and is now exposed as a structural unit referred to as the Santa Rosa Mylonite Zone, or Shear Zone. This structural feature both parallels and cuts across the previously described plutonic igneous and metamorphic rock boundaries. The mylonite zone was formed by a process in which rocks are broken and granulated by stress during movement along faults. This movement, or shearing, had to have occurred at great depth along a large fault zone. The upper drainage in Deep Canyon and much of the drainage in Palm Canyon is controlled by this shear zone.

Several faults are present within the wilderness and scenic areas. Many of these faults are mapped as low angle reverse faults, or thrust faults. A number of strike-slip, normal and high angle reverse faults are also present. Some of these faults appear to be controlled by the shear zone, and in part define some of the igneous and metamorphic rock boundaries. Deep Canyon is controlled by a thrust fault and the action of another thrust fault may have aided in the development of the Martinez Mountain rock slide. There are also a number of fracture and joint patterns present, most of which appear to be directly associated with the faulting. These faults, fractures and joints play an important role in controlling surface water runoff and ground water recharge and retention.

A portion of the San Jacinto Fault Zone passes through the southwestern part of the area (Figure 3). This fault is among the most active in the southern California region.

GENERAL HYDROGEOLOGY

Ground water within the Santa Rosa Wilderness Area and Santa Rosa Mountain National Scenic Area can be found in two principal types of hydrogeologic settings. These include 1) ground water in fractured basement rocks, and 2) ground water filling void spaces in sedimentary rocks. Most of the world's known ground water reserves are contained in porous sedimentary rocks and most of the Wilderness and Scenic Area's ground water is contained in the fractured basement rocks. Thus, usable ground water reserves within the wilderness and scenic areas are scarce to non-existent.

Ground Water In Fractured Basement Rocks. Solid pieces of fresh plutonic igneous and metamorphic rocks commonly have porosities of less than 3% and most less than 1% (Davis and DeWiest, 1966, p. 319). The pore spaces that are commonly present are small and generally not interconnected, so permeabilities are so small that they can usually be regarded as zero. Appreciable porosity and permeability can be developed through fracturing and weathering. Aquifer tests have shown that permeability in fractured rocks can be as much as one thousand times that of a piece of the same rock that is unfractured or even intensely weathered.

Fractures associated with faulting appear to produce the best water sources. Fractures that are not associated with pronounced faulting tend to produce only small increases in overall porosity and permeability of the rock. Results from grout injection at dam sites, and observations of exposed rock in tunnels and road cuts indicate that the total volume of open space within fractures is only a small percentage of the total volume of the rock. The fractures, or cracks, are almost always nearly closed, especially with depth. Fracture widths in excess of 2 millimeters in fresh, undisturbed rock are not very common. Some fracturing is known to exist at great depths, as water flowing into mines at depths of over 1,000 feet can be large enough to supply water to wells. Where water charged fractures cross low points, such as in stream beds and canyon bottoms, springs are locally created. Springs of this type can be found in both the Wilderness and Scenic Areas.

Most water wells completed in fractured crystalline rock are low yield wells, commonly yielding less than 10 gallons per minute. These wells can provide enough water to handle domestic needs, but often cannot provide water for agricultural or commercial development needs. Wells of this low yield type are common to the Pinyon Flats area, immediately adjacent to portions of the Wilderness and Scenic Areas.

Effects of weathering may extend to depths of over 300 feet into bedrock regions, however, weathering to depths of 5 to 50 feet are more normally encountered. This weathered zone, where it occurs below the water table, can produce a ground water zone capable of sustaining water wells. This near surface weathered zone, however, usually acts more as a conduit for surface water to percolate down and recharge the underlying fractured basement rocks, as can be seen in the Pinyon Flats area. The presence and thickness of this weathered zone can be determined by refraction seismic surveys, like that commonly used for rippability studies at development sites. Such studies conducted at sites adjacent to the Santa Rosa Mountain National Scenic Area support the presence of a weathered zone that is commonly less than 100 feet thick, and usually less than 50 feet in thickness (personal observations).

Some metamorphic rocks, such as marbles, contain minerals which are subject to relatively rapid dissolution by circulating ground water. Even though numerous solution zones and cavities may be formed in rocks such as marble, making local permeability large, the pore space for most of these rocks is probably less than 5%. Because of the low pore space, dissolution is usually highly localized along fractures or faults and does not affect the rock as a whole. No large solution features have been reported or personally observed in any of the marbles present in the wilderness or scenic areas; however, there is a good chance that some do exist.

Ground Water In Sedimentary Rocks. The alluvial and sedimentary rocks in the Santa Rosa Wilderness and Santa Rosa Mountain National Monument are confined mainly to the canyon bottoms and canyon mouth fans. For the most part, these sediments are composed of medium- to coarse-grained detrital material with some thin fine-grained detrital rock interbeds. Porosities within the coarse alluvial material probably range between 10 and 25 percent. These rocks tend to become more consolidated, and therefore less porous, with depth. The alluvial sediments in the canyon bottoms tend to be thin and to commonly rest directly on the bedrock. Where the bedrock comes to the surface along the bottom of the canyons, ground water held within the alluvium can be forced to the surface to produce a seep or spring.

Even though porosities in the alluvial sediments may be high, the permeabilities may not. Some of the reduction in permeability within these sediments can be related to closer packing of the grains, but most is due to the restriction of pore spaces by the presence of cement, such as clay and calcium carbonate. The large number of variables influencing permeability make it impossible to predict permeabilities based on rock porosity or rock type. However, usually the more porous the rock the better the permeability. The best wells in this alluvial material would result from permeable zones found some distance below the static ground water table. Ground water in the area of the canyon mouth fans tends to be rather deep, commonly deeper than 100 feet below ground level. These alluvial fan areas commonly serve as the recharge areas for the Coachella Valley's main ground water basin. Precipitation that falls on the bordering highlands, from occasional heavy winter storms and/or locally heavy summer thunderstorms, drains off the highland areas and out onto the valley floor. These surface waters sink rapidly into the more porous alluvial sediments adjacent to the mountain front, especially the canyon mouth alluvial fans, and recharge the deep aquifers under the adjacent valleys.

Formation of Springs. Most of the springs within the Santa Rosa Wilderness Area and the

Santa Rosa Mountain National Monument occur in low spots along major fracture systems. Many of these fracture systems can be related to adjacent faulting and some are associated with mappable fault zones. Some seasonal springs in the bottoms of the large canyons like Deep Canyon and Palm Canyon, appear to be related to ground water surfacing from alluvial sediments at local basement highs. In the past, the regions of the two canyons contained more perennial springs than they do today. Many of the springs that over forty years ago ran water for eight to twelve months of the year are now dry, or only produce water for a short time during wet seasons. Most of these springs dried up prior to the drilling of wells for domestic water, especially in the Pinyon Flats area. Many of the springs in the Pinyon Flats-Asbestos Mountain area were used by Mr. Jim Wellman to water his cattle. Some of these springs had been developed as water sources before Mr. Wellman bought the ranch from Mr. Frank Paul in 1934. When the "Wellman 101 Ranch" was sold to Mr. H. B. Chesser in late 1945 or early 1946, many of these springs had already dried up or become unreliable as year-round water sources.

APPRAISAL OF IDENTIFIED MINERAL RESOURCES

The Santa Rosa Wilderness Area and the Santa Rosa Mountain National Scenic Area both contain known potentially economic mineral deposits. Most of these potential mineral deposits contain nonmetallic minerals, such as dolomite, limestone, garnet, quartz, and feldspar. In addition, there are a number of scattered areas that contain potential aggregate deposits. Within the Santa Rosa Mountain National Scenic Area are a number of locations with potential economic metallic mineral deposits. Even though the region has been prospected off and on for the past hundred years without much success, the need for new minerals, metals, and new mining techniques could turn some of the known mineralized zones into economic deposits. Both the wilderness and scenic areas appear to have been formed and dedicated without having been adequately evaluated for their actual mineral potential, especially their nonmetallic mineral potential. A number of responses to the United States Forest Service Draft Plan covering the Environmental Impact Statement for the San Bernardino National Forest, Land and Resource Management Plan, suggested that the plan did not give adequate consideration to the economic and strategic values of the minerals on the Forest lands (Record of Decision, Final Environmental Impact Statement, 1989, p. 19).

Even though there are some warm water springs within and adjacent to the Santa Rosa Mountain National Scenic Area, the potential for geothermal resources within the two area appears to be low. The low potential is mainly based on the lack of permeable rocks in the subsurface commonly needed for the storage and production of the hot fluids required for present day geothermal operations.

The wilderness and scenic areas appear to have a very low to nonexistent potential for commercial hydrocarbon (oil and gas) deposits. The rocks within these two areas are mainly igneous and metamorphic rocks, neither of which are commonly associated with commercial petroleum deposits. The area also lacks the potential for petroleum source rocks" reservoir rocks and structural traps needed for a commercial petroleum producing province.

Mineral Resource Potential. The mineral resource potential for the Santa Rosa Wilderness Area and the Santa Rosa Mountain National Scenic Area can be subdivided into two main types: 1) metallic mineral deposits and 2) nonmetallic mineral deposits. Each of these two types of mineral deposits are discussed in the following paragraphs.

Metallic Mineral Potential: Geologic and geochemical data indicate that some areas of the gneiss, schist, marble and mylonite rocks, where intruded by granitic igneous rocks, have a high potential for tungsten resources (Calzia et al, 1988). Tungsten skarns often occur along the contact between carbonate and intrusive 'rocks and typical skarn deposits are often characterized by anomalous concentrations of tungsten, beryllium, bismuth, copper, molybdenum, tin, and zinc (Bateman, 1950; Cox, 1986).

The Mesozoic intrusive rocks include tonalities, monzogranites, and granodiorites. The medium-grained tonalites contain variable amounts of hornblende, biotite, and sphene. The medium- to coarse-grained granodiorites are commonly foliated and are characterized by large subhedral to euhedral hornblende, magnetite, and sphene crystals. While the amounts of magnetite and sphene are probably too small to be economically mined by today's standards, both are potential ores of needed metals. Some of the larger sphene crystals make good mineral specimens for collectors.

The quartz veins and veinlets intruded into the Cretaceous quartz monzonites and pre-Cretaceous gneisses and schists have potential for gold resources. These potential gold-bearing veins are commonly characterized by anomalous concentrations of arsenic, bismuth, copper, lead, silver, tungsten, and zinc (Calzia et al, 1988). The veins typically have an absence of visible sulfide mineralization. Gold was economically mined in the Butterfly Mountain region, westerly of the scenic area, from the same age and type of rocks found in both the Scenic and Wilderness Areas (Unruh and Ruff, 1981, pp. 68–75).

Both the Wilderness and Scenic Areas have been shown to contain major mylonite zones that may be related to detachment faulting. During the 1980s, mineral exploration has focused its attention along detachment faults for large-tonnage, low-grade gold deposits (Pridmore and Frost, 1992, p. 3–17). The Mesquite mine is considered to be one of these mineralized zones associated with a detachment fault.

Placer deposits, mainly for gold, may exist In alluvial deposits down gradient of the areas containing gold-bearing quartz veins. Though the potential is low, placer deposits for sphene and magnetite may also exist.

Nonmetallic Mineral Potential. The pre-batholithic metamorphic rocks locally have a high resource potential for marble (altered limestone) that may be used in the making of portland cement and construction materials. Dolomite has been mined in the Pinyon Flats area for roof granules, decorative stone and as a concrete additive to make a white concrete for swimming pools. One of these presently inactive dolomite mines is adjacent to the Santa Rosa Wilderness Area and it is located in a mappable zone that extends easterly into the Wilderness Area.

Some of the coarse-grained monzogranites contain muscovite and garnet. The metasedimentary rocks commonly include pelitic schist, gneiss, slate, and marble and locally contain amphibolites and calc-silica hornfels. The gneiss, schist and calc-silica hornfels locally contain garnet. Some of the recent alluvial deposits, especially in Palm Canyon, contain high amounts of garnet bearing sand. Garnet is commonly used as an abrasive, such as for sand paper, and some of the larger crystals make good mineral specimens.

Muscovite is present In the monzogranites and some of the metasedimentary rocks, mainly the schists. This muscovite is usually present as small flakes, so presently has limited economic uses. While the amount of known muscovite in the Wilderness and Scenic Areas is sub-economic at this time, it could become a viable potential future resource.

The quartz and feldspar rich dikes and veins that are common to both areas, though not considered economic at this time, could become a potential future resource. Silica (quartz) and feldspar are used as refractories, being common constituents in ceramic glazes.

Some of the quartz, feldspar and pegmatite dikes locally contain minerals considered to be gems and/or mineral collection material. Such minerals as tourmaline, beryl and microcline are known from the Santa Rosa Mountain area.

Narrow seams of slip-fiber asbestos occur in the amphibolite rocks at Pinyon Flats, just outside the scenic area. Further geologic mapping is needed to determine if slip-fiber asbestos deposits occur within the scenic area. Asbestos was an economic mineral commodity until it was classified as a hazardous substance by the U.S. Environmental Protection Agency.

The alluvial deposits in the bottom of the canyons and forming the alluvial fans at the mouth of the canyons offer a potential source for aggregate (sand and gravel). The canyon bottom deposits appear to be thin and of limited areal extent, however, visual inspection indicates that some of the material is of usable quality. The alluvial fan deposits appear to be thicker, of larger areal extent, and visually appear to be of usable quality.

Some of the granitic rocks offer a potential source for quarry material, rip-rap, and construction aggregate. Most of the surface exposed granitic rock is weathered and probably not usable as construction aggregate, but even some of it may be usable as rip-rap. The deeper, less weathered intrusive granitic rocks, like those exposed in road cuts, construction sites, and along some of the near vertical canyon walls, appear to be usable as a source for crushed aggregate construction material.

HISTORICAL MINING ACTIVITY

Some mineralization has been found associated with quartz-feldspar dikes that can be seen from and along Highway 74 near the base of the mountain, between Dead Indian-Carrizo Creeks and Black Hill. These dikes have been prospected by a number of small pits and at least one large adit. The remains of the adit can be seen near the base of the hill, west of the highway behind the Coachella Valley Water District flood control basin.

Up on the ridge behind the Bighorn development, is at least one small pit dug to evaluate a potential vanadium prospect. This prospect appears to be inactive at this time.

There are some rusty brown gossan zones associated with faults between the passing lane on Highway 74 and Black Hill. These have been prospected by some small pits and possibly one shaft.

There is an old inactive tungsten claim at the Miller Ranch tungsten prospect, in the upper end of Martinez Canyon. This claim contains some pits, a trench and at least one large adit.

There is reported to be an old prospect for gold over in the eastern Rabbit Peak area (personal communication). This may be in the area of the mineralized fault zone reported by Calzia et al. (1981, p. D7), in which the presence of an un-named prospect north of Rabbit Peak is also reported.

A number of areas northeast of Palm Springs have been prospected for limestone. The area contains pits and at least one adit. This area could contain potentially commercial limestone deposits.

There is one shear zone containing fibers of asbestos shown by Calzia et al. (1981, p. D6) to be in the southwest portion of the Santa Rosa Wilderness Area. No additional information has been found on this deposit.

Some mineralized zones near Potrero Spring-Potrero Flats have been prospected by a couple of shallow pits and one small adit. Based on the materials present, one of the pits was probably for gold and the adit may have been for uranium.

One small prospect pit can be seen up on a hill south of Highway 111 within the city limits of Cathedral City. This pit appears to be on a quartz dike that cuts through the metasedimentary rock complex.

Numerous borrow pits occur along Highway 74 where California Department of Transportation (Caltrans) has obtained construction materials for road building and maintenance. Some of these are within the scenic area and they include both pits and quarries.

EDUCATIONAL RESOURCE POTENTIAL

The Santa Rosa Mountain National Scenic Area contains at least five geologic features of significant educational value. These include the Martinez Mountain Rockslide, ancient Lake Cahuilla wave cut shoreline, the Santa Rosa Mylonite Zone, a series of westward directed late Cretaceous thrust faults, and some large detachment faults. The Santa Rosa Mountain Mylonite Zone passes through, and occupies much of, the Santa Rosa Wilderness Area.

The Martinez Mountain Rockslide, or Rock Avalanche, is considered to be the second largest such feature on the North American continent and the fifth largest in the world (Baldwin, 1986). The rock avalanche occurred on the eastern flank of Martinez Mountain during prehistoric time, probably during the Pleistocene, and the rocks slid all the way down and out onto the Coachella Valley floor. Baldwin (1986) estimates this Rock Avalanche to contain approximately 240,000,000 cubic meters of material.

The old beach line, or terrace, is believed to have formed during the last high stand of Ancient Lake Cahuilla, about 450 to 500 years ago. Although other such wave formed features occur along the same slopes, none of them are as prominent as the high stand beach line. From a distance, this wave formed shoreline, which is locally cut into the granitic intrusive rocks, appears to be a road or canal cut along the side of the mountain front. Wave formed features occur along other high water stands of large inland Pleistocene lakes, such as Lake Lahonton and Lake Bonneville in Nevada and Utah. However, no other such prominent inland lake shoreline feature is known to occur in the southern California area.

This shoreline feature is still in need of investigation. Detailed investigations of this feature, as well as some associated features, could lead to some insight into the ancient lakes' current patterns, paleo-wind directions, and to when the lakes were actually there. Such information could help to understand the paleoclimate of the region.

The Santa Rosa Mylonite Zone can be traced up Palm Canyon, across the Pinyon Flats area, around the east end of Santa Rosa Mountain and then south to the Borrego Valley area (Brown, 1981, p. 121; Parcel, Jr., 1981, pp. 139–150; Calzia et al., 1988, pp. D4–D5). This mylonite zone passes through, nearly bisects, and occupies approximately half of the Santa Rosa Wilderness Area.

A series of late Cretaceous age westerly directed thrust faults may play an important role in helping to understand the tectonic history of western North America. As an aid in understanding plate movement for western North America, it became apparent that some large scale westward crustal shortening should have occurred during late Cretaceous time. A search of the western United States for such westerly directed movement has now been centered on the Santa Rosa Mountain area.

Some large scale detachment faulting has also been mapped within the Santa Rosa Mountain National Scenic Area (Pridemore and Frost, 1992, p. 13). Studies of detachment faulting are presently under way in an effort to better understand California's mid-Tertiary geologic history.

POTENTIAL GEOLOGIC HAZARDS

There are a number of potential geologic hazards that can be faced by people entering the Santa Rosa Mountain Area. Some of these potential hazards are described below.

Ground Shaking: Ground shaking results from movement along faults during earthquakes. The Santa Rosa Mountain National Scenic Area and the Santa Rosa Wilderness Area are located about midway between the San Andreas and San Jacinto Fault Zones. These fault zones are both considered to be active, with the San Jacinto Fault Zone considered by many to be the most active fault zone in the southern California area. Movement·along either of these fault zones between San Bernardino and the Salton Sea region could result In strong ground shaking within the wilderness and scenic areas.

Rockfalls: Any steep slope has the potential for isolated rockfalls. Larger, local rockfalls can be triggered by wind, rain, snow, freezing, hikers, or animals. However, the greatest potential for major rockfalls would result from ground shaking during an earthquake.

Landslide/Rockslides: Landslides and/or rockslides will be mainly confined to steep, unstable slopes. Minor landslides or rockfalls will be most frequent in areas of over steepened slopes cut into alluvial material. Some will also occur in areas where steep slopes have been cut into highly fractured basement material. Such slides can be triggered by wind, rain, snow, freezing, hikers, or animals; however, the greatest potential for such slides would result from ground shaking during an earthquake.

Waterfalls: Waterfalls, especially in Deep Canyon, have stranded a number of hikers. These steep rock slopes are commonly water polished, so are easier to descend than to climb. In areas with more than one waterfall, a person can easily become trapped between them.

Flooding: One of the more common hazards to be encountered within the wilderness and scenic areas is that of flooding. The most dangerous time for flooding is during the summer months when the less predictable flash floods, or thunderstorms, occur. The larger canyons tend to be steep walled and to have rather large drainage areas. During storms, these canyon bottoms can go from dry to raging torrents in a matter of minutes and the steep canyon walls can prevent climbing to safety. Flooding can occur in areas where it is not raining, from storm waters dropped in the upper part of the drainage basin, which can be some miles away.

Prospect Pits And Adits: Since mining regulations require annual assessment work to be done on valid mining claims, past and present areas under mineral claim will commonly contain pits, adits, or both. Most of the prospect pits will be less than ten feet across and six feet deep and should not represent a serious hazard to the experienced hiker. There may, however, be some unmapped pits, shafts, or adits that could pose a fatal hazard.

Other: There are some muddy and pseudo-quicksand areas present in the alluviated areas of the canyon bottoms and canyon mouth fans. While not commonly encountered because of the dry climate, they can be found during wet years, winter months, and following heavy storms. One can become stuck in these areas, but most are only a couple of feet deep and getting out of them should not present a serious problem for adults and most older children.

SENSITIVE GEOLOGIC AREAS

The scenic value of any portion of the Santa Rosa Mountain National Scenic Area could be damaged by development. However, properly regulated development, such as roads, trails, mining, and even limited housing development, should result in only minimal visual damage occurring to the area. While access to some portions of the scenic area may need to be limited, mainly from non-geologic factors, most of the scenic area should be kept open to free public access. There are a few geologic features within the scenic area that could face damage or destruction if not protected. Two of these features are listed below.

Old Beach Line: The old beach line, which is attributed to ancient Lake Cahuilla, could be severely damaged or destroyed by development. As much as possible of this old shore line, including where it crosses alluvial fans, should be incorporated into the scenic area for protection. Access roads and trails into the scenic area that cross the old shoreline need to be kept to a minimum.

Martinez Mountain Rock Slide: The Martinez Mountain Rock Slide (Rock Avalanche) could be severely damaged by development. All of the lower portion of the slide should be incorporated into the scenic area for protection. Access roads and trails in the area of the slide should be kept to a minimum.

II. GLOSSARY

Adit: A tunnel dug horizontally into a hill for exploration or production of ore.

Aggregate: Sands and gravels used in the production of concrete or construction materials.

Alluvial fan: A fan-shaped deposit of sediment formed near the mouth of a canyon when a stream's slope is abruptly reduced.

Alluvium: Unconsolidated sediment deposited by a stream.

Aplite: A fine-grained igneous rock composed of quartz and feldspar, commonly found as dikes and veins.

Aquifer: Rock or soil through which ground water passes easily.

Arkose: A feldspar-rich sandstone.

Batholith: A large mass of igneous rock that formed when magma was emplaced at depth, cooled, crystallized and was subsequently exposed by erosion.

Cenozoic Era: A geologic time span following the Mesozoic Era, or a time span from about 66 million years ago until the present.

Detachment fault: A low angle normal fault, or gravity slide block, commonly associated with a shear zone or mylonite zone.

Dike: A tabular-shaped intrusive igneous feature that cuts through the surrounding rock.

Erosion: The removal and transportation of lithic material by a mobile agent, such as water, wind, or ice.

Fault: A break in a rock mass along which movement has occurred.

Foliated: A texture of metamorphic rocks that gives the rock a layered appearance.

Fossil: The remains or traces of organisms preserved from the geologic past.

Fracture: A break in a rock mass along which no movement has occurred.

Geothermal energy: Natural subsurface heat, including steam, used to power generators.

Gneiss: A metamorphic rock with alternating layers of different composition, usually parallel-to-subparallel foliation of medium to-coarse-grained platy minerals; product of intermediate-to-high grade metamorphism.

Granodiorite: A granular intrusive quartzose igneous rock intermediate between granite and diorite.

Ground water: Water present in the zone of saturation.

Hornblende: The most common of the amphibole group minerals, usually dark green to black; a common accessory mineral in granitic rocks.

Hydrogeology: The study of ground water conditions.

Igneous rock: A rock formed by the crystallization of molten magma.

Infiltration: The downward movement of surface waters into rock or soil through cracks and pore spaces.

Leucogranite: A granular intrusive igneous rock with a potash rich granite composition.

Magma: A body of molten rock found at depth.

Magnetite: An oxide of iron mineral, an ore of iron.

Mesozoic Era: A geologic time span between the Paleozoic and Cenozoic Eras, or a time span from about 252 to 66 million years ago.

Metamorphism: The change in mineral composition and texture of a rock when subjected to high temperature and pressure within the earth.

Mineral: A naturally occurring, inorganic crystalline material with a unique chemical composition.

Mylonite: A rock material smeared or milled by strong differential movement, commonly formed in association with deep seated faulting.

Normal fault: A fault in which the rock above the fault plane (hanging wall) has moved down relative to the rock below (footwall).

Paleozoic Era: A geologic time span between the Precambrian and Mesozoic Era, or a time span from about 541 to 252 million years ago.

Pegmatite: A coarse- to very coarse-grained variety of granite occurring as dikes or veins.

Permeability: A measure of material's ability to transmit a fluid.

Pit: 1) A hole or small shaft dug to evaluate or remove ore from the ground, and 2) A hole excavated into sedimentary rock for the extraction of aggregate.

Pleistocene: A geologic time span commonly referred to as the "Ice Age," or a time span from about 2.6 million to 11,000 years ago.

Porosity: The volume of open space within a rock or soil.

Precambrian: All the geologic time prior to the Paleozoic Era, or a time span from about 4.6 billion years ago until 541 million years ago.

Quarry: An open excavation for obtaining building stone, slate, limestone, granite or crushed aggregate; usually requires blasting.

Quartzite: A compact granular quartzose rock commonly derived from a sandstone.

Quartz monzonite: A granular intrusive igneous rock with a pyroxene-rich granodiorite-like composition.

Regional metamorphism: Metamorphism associated with large-scale mountain-building processes.

Reverse fault: A fault in which the rock above the fault plane (hanging wall) has moved up relative to the rock below (footwall).

Rock: A consolidated mixture of minerals.

Runoff: Water that flows at the surface rather then infiltrating into the ground.

Schist: A mica-rich crystalline metamorphic rock having a closely foliated structure, product of intermediate-grade metamorphism.

Sedimentary rock: Rocks formed from weathered products of preexisting rocks that have been eroded, transported, deposited, and lithified.

Sill: A tabular igneous body that was intruded parallel to the layering of preexisting rock.

Slate: A dense fine-grained metamorphic rock produced from an altered shale, product of low-grade metamorphism.

Sphene: A mineral composed of calcium, titanium, silica, and oxygen; mined in Russia as a source of titanium metal.

Spring: A flow of ground water that emerges naturally at the ground surface.

Strike-slip fault: A fault along which the principle movement is horizontal.

Terrace: A flat, bench like structure produced by a stream, which has been left elevated as the stream continues to cut downward.

Thrust fault: A low-angle reverse fault, usually less than 45 degrees.

Tonalite: A granular intrusive quartzose igneous rock similar to a diorite.

Unconformity: A surface that represents a break in the rock record, caused by erosion or nondeposition.

Water Table: The top of the ground water zone.

Weathering: The disintegration and decomposition of rock at or near the earth's surface.

Zone of saturation: The zone where all open pore spaces in sediment and rock are completely filled with water.

III. FIELD TRIP

Before starting this field trip, be sure that your vehicle is in good running condition and that you have an adequate supply of water and other drinks with you. It is recommended that you take along binoculars and cameras. You will need at least two cameras if you want to take photos and slides during the same trip. Always remember that you are not the only vehicles on the road, so be careful and above all, courteous to others. This field trip can probably best be taken during the week or on non-holiday weekends. Hopefully you will find this field trip as interesting as I have.

ROAD LOG

Starting at the intersection of Highway 111 and Highway 74, head up Highway 74 (south) toward the mountains. As you proceed up Highway 74, you are traveling up a large alluvial fan complex developed from the past flow of water and sediment out of Carrizo and Dead Indian Canyons. Just as you pass the last church on the right, or about where Homestead Drive meets the highway, you will see a small outcrop (knob) of folded, foliated metamorphic rocks adjacent to the highway on your right. If you look beyond this knob, you will see that the lower portion of the mountain is also composed of foliated metamorphic· rocks. There is a white to light gray outcropping of metamorphic rocks toward your right, across the channel behind the church play yard. This white material is a dolomitic marble, or metamorphosed dolomitic limestone. These metamorphic rocks were once sediments deposited in a late Precambrian or early Paleozoic sea that have since been metamorphosed to their present rock types by deep burial and the late Cretaceous intrusions of igneous granitic rock material.

You will reach a turn in the highway as you approach the base of the mountain. There is a turn off (parking area) with a gate on the right hand side of the road where you can pull off and park. There may be a number of cars parked here, as this is a parking place for people using the hiking trails that start from this area. You will be across the highway from the Bureau of Land Management Visitor Center for the Santa Rosa Mountain National Scenic Area. Walk up onto the flood control dike and look to the northwest (over toward the water tank). You will notice a sharp change in the rock types, with intrusive granitic rocks setting atop foliated metamorphic rocks. This change runs diagonally up the south side of the ridge and tends to flatten near the top of the ridge. This sharp contact marks the slide plane of a Detachment Fault Block. As you turn towards your vehicle, take a look at the Detachment Fault Block that forms the ridge behind (easterly) the new visitor center. If you look carefully, you can see some of the high-angle normal faults that cut through this Detachment Fault Block. This large block of granitic rock extends from the Ironwood Country Club on the north to the exposed slide plate just across the highway from the dirt pull off with the emergency phone just above the beginning of the passing lane. This is one large Detachment Fault Block. This block apparently slid off the side of the mountain about mid-Tertiary time, as the mountain range was being up lifted.

Instead of returning to your vehicle, continue on up along the right side of Highway 74 to the old road cut. This road cut was where the highway ran before the flood control dam and dike were installed. Here you can see a number of milky (white) quartz dikes cutting through the metamorphic rocks, primarily gneiss and schist. At one point toward the highway, you will see a white quartz dike that has been offset by a small fault. This is a good area to view white quartz veins filling fractures in a series of metamorphic rocks. By the geologic principle of cross-cutting, the metamorphic rocks had to be present before the quartz dike could be intruded into them, and the quartz had to in place before the faulting could offset it. Therefore, the faulting is the youngest and the metamorphic rocks are the oldest. Return to your vehicle from here.

As you continue up Highway 74 to look at other geologic features, remember that you are not the only one using this highway. For some of these features, the driver should just let his passengers observe and take their turn later when they are a passenger in someone else's vehicle or with someone else driving on the return trip. As you pass through the first outside curve and into the first inside curve, you will notice that the road cut on your right side is composed of metamorphic rocks similar to those viewed earlier in the old road cut. Just as you are coming out of the inside turn, you will notice a sharp change in which the foliated metamorphic rocks change to a massive weathered, well-indurated, medium gray, granitic fanglomerate. This is a very old sequence of sedimentary rocks that has been preserved beneath the slide plane of the Detachment Fault Block. As you continue up the highway, you will see this old fanglomerate exposed in the first highway cut on your left and again in a side road cut on the left. Also exposed in this side road cut is a linear outcrop of a pinkish material, yet to be studied.

As you continue up the highway, you will approach a turn out with an emergency phone on your right. If you look to your left just as you reach this turnout, you will see a sharp contact between the foliated metamorphic rocks in the highway cut and the overlying granitic rocks of the Detachment Fault Block. You may want to stop here to get a closer look. Again, be very careful getting back onto the highway if you stop here. As you round the first corner above this turnout, you will get a better look at this detachment fault plane and can see that it continues up the ridge to near the highway at the passing lane. This sharp contact, or slide plane, will be more pronounced on your trip back down the mountain.

You will pass through a wide highway turn above the referenced turn out and then proceed through some highway cuts until you reach a rather sharp turn within a highway cut. On the right hand side (west wall) the rocks have been folded into an antiformal feature, that is, the rocks have been bent so that they are high near the center and dip downward on both sides. There is also a fault cutting through this exposed wall, but it and the antiform are better viewed on the trip back down.

As you approach the first turn after reaching the passing lane, you can see on the left side of the highway the gray-green fault gouge (ground up rock) that marks the slide plane at the base of the Detachment Block. There is a large turnout here that you might want to take advantage of. A stop here allows you to look at the Detachment Fault Block. It can also show you some folded metamorphic rocks by walking to the south end (up hill) of the turnout and looking east across the highway.

You will continue to see exposures of the foliated metamorphic rocks in the Highway cuts as you continue up the Highway. There are also some additional gray-green fault gouge lenses exposed along the road cuts to your left as you continue up the highway toward the end of the passing lane. You can also see a waviness to the foliated metamorphic rocks exposed in the highway cuts. These rocks have actually flowed under intense heat and pressure, like toothpaste, to form one layer atop another. As you make the turn at the end of the passing lane, you can see a nearly horizontal zone of rusty looking fractured rock exposed in the highway cut on the right. This zone marks the plane of a flat thrust fault. You can stop at the parking area across from this flat thrust on the way back down, if you want to get a better look or take pictures.

The highway will make a right bend and then a left before nearly straightening out. You will again notice the outcrops of foliated metamorphic rocks in the highway cuts. This near straight road will make a sharp right turn, know as "Deadman's Curve" and then enter a highway cut. At this point there are granitic rocks exposed in the highway cut, along with a series of faults that parallel the highway through the cut. These faults are better seen on the trip back down. A little farther up the highway you will see a recumbent fold (a fold lying on its side) in with a series of iron stained rocks. This is an extension of the faulting described in the first part of this highway cut. You can view additional iron stained rocks associated with the same, or similar fault zones. About four turns later, you can see a small prospect pit dug into a lens of the iron stained rock, or gossen. A little further up, the road straightens out and in the last highway cut before the right angle bend, there is a well exposed normal fault. This fault is down to the north, and has foliated metamorphic rocks on the down thrown side. Remember to look at this fault in the first highway cut on the way back down. This fault appears to control a spring in Carrizo Creek, about two hundred yards to the west of the highway. As you make the turn, you will notice the exposures of foliated metamorphic rocks on your right (south) and then on both sides of the highway. This appears to be a window within a thrust fault that exposes the metamorphic rocks below the granitic rocks. If you look at the top of the ridge directly above and on the right, you will see that a block of granitic rock forms the peak and the foliated metamorphic rocks form the flanks of the ridge. You will come back into the granitic rocks on both side of the highway before you reach the Vista Point (stop here on the way back) and will continue in them until shortly after making the turn. About where the dirt road comes in on the right, the foliated metamorphic rock begin again. You will remain in the metamorphic rock all the way to Black Hill. As you approach Black Hill, that high mound of granitic rock on the right that blocks your view of Santa Rosa Mountain, you will notice that there are granitic rocks exposed in the highway cuts on the right and foliated metamorphic rocks exposed in outcrop to your left. As you make the right turn at the southeast corner of Black Hill, you will find granitic rocks exposed in the highway cuts on both sides. You will remain in granitic igneous rocks along both sides of the highway until just before reaching the Pinyon Fire Station (Station 30) and the Sugarloaf Cafe. About where Big Horn Drive intersects the highway, you will be leaving the Santa Rosa Mountain National Scenic Area.

Continue on up to the Sugarloaf Cafe area and turn around for the trip back down the hill. You may want to change drivers and take a rest and food break at this point. Make a stop at the café parking area and look back across the highway toward the Fire Station. Behind the station you will see granitic rocks in outcrop, but if you look at the highway cut over to the left (west) of the station parking lot you will see foliated metamorphic rocks. The fault that separates these two rock types passes just behind the station and is responsible for the small spring located in the gully behind the fire station. This spring used to supply the water for the store when I was a kid.

As you start the trip back down the hill, be sure to enter traffic safely. You will pass through a number of highway cuts that expose granitic rocks on both sides of the highway. About a mile down from the cafe you will come to a large dirt pullout on your right side. Be careful at this stop, it is a long way to the bottom of the canyon and the side is very steep. You will need to stop here to look at the exposures of foliated metamorphic rocks in Deep Canyon and along the base of Santa Rosa Mountain. You will notice a cut into the white rock over in the upper Deep Canyon Area. This was a "Dolomite Mine" where they quarried a dolomitic marble, similar to that you saw earlier along Highway 74 down behind the church playground. Most of the rock removed from this mine was used as roofing granules on homes in the Palm Desert area. The area between you and the base of Santa

Rosa is known as the "Santa Rosa Mountain Shear Zone." This shear zone was formed as a deeply buried fault zone. Because of the depth of burial, the high pressure and temperature allowed the rock to flow, rather than break. This is the reason you see the flow structures, such as thickening and thinning along the planes of foliation, in the rocks exposed in this zone. This is probably the same group of foliated rocks you were looking at during the trip up the hill. More geologic mapping is needed to prove or disprove such a relationship.

Again, be careful getting back onto the highway. About a half mile down the road is the Tewanet Overlook, you may want to stop here to get a different look at the shear zone, or just to look at the scenery.

About another half mile down the highway is another large dirt turnout on your left. There is also an emergency phone at this location. You will need to stop here to look at a fault. Be sure to pullover to the right as far as you can, as this area is sometimes used by slower traffic to let other cars pass and you do not want to get hit. If you look back across the highway (west) toward Sugarloaf Mountain, you will see a ridge composed of a light colored feldspathic dike material. You will notice that this material abruptly stops at the base of the ridge. Now turn completely around and look to the east. You will notice that there is the continuation of the light colored feldspathic dike material. This dike has been offset over a hundred yards by a fault that comes up out of Deep Canyon and continuous over to Asbestos Mountain. Now walk back up the highway, staying out of the road and away from traffic. About 100 feet up the road (south) from the end of the turnout, look back across the highway (west) at the highway cut. In this cut you will see a zone of pinkish fault gouge about 20 feet wide. This is the fault that offset the dike, and as you walk back toward your car you will notice that this fault zone creates a low area that cuts through the ridges as it continues over toward Asbestos Mountain.

As you continue on back down the highway, look at the highway cuts for additional faulting, jointing and fracturing. You will notice that many of these smaller faults, joints, and fracture zones can be recognized by iron oxide (rust colored) staining. Continue on down to the "Coachella Valley Vista Point" and stop here. This stop gives you the opportunity to look at many geologic features and to look at most of the Coachella Valley, provided the weather cooperates. Across the valley you can get a good look at the Indio Hills and the San Andreas Fault System that has created them. On a clear day, you will be able to see a number of the palm oases, both in and along the base of the hills, that are growing along the faults. Down in front of you is the Detachment Fault Block that you were looking at on the way up and over to your right is a much larger Detachment Fault Block. This larger block, referred to as the "Palm Desert Detachment Fault Block" (Pridemore and Frost, 1992), extends from the base of Sheep Mountain all the way out to "Point Happy" near La Quinta. These two detachment blocks were probably one when they slid off the side of the mountain and have now been separated by the erosion of water and sediment coming out of Deep Canyon.

As you continue on down the highway, remember to look for the features described in this trip log when you were coming up the hill. These include such features as faults, the detachment block slide plane, and the old fanglomerate that was preserved under the detachment block. Your last stop should be the visitor center, which is on your right as you reach the bottom of the hill. I believe it is open on weekends, Friday, and Monday. If you have not stopped here before, I think you will find the exhibits educational and interesting. As you head back to Palm Desert, do not forget to look at the Dolomitic Marble again, this time on the left, as you recall, this may be related to the dolomitic marble quarried in the mine you looked at in the Santa Rosa Mountain Shear Zone.

SPECIFIC GEOLOGIC REFERENCES

Baldwin, J. E., 1980, The Martinez Rock Avalanche: Unpublished Masters Thesis, California State University, Los Angeles, 58 p.

Baldwin, J. E., 1986, Martinez Mountain Rock Avalanche: in Guptil, P. D., et al., eds., Geology of the Imperial Valley, California: South Coast Geological Society Annual Field Trip Guidebook Number 14, p. 37–48.

Bock, C. G., 1967, The Martinez Mountain Rockslide: unpublished Bechtel Corporation Report to Coachella Valley County Water District, 6 p.

Bock, C. G., The Martinez Mountain Rock Avalanche: in Coates, D. R., ed., Reviews in Engineering Geology, Landslides: Geological Society of America, Volume 3, p. 155–168.

Brown, A. R., 1966, Geologic Report on the Whitlock Mining Claims: Unpublished Report for Mr. Robert M. Harris, San Marcos, California.

Brown, A. R., 1968, Geology of a Portion of the Southeastern San Jacinto Mountains, Riverside County, California: Unpublished Master's Thesis, University of California, Riverside.

Brown, A. R., 1980, Limestone Deposits of the Desert Divide, San Jacinto Mountains, California: in Fife, D. L. and Brown, A. R., eds., Geology and Mineral Wealth of the California Desert: South Coast Geological Society Publication, p. 284–293.

Brown, A. R. 1981, Structural History of the Metamorphic, Granitic, and Cataclastic Rocks in the Southern San Jacinto Mountains: in Brown, A. R. and Ruff, R. W., eds., Geology of the San Jacinto Mountains: South Coast Geological Society Annual Field Trip Guidebook Number 9, p. 100–138.

Brown, A. R. (Editor), 1981, Geologic Map of the Idyllwild (15 Minute) Quadrangle, California Compiled by Thomas W. Dibblee, Jr.: South Coast Geological Society Map SCGS-5.

Brown, A. R. (Editor), 1981, Geologic Map of the Palm Springs 15 Minute Quadrangle, California Compiled by Thomas W. Dibblee, Jr.: South Coast Geological Society Map SCGS-3.

Brown, A. R. and Ruff R. W., eds., Geology of the San Jacinto Mountains: South Coast Geological Society Annual Field Trip Guidebook Number 9, 219 p.

Brown, V., Allan, D., and Stark, J., 1972, Rocks and Minerals of California: Naturegraph Publishers, 200 p.

Buono, A., Moyle, W. R., Jr., and Dana, T., 1979, Water Resources of the Santa Rosa Indian Reservation and Vicinity, Riverside County, California: US Geological Survey Open-File Report 79-1172, 32 p.

Bushee, Jo, Holden, J., Geyer, B., and Gastil, G., 1963, Lead-Alpha Dates for Some Basement Rocks of Southwestern California: Geological Society of America Bulletin, Volume 74, p. 803–806, 1 figure.

California Division of Mines and Geology, 1966, Mineral Resources of California: Bulletin 191, 450 p., 1 plate.

California State Mining Bureau, 1916, Riverside County Field Report of the State Mineralogist XV, page 535.

California State Mining Bureau, 1932, Riverside County Field Report of the State Mineralogist XXVIII, p. 5 and 6.

Calzia, J. P., Crowley, J. A., Dockel R. D., Somoni, T. R., Jr., and Server, G. T., 1979, Leasable Mineral Resources of the California Desert Conservation Area: U.S. Geological Survey Administrative Report, 183 p.

Calzia, J. P., Madden-McGuire, D. J., Oliver, H. W., and Schreiner, 1988, Chapter D: Mineral Resources of the Santa Rosa Mountain Wilderness Study Area, Riverside County, California: US Geological Survey Bulletin 1710-D.

Cox, D. P., 1986, Descriptive Model of W Skarn Deposits: In Cox, D. P. and Singer, D. A., Mineral Deposit Models: U.S. Geological Survey Bulletin 1693, page 55.

Crowell, J. C., ed., 1975, San Andreas Fault in Southern California, A Guide to San Andreas Fault from Mexico to Carrizo Plain: California Division of Mines and Geology Special Report 118, 272 p.

Damiata, B. N., Park, S. K., and Lee, T-C, 1986, Geophysical Overview: Imperial Valley, Southern California: In Guptil, P. D., et al.,, eds., Geology of the Imperial Valley, California: South Coast Geological Society Annual Field Trip Guidebook Number 14, 1986, p. 116–133.

Dibblee, T. W, Jr., 1954, Geology of the Imperial Valley Region, California: in Jahns, R. H., ed., Geology of Southern California: California Division of Mines Bulletin 170, p. 21–28.

Dibblee, T. W. Jr., 1981, Geology of the Idyllwild (15-Minute) Quadrangle, California: South Coast Geological Society, Geologic Map SCGS-5, Scale 1:62,500.

Dibblee, T. W., Jr., 1981, Geology of the San Jacinto Mountains and Adjacent Areas: in Brown, A. R. and Ruff, R. W.,, eds., Geology of the San Jacinto Mountains: South Coast Geological Society Annual Field Trip Guidebook Number 9, p. 1–47.

Diment, T. C., Urban, T. C., Sass, J. H., Marshall, V. V., Monroe, R. J., and Lachenbruch, A. H., 1975, Temperatures and Heat Contents Based on Conductive Transport of Heat: In Assessments of Geothermal Resources of the United States: U.S. Geological Survey Circular 726.

Dikka, R. K., 1984, Fission-Track Geochronologic Evidence for Late Cretaceous Mylonitization and Early Paleozoic Uplift of the Northern Peninsular Range, California: Geophysical Research Letters, Volume 11, p. 46–49.

Dutcher, L. C. and Bader, J. S., 1963, Geology and Hydrology of Agua Caliente Spring, Palm Springs, California: US Geological Survey Water Supply Paper 1605, 43 p., 2 plates.

Dutcher, L. C., Hardt, W. F., and Moyle, Jr., W. R., 1972, Preliminary Appraisal of Ground Water In Storage With Reference to Geothermal Resources in the Imperial Valley Area, California: US Geological Survey Circular 649, 57 p.

Elders, W. A., Cohen, L. H., and Mehegan, J. M., 1986, Magmatism and Volcanism In the Salton Trough: In Guptil, P. D., et al., eds., Geology of the Imperial Valley, California: South Coast Geological Society Annual Field Trip Guidebook Number 14, p. 134–143.

Envicom Corporation and the County of Riverside Planning Department, 1976, Seismic Safety and Safety Elements Technical Report: County of Riverside.

Erskine, B. G. and Wenk, H. R., 1985, Evidence for Late Cretaceous Crustal Thinning in the Santa Rosa Mylonite Zone, Southern California: Geology, Volume 13, p. 274–277.

Foster, J. H., 1980, Geology: Deep Canyon, a Desert Wilderness for Science, p. 89–97.

Fox, R. C., 1980, Investigation for an Additional Water Supply, Pinyon Pines and Vicinity: unpublished Report for the Pinyon Pines County Water District, Riverside County, California, 14 p. plus maps.

Fraser, D. M., 1931, Geology of the San Jacinto Quadrangle South of the San Gorgonio Pass, California: California Journal of Mines and Geology, Volume 27, p, 494–540.

Gath, E. M., Ruff, R. W., and Raub, M. L., 1986, Geology of the Salton Trough Field Trip Road Log: in Guptil, P. D., et al., eds., Geology of the Imperial Valley, California: South Coast Geological Society Annual Field Trip Guidebook Number 14, 225 p.

Gayer, B. L., 1962, Geology of the Palm Desert Region, Riverside California: San Diego State College, unpublished senior thesis.

Goudarzi, G. H., 1984, Guide to Preparation of Mineral Survey Reports on Public Lands: US Geological Survey Open File Report 84-787, 51 p.

Greensfelder, R. W., 1974, Maximum Credible Rock Acceleration from Earthquakes in California: California Division of Mines and Geology Map Sheet 23, Scale 1:2,500,000.

Guptil, P. D., Gath, E. M., and Ruff, R. W.,, eds., 1986, Geology of the Imperial Valley, California: South Coast Geological Society Annual Field Trip Guidebook Number 14, 225 p.

Hart, E. W., 1974, Zoning for Surface Fault Hazards In California: The New Special Studies Zones Maps: California Geology, Volume 27, Number 10, p. 227–230.

Higgens, C. T., ed., 1980, Geothermal Resources of California: California Division of Mines and Geology Data Map Series Number 4.

Hill, R. I., 1981, Field, Petrological and Isotopic Studies of the Intrusive Complex of San Jacinto Mountains: in Brown, A. R. and Ruff, R. W., eds., Geology of the San Jacinto Mountains: South Coast Geological Society Annual Field Trip Guidebook Number 9, p. 76–89.

Hill, R. I.,1984, Petrology and Petrogenesis of Batholithic Rocks, San Jacinto Mountains, Southern California: Unpublished PhD Thesis, California Institute of Technology.

Hodgson, S. F., ed., 1982, Oil and Gas Prospect Wells Drilled in California through 1980: California Division of Oil and Gas Publication Number TR01, Second Edition.

Hood, R. G. and Russell, G., 1966, Palm Springs Aerial Tramway: in Lung, R. and Proctor, R. J., eds., Engineering Geology in Southern California: Association of Engineering Geologists, Special Publication, p. 370–373.

Jahns, R. H., 1954, Geology of the Peninsular Range Province, Southern California and Baja California: in Jahns, R. H., ed., Geology of Southern California: California Division of Mines Bulletin 170, p. 29–52, 1 plate.

Jahns, R. H., 1954, Northern Part of the Peninsular Range Province, Geologic Guide Book Number 5: in Jahns, R. H., ed., Geology of Southern California: California Division of Mines Bulletin 170.

Jennings, C. W., 1973, Preliminary Fault and Geologic Map of California: California Division of Mines and Geology Preliminary Report 13, Scale 1:750,000.

Larson, E. S., Jr., Gottfried, D., Jaffe, H. W., and Waring, C. L., 1958, Lead-Alpha Age Determinations of the Mesozoic Batholiths of Western North America: U.S. Geological Survey Bulletin 1070-B, p. 35–62.

Leadabrand, E. S., Jr., 1971, Guidebook to the San Jacinto Mountains of Southern California.

Lockwood, J. P. 1961, Geology of an Area in the Western Santa Rosa Mountains of Southern California, with Special Emphasis on a Previously Unreported Mylonite Zone: Unpublished Senior Theses, University of California, Riverside.

Mariano, J. C., Roberts, C. W., Erdman, C. F., and Biehler, S., 1987, Isostatic Residual Gravity Map of the Santa Rosa Mountains Wilderness Study Area and Vicinity, East-Central Santa Ana 1° × 2° Quadrangle, California: U.S. Geological Survey Open-File Report 87-88, Scale 1:62,500.

McKelvey, V. E., 1972, Mineral Resource Estimates and Public Policy: American Scientist, Volume 60, p. 32–40.

Miller, W. J., 1944, Geology of the Palm Springs-Blythe Strip, Riverside County, California: California Journal of Mines and Geology Volume 40, p. 11–72.

Murdock, J. and Webb, R. W., 1958, Minerals of California: California Division of Mines and Geology Bulletin 173, 452 p.

Murdock, J. and Webb. R. W., 1960, Supplement to Minerals of California for 1955 Through 1957: Supplement to Bulletin 173, California Division of Mines, 64 p.

Murdock, J. and Webb, R. W., 1964, Supplement 2 to Minerals of California for 1958 Through 1961: Supplement 2 to Bulletin 173, California Division of Mines, 28 p.

Oakeshott, G. B., Braun, L. T., Jennings, C. W., and Wells, R, 1950, Exploratory Wells Drilled Outside of Oil and Gas Fields In California to December 31, 1950: California Division of Mines and Geology Special Report 23, 77 p., 1 map.

Oliver, H. W., Chapman, R. H., Biehler, S., Griscom, A., Silver, E. A., Beyer, L. A., and Robbins, S. L., 1981, Gravity Map of California and Its Continental Margin: California Division of Mines and Geology Special Map, Scale 1:750,000.

Parcel, R. F., Jr., 1972, Geology of a Part of the Santa Rosa Mountain Shear Zone, Riverside County, California: Unpublished master's thesis, University of California, Riverside.

Parcel, R. F., Jr., 1981, Structure and Petrology of the Santa Rosa Mountain Shear Zone in the Pinyon Flats Area, Riverside County, California: in Brown, A. R. and Ruff, R. W., eds., Geology of the San Jacinto Mountains: South Coast Geological Society Annual Field Trip Guidebook Number 9, p. 139–150.

Pridmore, C. L. and Frost, E. G., 1992, Detachment Faults California's Extended Past: California Geology Volume 45, Number 1, January/February issue, p. 3–17.

Proctor, R. J., 1968, Geology of the Desert Hot Springs Upper Coachella Valley Area, California: California Division of Mines and Geology Special Report 94, 50 p., 1 map.

Real, C. R., Toppozada, T. R., and Parks, D. L., 1978, Earthquake Epicenter Map of California, 1900 Through 1974: California Division of Mines and Geology Special Map.

Rogers, C. R. (comp.), 1965, Geologic Map of California Santa Ana Sheet: California Division of Mines and Geology Map, Scale 1:250,000.

Rogers, T. H. and Beach, D. H., 1963-9164, Reconnaissance Geologic Mapping and Photogeologic Interpretation in Parts of Palm Desert, Coachella, Clark Lake, and Rabbit Peak Quadrangles, Scale 1:62,500: California Division of Mines and Geology, Reconnaissance Mapping for the State Geologic Map.

Ruff, R. W., Bogseth, P. A., and MacGregor, B. M., 1981, Geology of the San Jacinto Mountains Field Trip Road Log: in Brown, A. R. and Ruff, R. W., eds., Geology of the San Jacinto Mountains: South Coast Geological Society Annual Field Trip Guidebook Number 9, p. 175–219.

Saul, R. B., Gray, C. H., Jr., and Evans, J. R., 1968, Mines and Mineral Resources of Riverside County, California:

California Division of Mines and Geology Open-File Report, 1968-67, p. 438–440 and 486–487.

Schnabel, P. and Seed, H. B., 1973, Acceleration in Rock for Earthquakes in the Western United States: Seismic Society of America Bulletin, Volume 63, Number 2, p. 501–516.

Schreiner, R. A., 1983, Mineral Investigation of the Santa Rosa Mountains Wilderness Study Area, California Desert Conservation Area, Riverside County, California: U.S. Bureau of Mines Open-File Report MLA 100-83, 9 p.

Sharp, R. V., 1966, Ancient Mylonite Zone and Fault Displacement in the Peninsular Ranges of Southern California: Program, 1966 Annual Meeting, Cordilleran Section, Geological Society of America, (abs.), p. 65.

Sharp, R. V., 1967, San Jacinto Fault Zone in the Peninsular Ranges of Southern California: Geological Society of America Bulletin, Volume 79, Number 6, p. 706–730.

Sharp, R. V., 1975, En Echelon Fault Patterns of the San Andreas Fault in Southern California: in Crowell, J. C., ed., San Andreas Fault in Southern California, A Guide to San Andreas Fault from Mexico to Carrizo Plains: California Division of Mines and Geology Special Report 118, p. 147152.

Sharp, R. V., 1979, Some Characteristics of the Eastern Peninsular Ranges Mylonite Zone: in Speed, R. and Sharp, R. V., (org.), Analysis of Actual Fault Zones In Bedrock: Proceedings of Conference 8, National Earthquake Hazards Reduction Program, U. S. Geological Survey Open-File Report 79-1239.

Silva, M. A. and Rapp, J. S., 1991, Mines and Mineral Prospects of the California Desert: California Division of Mines and Geology Open-File Report 91-18, map and letter.

Simpson, C., 1984, Borrego Springs -Santa Rosa Mylonite Zone: A Late Cretaceous West-Directed Thrust In Southern California: Geology, Volume 12, p. 8–11.

Sims, S. J., 1960, Geology of Part of the Santa Rosa Mountains, Riverside County, California: unpublished PhD Thesis, Stanford University.

Slosson, J. E., 1973, Review of Special Studies Zones Maps that Delineate Zones Encompassing Potentially Active Faults: Letter to Concerned Cities and Counties from the State Geologist, California Division of Mines and Geology, December 26, 1973.

Stanley, G. M., 1963, Prehistoric Lakes in Salton Sea Basin (abs.): Geological Society of America Special Paper 73, p. 249–250.

Stanley, G. M., 1966, Deformation of Pleistocene Lake Cahuilla Shoreline, Salton Sea Basin, California (abs.): Geological Society of America Special Paper 87, page 165.

Snydor, R. H., 1975, Geology of the Northern Lodes of the San Jacinto Pluton, Palm Springs, California: unpublished Masters Thesis, University of California, Riverside.

Theodore, T. G., 1966, The Fabric of a High-Grade Mylonite

Zone In Southern California: American Geophysical Union Transactions, Volume 47, Number 3, p. 491–192.

Theodore, T. G., 1970, Petrogenesis of Mylonites of High Metamorphic Grade In the Peninsular Ranges of Southern California: Geological Society of America Bulletin, Volume 81, p. 435–350.

Thomas, R. G., 1963, The Late Pleistocene 150-foot Fresh Water Beachline of the Salton Sea: Southern California Academy of Sciences Bulletin Volume 82, Part I, page 9.

Tucker, W. B. and Sampson, R. J., 1945, Mineral Resources of Riverside County: California Journal of Mines and Geology, Volume41, p. 121–182.

U.S. Bureau of Mines and U.S. Geological Survey, 1980, Principles of a Resource/Reserve for Classification Minerals: U.S. Geological Survey Circular 831, 5 p.

U.S. Department of Agriculture, 1979, Soil Survey of Riverside County, California; Coachella Valley Area: USDA Soil Conservation Service.

U.S. Department of Agriculture, 1988, San Bernardino National Forest Land and Resource Management Plan: USDA Forest Service, Pacific Southwest Region, 3 Volumes plus maps.

Unruh, M. E. and Ruff, R. W., 1981, History of Gold Mining in the Southern San Jacinto Mountains, Riverside County, California: in Brown, A. R. and Ruff, R. W., eds., Geology of the San Jacinto Mountains: South Coast Geological Society Annual Field Trip Guidebook Number 9, p. 68–75.

Wallace, R. D., 1982, Evaluation of possible Detachment Faulting West of the San Andreas Fault, Southern Santa Rosa Mountains, California: Unpublished Master's Thesis, California State College, San Diego.

Waters, M. R., 1983, Late Holocene Lacustrine Chronology and Archaeology of Ancient Lake Cahuilla, California: Quaternary Research, Volume 19, p. 373–387.

Webb, R. W., 1939, Evidence of the Age of a Crystalline Limestone in Southern California: Journal of Geology, Volume 47, p. 198–201.

Wilke, P. J., 1978, Late Prehistoric Human Ecology at Lake Cahuilla, Coachella Valley, California: Contents of the University of California Archaeological Research Facility, Number 38.

Wright, L. B., 1946, Geology of the Santa Rosa Mountain Area, Riverside County, California: California Journal of Mines and Geology, Volume 42, p. 9–13.

TEXTBOOK REFERENCES

Bateman, A. M., 1950, Economic Mineral Deposits: J. Wiley and Sons, Inc., 768 p.

Berry, L. G. and Mason, B., 1959, Mineralogy; Concepts, Descriptions, and Determinations: W. H. Freeman and Company, 630 p.

Bowen, N. L., 1956, The Evolution of the Igneous Rocks: Dover Publications Inc., 332 p.

Dalrymple, G. B. and Lanphere, M. A., 1969, Potassium-Argon Dating; Principles, Techniques, and Applications to Geochronology: W. H. Freeman and Company, 258 p.

Dana, E. S. and Ford, W. E., 1952, Mineralogy: Revised by Hurlbut, C. S., Sons, Inc., 530 p.

Dolbear, S. H. and Bowles, O., eds., 1949, Industrial Minerals and Rocks; Nonmetallics other than Fuels: American Institute of Mining and Metallurgical Engineers, 1156 p.

Fetter, C. W., 1988, Applied Hydrogeology: Merrill Publishing Company, 592 p.

Fritzen, D. K., 1959, The Rock-Hunter's Field Manual; A Guide to Identification of Rocks and Minerals: Cornerstone Library, 191 p.

Lewis, R. S., 1941, Elements of Mining: J. Wiley and Sons, Inc., 579 p.

Lewis, R. S. and Clark, G. B., 1964, Elements of Mining: J. Wiley and Sons, Inc., 768 p.

Low, J. W., 1957, Geologic Field Methods: Harper and Brothers, 489 p.

Mason, B. and Berry, L. G., 1968, Elements of Mineralogy: W. H. Freeman and Company, 550 p.

Pearl, R. M., 1956, Rocks and Minerals: Barnes and Noble, Inc., 275 p.

Pearl, R. M., 1973, Handbook for Prospectors: McGraw-Hill, 495 p.

Pough, F. H., 1976, A Field Guide to Rocks and Minerals: Houghton Mifflin Company, 317 p.

Thornbury, W. D., 1954, Principles of Geomorphology: J. Wiley and Sons, Inc., 618 p.

Turner, F. J. and Verhougen, J., 1968, Igneous and Metamorphic Petrology: McGraw-Hill, 694 p.

Wager, L. R. and Brown, G. M., 1967, Layered Igneous Rocks: W. H. Freeman and Company, 588 p.

Wiley, Jr., J. and Davis, S. N. and DeWiest, R. J. M., 1966, Dana's Manual of Hydrogeology: J. Wiley & Sons, 447 p.

Revised: October 18, 1992

Oasis in the Desert: Geology, Hydrogeology, and History of Groundwater in the Coachella Valley, California

Scott Snyder PG, CHG Snyder Geologic
858-412-9848

INTRODUCTION

The Coachella Valley region is geologically dynamic, located predominantly between the San Jacinto and San Andreas Fault systems. Two primary features of interest are the highlands of the San Jacinto and Santa Rosa Mountains, and the lowlands of Coachella Valley ("Indio" on Figure 1). The striking contrasts of these two features resulted from historical movement along the fault systems.

Mount San Jacinto has among the greatest topographic relief in North America; from the floor of the Coachella Valley to the top of the mountain, the prominence of the San Jacinto is more than 8,300 feet in less than a couple of miles. The biological diversity of the mountain is equivalent to traveling from southern California to Alaska.

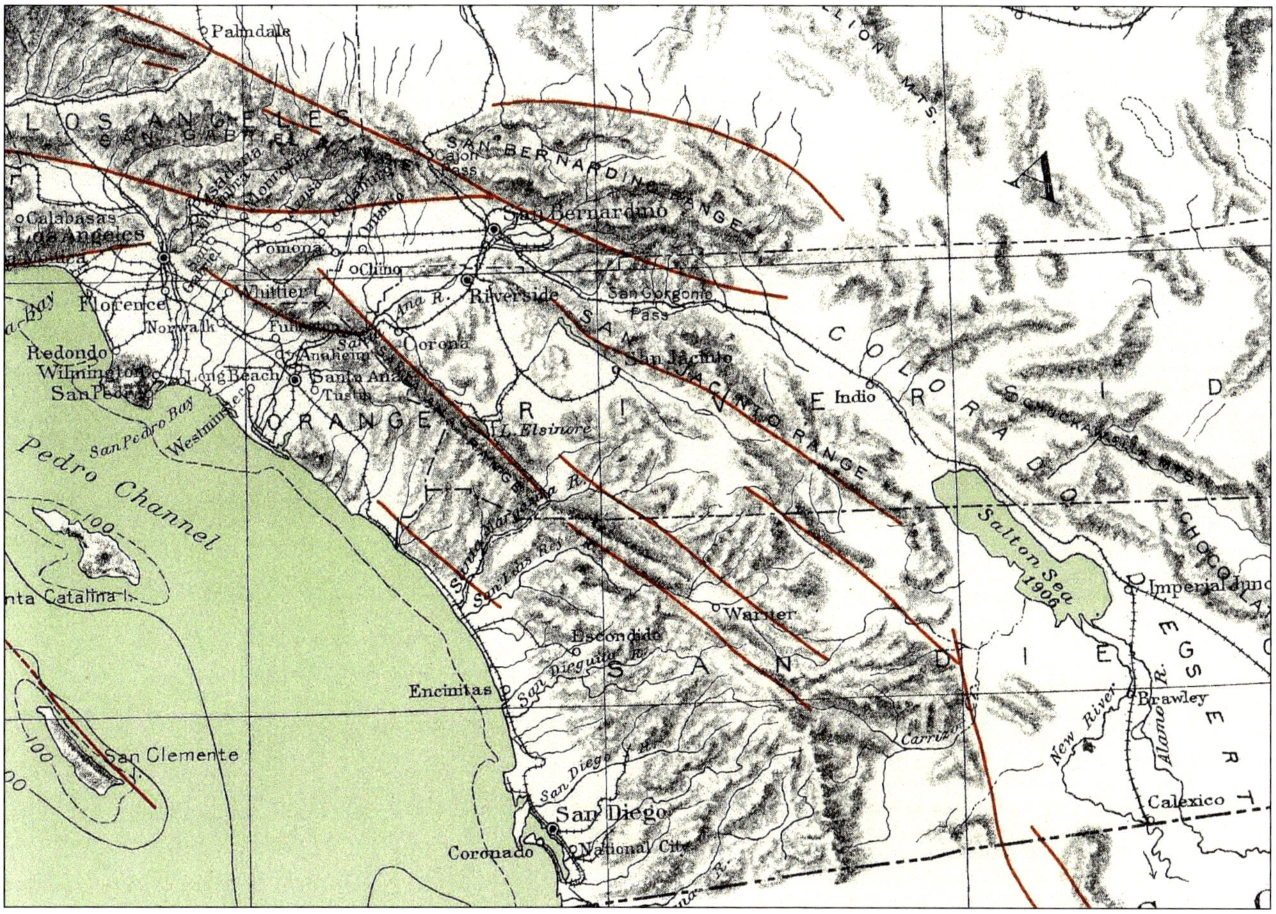

Figure 1. Geomorphic map of Coachella Valley and vicinity (Excerpt from California State Earthquake Investigation Commission, 1908, Geomorphic map, California, Nevada, portions of Oregon, Idaho. Source: David Rumsey Historical Map Collection.)

Figure 2. Coachella Valley from Vista Point along Pines to Palms Scenic Byway, Riverside County.

The Santa Rosa and San Jacinto Mountains trend northwest-southeast along the southwestern edge of the Colorado Desert Geomorphic Province. The mountains are part of the Peninsular Range Geomorphic Province that consists of a chain of mountains beginning with the northernmost Mount San Jacinto, and trending southward through southern California into Baja California, Mexico. The Santa Rosa and San Jacinto Mountains consist largely of pre-late Cretaceous meta-sedimentary and meta-volcanic rocks that have been intruded by late Cretaceous age plutonic rocks, largely of granitic composition. Much of these highlands are managed as the San Jacinto–Santa Rosa Mountains National Monument.

The summit of Mount San Jacinto is the highest point along the Peninsular Range Province, at an elevation of 10,805 feet above mean sea level. In the Coachella Valley, elevations range from below mean sea level to several hundred feet above sea level, resulting in an abrupt vertical relief of more than 8,000 feet on the steep eastern face of Mount San Jacinto, and exceeding the vertical relief in most other parts of the contiguous United States except Death Valley.

The elevation of the mountain range captures significant amounts of rain and snow primarily from winter storms of the Pacific Ocean. The drainage from these storms and spring snowmelt provide valuable water resources to the Coachella Valley in an otherwise dry desert region. Mount San Jacinto is also the site of the Palm Springs Aerial Tramway, which takes tourists by cable car from the desert up to an elevation over 8,000 feet above sea level in 15 minutes. The higher elevations of the San Jacinto Mountains are managed by the Mount San Jacinto State Park and Wilderness (BLM, 2012).

The Coachella Valley, part of the Colorado Desert Geomorphic Province, is bounded to the west by the Santa Rosa and San Jacinto Mountains. The San Andreas Fault Zone (SAFZ) trends roughly parallel to the mountains on the east side of the Coachella Valley; however, there are faults along the base of the west side of the San Jacinto and Santa Rosa Mountains which are part of the San Jacinto Fault Zone (SJFZ) and are partly responsible for the sharp contrast in elevation and the groundwater seeps and springs along the base of the mountains.

Figure 3. Coachella Valley is arid, receiving less than 4 inches of precipitation per year. This valley is deep in the rain-shadow of the San Jacinto and Santa Rosa Mountains of the Peninsular Ranges.

WHY VISIT HERE?

A major reason for the visiting the Coachella Valley is water. Desert oases are some of the most fascinating human and geologic localities. In areas that receive less than 4 inches of rain a year, the geology of the area has developed a plumbing system to deliver fresh, life-sustaining water in an otherwise inhospitable and deadly environment. The oases are interesting geologically because they are formed in the Coachella Valley by the presence of faults, some of which are known throughout the world. These faults act as barriers to the flow of groundwater that travels through the alluvial fans present throughout the valley, at the bases of the mountains that surround it. Groundwater flows through the sediments, hits the fault plane with its low permeability sediments, and rises up the plane, in some cases reaching the surface. Where it reaches the surface, oases are formed (in some cases containing streams and ponds). These watering holes support life of all kinds, including human. Where the groundwater doesn't reach the surface along these faults, it often rises high enough to support flora along the upgradient side of the fault, but on the downstream side, very little life is supported in the absence of this precious resource.

GEOLOGIC DESCRIPTION OF THE COACHELLA VALLEY

The Coachella Valley is within the Colorado Desert Geomorphic Province, and is located primarily in Riverside County, with small portions at each extreme end in San Bernardino and San Diego Counties. The valley is about 65 miles long and encompasses about 400 square miles (California Department of Water Resources, 1964). It is an alluvial valley filled near surface with sediments derived locally from igneous and metamorphic rocks of the mountain ranges that bound it including the San Jacinto, Santa Rosa, San Bernardino, and Little San Bernardino mountain ranges. At its deepest, it is estimated that the valley may contain sediments up to 12,000 feet thick (2.3 miles), deepest on the eastern side. The valley is drained by the Whitewater River (into the Salton Sea, rarely) and ranges in elevation from over 11,000 feet at San Gorgonio Mountain, to almost 200 feet below sea level at the Salton Sea.

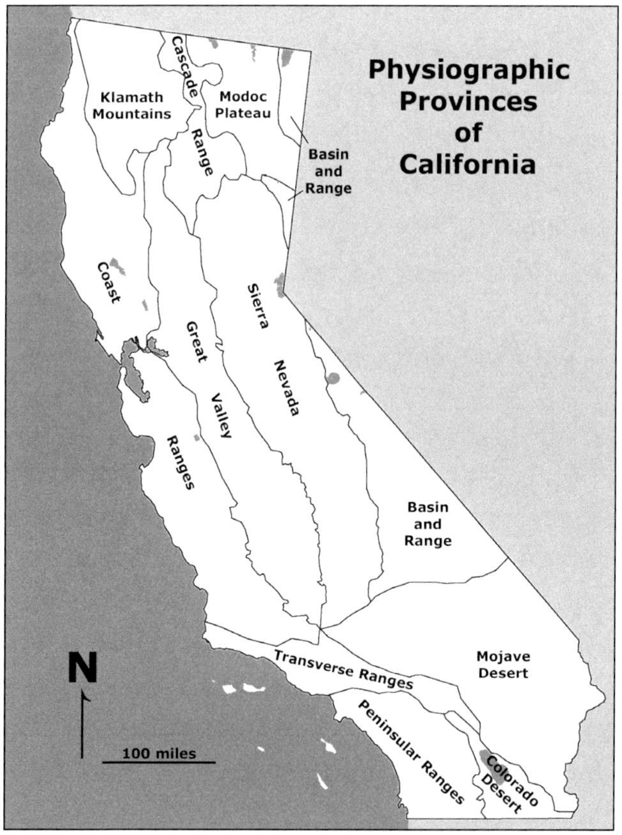

Figure 4. Physiographic provinces of California.

GEOLOGY OF THE COACHELLA VALLEY AND SALTON TROUGH REGION

Extracted from U.S. Geological Survey (2006)

The northwestern Coachella Valley is the alluviated lowland that extends southeastward from the San Gorgonio Pass region to the north end of the Salton Sea. The lowland is traversed by multiple strands of the San Andreas fault, and is punctuated by localized compressional squeeze-ups that form dome-shaped hills of uplifted sand and gravel. Current geologic understanding suggests that the lowland is a contractional region that has developed over the last 1 million years or so in response to left steps among various strands of the San Andreas fault. This left step still is taking place today, and provides the background for current earthquake activity in the northern Coachella Valley region. Sedimentary materials that are filling up the basin record information about interactions among the various fault systems, including their origin and movement history. The sediments also store ground water for the Coachella Valley region, and have been utilized for sand-and-gravel resources. Although the region does not have the population density of the more urbanized Inland Empire to the west, geologic studies in the Coachella Valley are relevant for several reasons:

- the Quaternary displacement history of the San Andreas fault strands is not well documented;
- ground-water withdrawal and recharge are occurring in an arid setting where the shape, thickness, and physical stratigraphy of subsurface basins are not completely understood;
- extensive luxury-resort growth has triggered extensive infrastructure development and service communities that are growing at a rapid rate;
- extensive tracts of Quaternary and Tertiary sand-and-gravel deposits are incompletely understood in terms of their stratigraphy, physical properties, and age, and their potential for development as industrial resources has not been systematically evaluated;
- Mission Creek and Whitewater River, two major drainages in the Coachella Valley, are significant sources for streamflows that historically have recharged the Salton Sea and its geologic predecessors (Lake Cahuilla). The Quaternary evolution of the Coachella Valley sedimentary fill thus is relevant to environmental and ecosystem studies of the Salton Sea.

In the northwest end of the Coachella Valley, multiple strands of the San Andreas fault zone appear to have stepped left into the San Gorgonio Pass structural knot in the San Andreas, thereby bypassing the southeastern San Bernardino Mountains and creating several dome-like squeeze-ups that have brought older Quaternary sedimentary materials to the surface. The domal uplifts are being dissected while late Quaternary surficial deposits simultaneously are being deposited around them on the Coachella Valley lowland.

The valley is part of the Salton trough, the landward extension of the Gulf of California (Sea of Cortez), which includes the Imperial Valley and Salton Sea. The valley lies within a structural trough because of the presence of several prominent fault zones including the San Andreas and San Jacinto. The SAFZ includes the Banning, Mission Creek, Garnet Hill, and Indio Faults; all of these faults control the flow of groundwater and its occurrence at the surface.

The climate of the valley ranges from below freezing in the winter at the highest peaks, to over 120°F in the valley during summer. Precipitation ranges from less than 4 inches in the valley to over 40 inches in the mountains. The source of groundwater in the valley is primarily from runoff from the rivers and streams that drain the mountains. At one time, it was commonly thought that the source of groundwater in the valley was from infiltration from the Colorado River, over 70 miles away!

Figure 5. Thousand Palms and oasis, and San Andreas Fault (Bruce Perry, Department of Geological Sciences, CSU Long Beach).

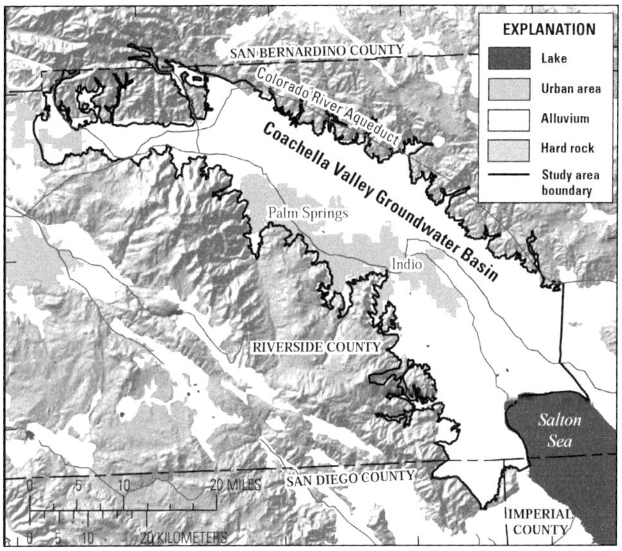

Figure 6. Coachella Valley Groundwater Basin (Dawson and Belitz, 2012).

HYDROGEOLOGIC DESCRIPTION OF THE COACHELLA VALLEY AQUIFER SYSTEM

The Coachella Valley Aquifer is subdivided into four subbasins, of which the Indio Subbasin is by far the largest (77%). The other 3 subbasins include the Mission Creek (10.6%), Desert Hot Springs (6.7%), and San Gorgonio Pass [7-21.04] (5.7%) Subbasins.

The groundwater storage capacity of the aquifer has been estimated by the Department of Water Resources and the Coachella Valley Water District (CVWD) at 39 million acre-feet, assuming a usable aquifer thickness of only 1,000 feet. By comparison, the total annual groundwater production by the Metropolitan Water District of Southern California (MWD) is 1.56 million acre-feet (MWD, 2007); the total annual water demand for California on an average year is 78 million acre-feet. Therefore, at full capacity, the basin could provide half of the water for all of California. This gives some perspective to the size of the basin.

The following descriptions by the California Department of Water Resources Bulletin 118 *California's Groundwater* (DWR, 2003) provide an excellent summary of the three groundwater basins of the Coachella Valley that we visit in this guidebook.

Indio (Subbasin Unit 7-21.01)

Basin Boundaries and Hydrology. Indio Subbasin is located northwest of the Salton Sea and receives low precipitation, averaging about 6 inches per year, and a wide range of temperatures. The Banning fault bounds the subbasin on the north and the semi-permeable rocks of the Indio Hills mark the northeast boundary. Impermeable rocks of the San Jacinto and Santa Rosa Mountains bound the subbasin on the south. A bedrock constriction separates the Indio Subbasin from the San Gorgonio Pass Subbasin on the northwest. The Salton Sea is the eastern boundary and the subbasin's primary discharge area. A low drainage divide forms a short boundary with the West Salton Sea Groundwater Basin in the southeast.

The Indio Subbasin is drained by the Whitewater River and its tributaries. The Whitewater River rarely flows throughout the year and flow in tributaries such as San Gorgonio River is intermittent. Surface flow is southeastward to the Salton Sea. The Colorado River Aqueduct and the Coachella Branch of the All-American Canal convey imported surface water into the Coachella Valley which overlies the subbasin.

Restrictive Structures. The west-trending Banning and Garnet Hill faults, along with the northwest trending San Andreas fault, are effective barriers to groundwater because of folded or displaced water bearing deposits and impermeable fault gouge (DWR, 1964). The northwestern boundary of the subbasin is a bedrock constriction that projects from the flank of San Jacinto Peak and extends northward beneath the subbasin deposits. Water level data indicate that groundwater cascades over this bedrock constriction. A drop in water level of more than 500 feet over a distance of 8,000 feet was measured in the area in April 1961 (DWR, 1964).

Recharge Areas. Surface runoff and subsurface inflow are significant sources of recharge to the subbasin. In addition, the Whitewater River spreading grounds northwest of Palm Springs receives Colorado River Aqueduct water and has a maximum capacity of 300,000 acre-feet per year (af/year)* (CVWD, 2000a). Colorado River water is conveyed into the subbasin via the Coachella Canal, which also supplies a pilot recharge project facility located in the southeastern part of the subbasin (CVWD, 2000b).

Groundwater Level Trends. Prior to 1949, water levels steadily declined because of pumping. After 1949 and into the early 1980s, water levels in the central and southern subbasin area rose as imported Colorado River water begin to recharge parts of the subbasin. Elsewhere in the subbasin during this time water levels continued to decline. Since the 1980s, water levels in the central and southern areas have declined despite Colorado River imports. These declines are largely due to increasing urbanization and groundwater pumping (CVWD, 2000a).

The primary water bearing units are the Ocotillo Conglomerate of Pliocene or Lower Pleistocene age (Dibblee, 1954), Cabezon Fanglomerate of Pleistocene age, and Pleistocene and Holocene alluvium, (DWR, 1964). The thickness of the water bearing units is estimated to be at least 2,000 feet. In the northern portion of the valley, to which ancient Lake Cahuilla did not reach, the aquifer is mostly unconfined. Where Lake Cahuilla was present (as far north as La Quinta) there are alternating aquifer units representing alternating alluvial deposits consisting of sand and gravel and aquitard units representing fine-grained silts and clays of lacustrine (Lake Cahuilla) origin.

*One acre-foot is approximately 326,000 gallons.

Mission Creek (7-21.02)

Basin Boundaries and Hydrology. Mission Creek Subbasin is located northwest of the Salton Sea, within the Colorado Desert region which is characterized by low precipitation, averaging about 6 inches per year, and a wide range of temperatures. The subbasin underlies the northwest portion of the Coachella Valley and is bounded by the impermeable rocks of the San Bernardino Mountains on the west and the Banning fault on the south. The Mission Creek fault bounds the northern and eastern edges of the subbasin and the Indio Hills bound the subbasin on the southeast. Major surface water features in the area are the Whitewater River, Mission Creek, San Gorgonio River, Little and Big Morongo Washes. These drainages flow intermittently during high precipitation events.

Restrictive Structures. The Mission Creek Basin is in the northwestern part of a large structural trough that includes the Gulf of California. The west-trending Banning and northwest-trending Mission Creek faults are the major groundwater controls in the subbasin. Both act as barriers to groundwater movement as these faults have folded sedimentary deposits, displaced water bearing deposits, and caused once permeable sediments to become impermeable (DWR, 1964).

Recharge Areas. Runoff from the surrounding highlands drains into the subbasin from intermittent creeks and rivers supplying most of the recharge to the subbasin (DWR, 1964). Subsurface leakage occurs across the Mission Creek Fault approximately three miles southeast from the City of Desert Hot Springs, allowing groundwater of different quality to enter the subbasin from the neighboring Desert Hot Springs Subbasin (Slade, 1981).

Groundwater Level Trends. Water levels have been declining since the early 1950s due to scarce annual precipitation and groundwater extractions (DWR, 1964; Slade, 1981). Groundwater level data indicate that since 1952, water levels have declined at a rate of 0.5 foot to 1.5 feet per year (CVWD, 2000a). In 1971 the United States Geological

Survey (USGS) determined water levels within the subbasin and found a semi-flat gradient to exist making groundwater movement slow with general movement to the southwest. Current water levels vary in domestic wells from 140 to 721 feet below ground surface with an average depth to water being 372 feet (MSWD, 2000).

Desert Hot Springs (7-21.03)

Basin Boundaries and Hydrology. Desert Hot Springs Subbasin is located northwest of the Salton Sea and is within the Colorado Desert Region which is characterized by low precipitation (5.66 inches) and a wide range of temperatures. The subbasin underlies the northeastern portion of the Coachella Valley. The northeasterly boundary of the subbasin is the contact with non-permeable rocks of the Little San Bernardino Mountains from Little Morongo Canyon southeast to Thermal Canyon. The southwest boundary of the subbasin is defined by the contact with the semi-permeable rocks of the Indio Hills along with the Banning and Mission Creek faults.

The Mission Creek Fault forms the boundary from the Little Morongo Canyon southeast to Pushawalla Canyon in the Indio Hills. The semi-water bearing rocks of the Indio Hills border the subbasin from the Mission Creek Fault east to the Indio Hills Fault. The Banning-Mission Creek Fault separates the Desert Hot Springs Subbasin from the Indio Subbasin beneath the alluvial debris cone between the Indio Hills and Mecca Hills. Between the Indio Hills Fault and the Banning-Mission Creek Fault at the southeast end of the Indio Hills, the subbasin boundary is the contact between Holocene alluvium and Plio-Pleistocene deposits. The subbasin contacts with the Mecca Hills to the southeast and the boundary is the southeastern side of Thermal Canyon from the Banning-Mission Creek Fault. The boundary continues easterly along the southeast wall of a tributary wash to outcrops of crystalline basement rock of the Little San Bernardino Mountains near Interstate 10 (CVWD, 2000a).

Water Bearing Formations. The primary water-bearing materials in the subbasin are relatively undisturbed and unconsolidated late Pleistocene and Holocene alluvial fan deposits, principally the Ocotillo Conglomerate; a Pleistocene aged thick sequence of poorly-bedded, coarse sand and gravel (CVWD, 2000a). These deposits underlie the Dillon Road Piedmont Slope, which is higher in elevation than the central plain of the Coachella Valley and is actively being deposited by runoff and erosion from the Little San Bernardino Mountains. Thickness of the water bearing deposits is estimated to be in excess of 700 feet and the majority of groundwater is probably unconfined. (DWR, 1964). Due to lack of development within the subbasin, groundwater and other hydrologic data are sparse except for the Miracle Hill area where development is greater due to the thermal waters that supply resorts. These hot thermal waters occur near active faults such as the Mission Creek fault. In the Miracle Hill area, more than 130 water wells have been drilled and approximately half of these were active in 1961 and pumped water for the hot water spas (DWR, 1964). Many of these wells or replacement wells are still active (MSWD, 2000).

Restrictive Structures. The Desert Hot Springs Subbasin is in the northwestern part of a great structural trough that includes the Gulf of California. The northwest trending Mission Creek and Banning faults, along with the Indio Hills fault are the major groundwater barriers in the subbasin. Smaller, related faults parallel these larger faults and all act as barriers in various degrees to groundwater movement. Direction of groundwater movement varies within the subbasin because of these structural controls imposed upon it, however flow is generally southeastward toward the Mecca Hills-Thermal Canyon area (DWR, 1964).

Recharge Areas. Seasonal runoff draining from the Little San Bernardino Mountains recharges the subbasin by percolating through the underlying water bearing coalescing alluvial fan deposits. Surface runoff, from high precipitation or snow-melt, is contained by intermittent creeks that discharge into the subbasin (DWR, 1964).

Groundwater Level Trends. In 1961 depth to water ranged from 12 feet below ground surface near the Mission Creek Fault to over 300 feet southeast of Miracle Hill (DWR, 1964). Water level data was sparse in most areas within the subbasin except the Miracle Hills, where the water table is declining because of use by resorts. Water level data in the other areas of the subbasin suggest that the water table remained stable (DWR, 1964).

The 1964 synopsis of groundwater level trends is probably still relevant to current basin conditions. Presently, the subbasin is still underdeveloped due to high levels of TDS and most of the groundwater extraction is in the Miracle Hill Area where water levels are still in decline.

Impairments to the water quality include inorganics, organic contaminants, and radiologicals, in addition to salts being imported from the Colorado River water.

GROUNDWATER FLOW

Groundwater within the valley flows generally from topographic highs (mountains and alluvial fans) to topographic lows (Salton Sea), but is strongly controlled by the presence of several prominent faults such as the strong gradients along the faults that comprise the SAFZ. Of particular note are the water levels in the area of the junction of the Banning and Mission Creek Strands, as well as the Garnet Hill and Banning Strands (Tyley, 1974). Because of the control these faults exert on the groundwater flow, it becomes apparent how groundwater reaches the surface at locations along faults, resulting fan palm oases (see Cornett, this volume).

SUPPLIED WATER

Within the Coachella Valley are two primary water purveyors, the Coachella Valley Water District and the Desert Water Agency. These two agencies, along with several smaller water districts in the region (Mission Springs Water District, Coachella, Indio, and Myoma Dunes Water Company), supply water for domestic and agricultural use in an area that receives less than 4 inches of rain per year. The water supply comes from two primary sources, runoff from portions of the San Jacinto, Santa Rosa, San Bernardino, and Little San Bernardino mountain ranges, and water imported from the Colorado River via the Coachella Canal and Colorado River Aqueduct. Although CVWD is entitled to State Water Project (SWP) water that originates in northern California, the valley has no physical connection via canals or aqueducts to receive this water. Therefore, the Metropolitan Water District of Southern California has agreed to a water swap in which CVWD receives an amount of water from the Colorado River Aqueduct to which it is entitled from the SWP, and MWD receives an equal amount of SWP water to which CVWD is entitled (CVWD Urban Water Management Plan, 2008).

Runoff from rainfall and snowmelt travels down the mountain slopes over weathered and unweathered bedrock until it reaches the contact of the alluvial fans of the valley that lie against the mountains. Groundwater infiltrates into these alluvial fans, ultimately reaching the aquifer system of the Coachella Valley. In some locations, the precipitation collects in small drainages far up the mountain slope and converge to create larger streams and rivers that reach the valley. The Whitewater River, that drains a part of San Gorgonio and San Bernardino Mountains, is the primary example of such a condition. On average, the Whitewater River delivers approximately 12,000 acre feet of natural recharge to the valley annually (Tyley, 1974). The next largest source of natural recharge is only 2,700 af/year (Snow Creek). Clearly, the recharge from the Whitewater River (snow pack runoff) is a significant part of the natural recharge (33%), but falls well short of the annual valley demand of nearly 500,000 af (CVWD, 2008).

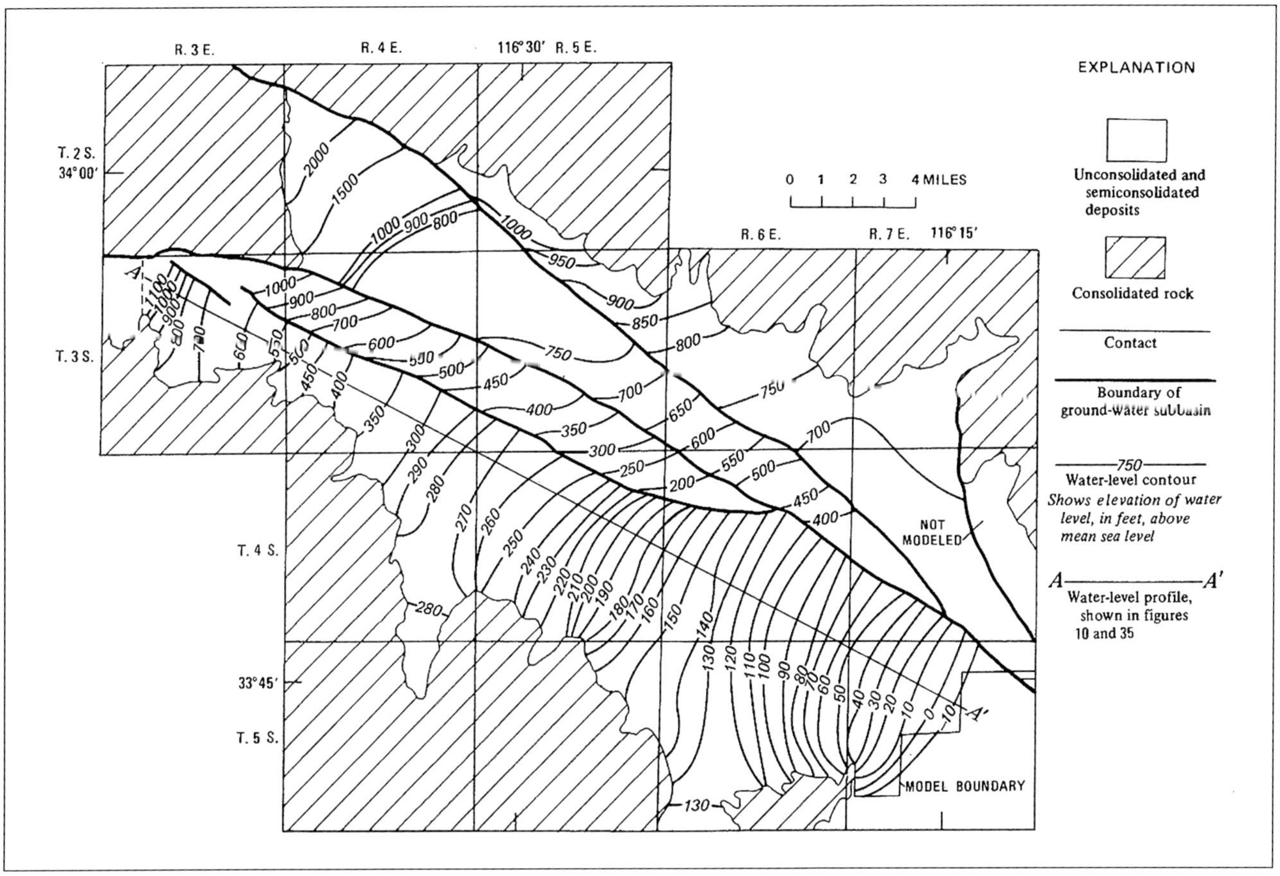

Figure 7. Water-level contours, autumn 1936 (Tyley, 1974, Figure 9)

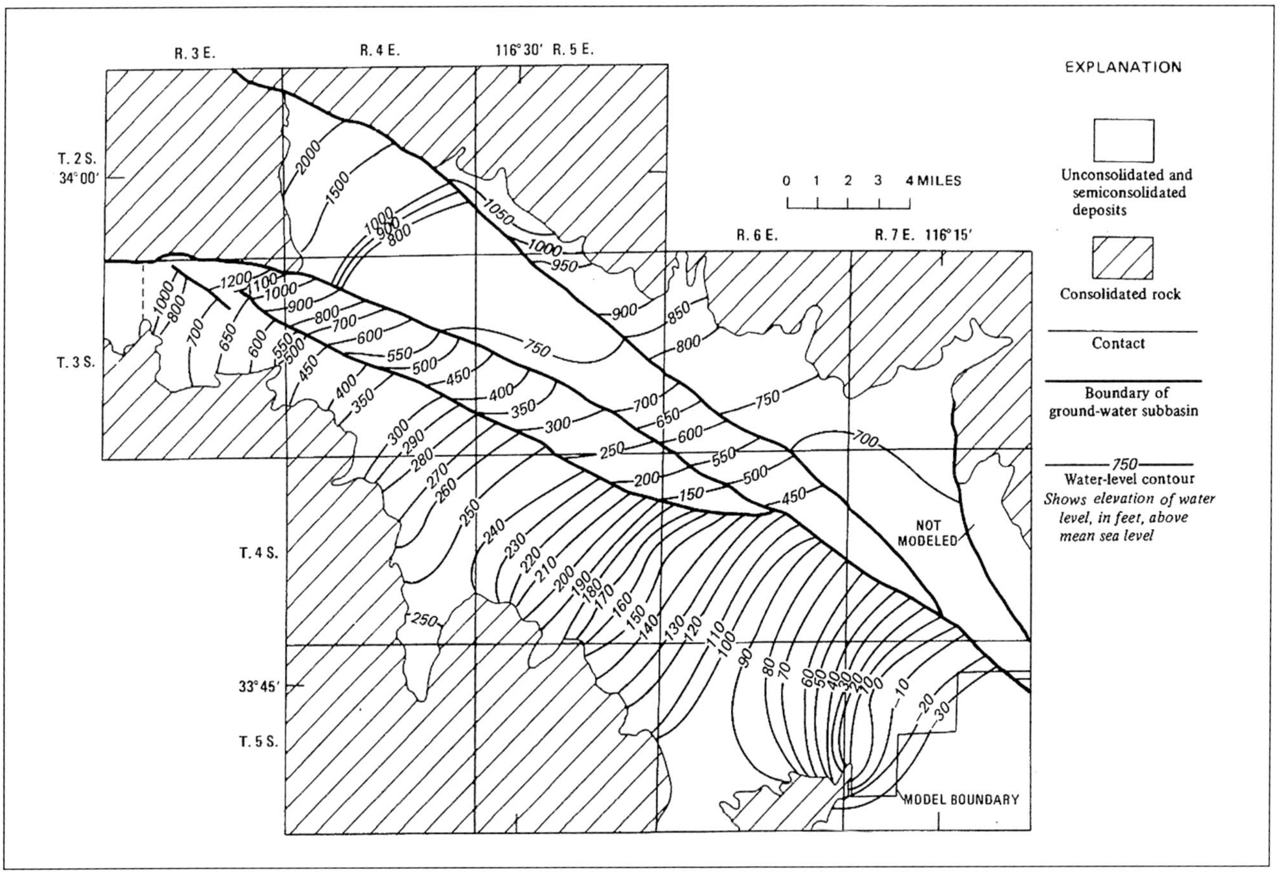

Figure 8. Water-level contours, autumn 1951 (Tyley, 1974, Figure 12)

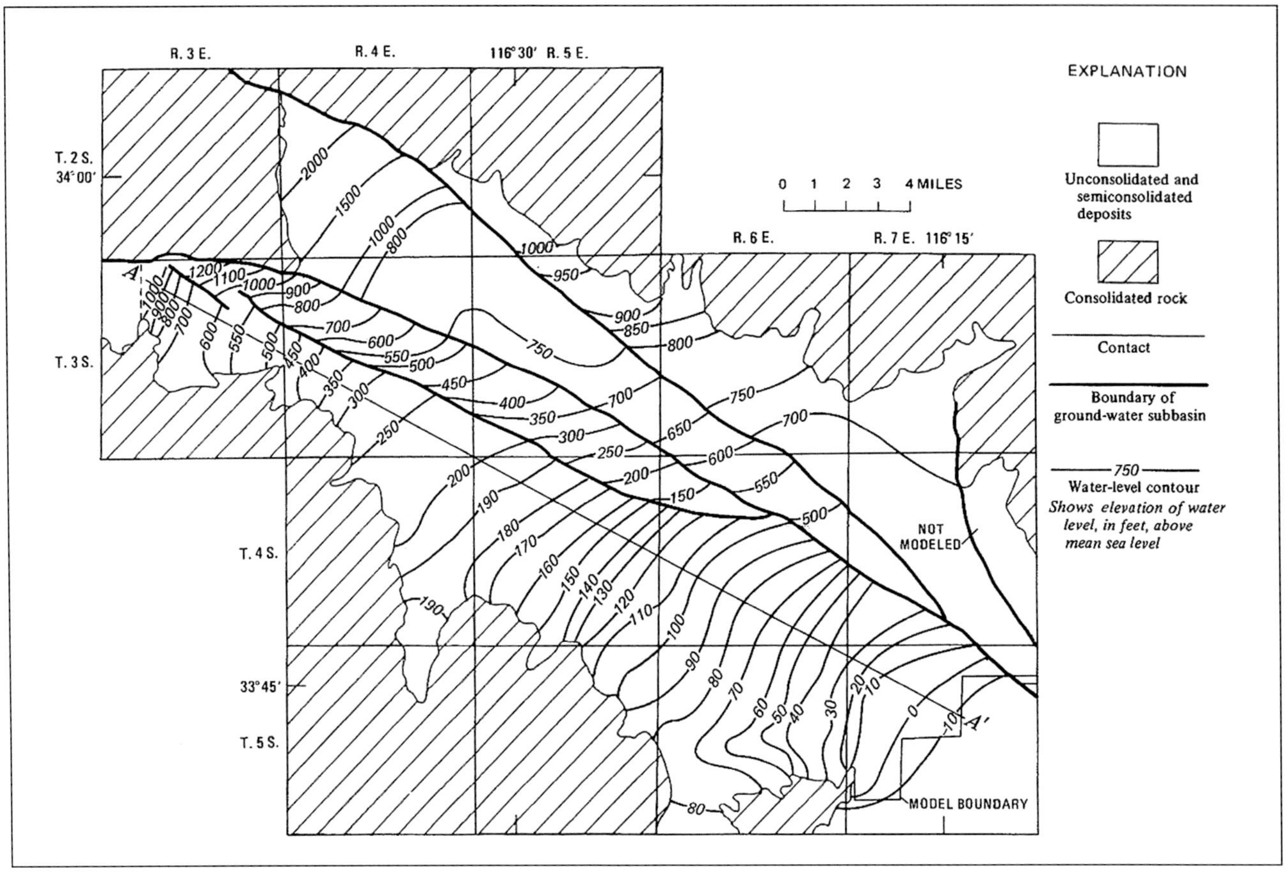

Figure 9. Water-level contours, autumn 1967 (Tyley, 1974, Figure 13)

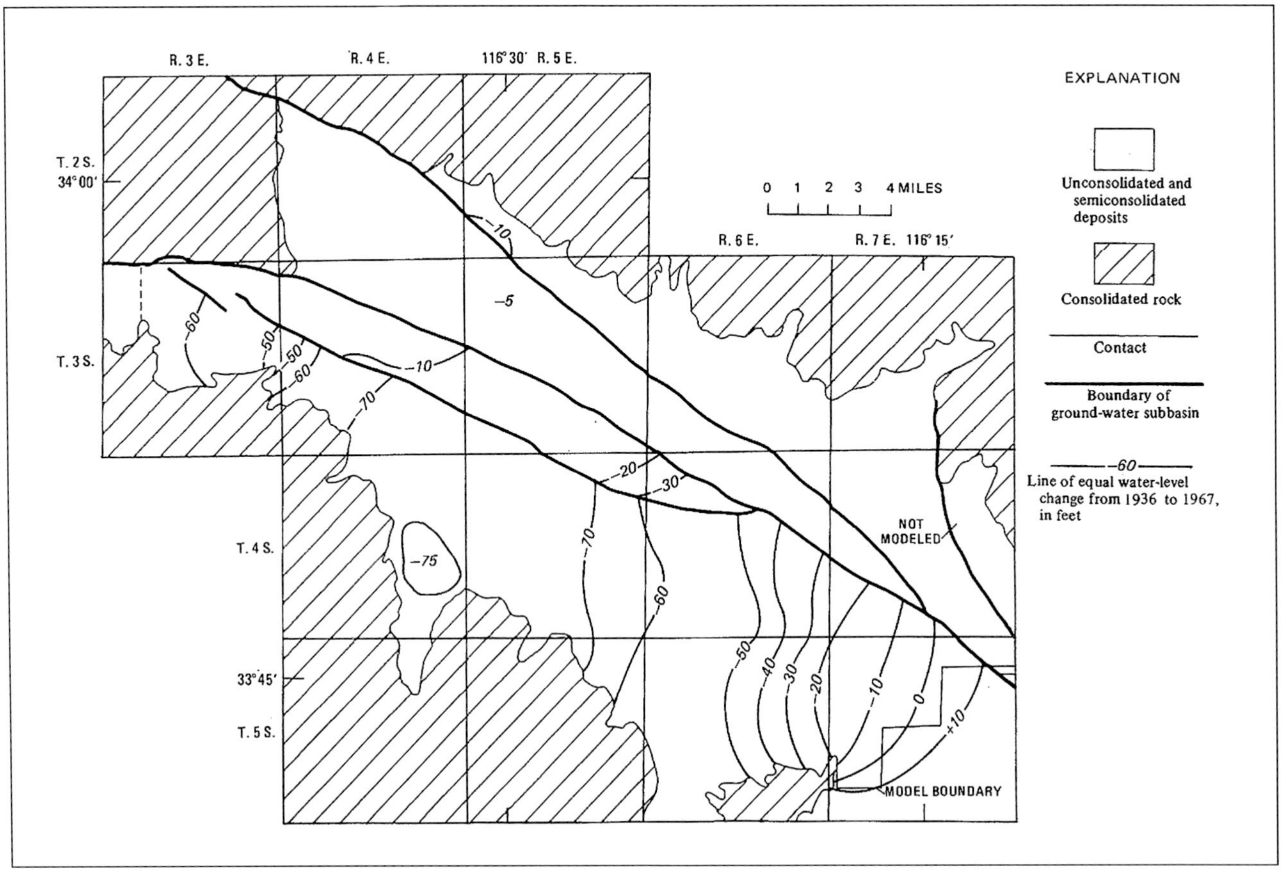

Figure 10. Changes in water-level elevation, 1936–67 (Tyley, 1974, Figure 14)

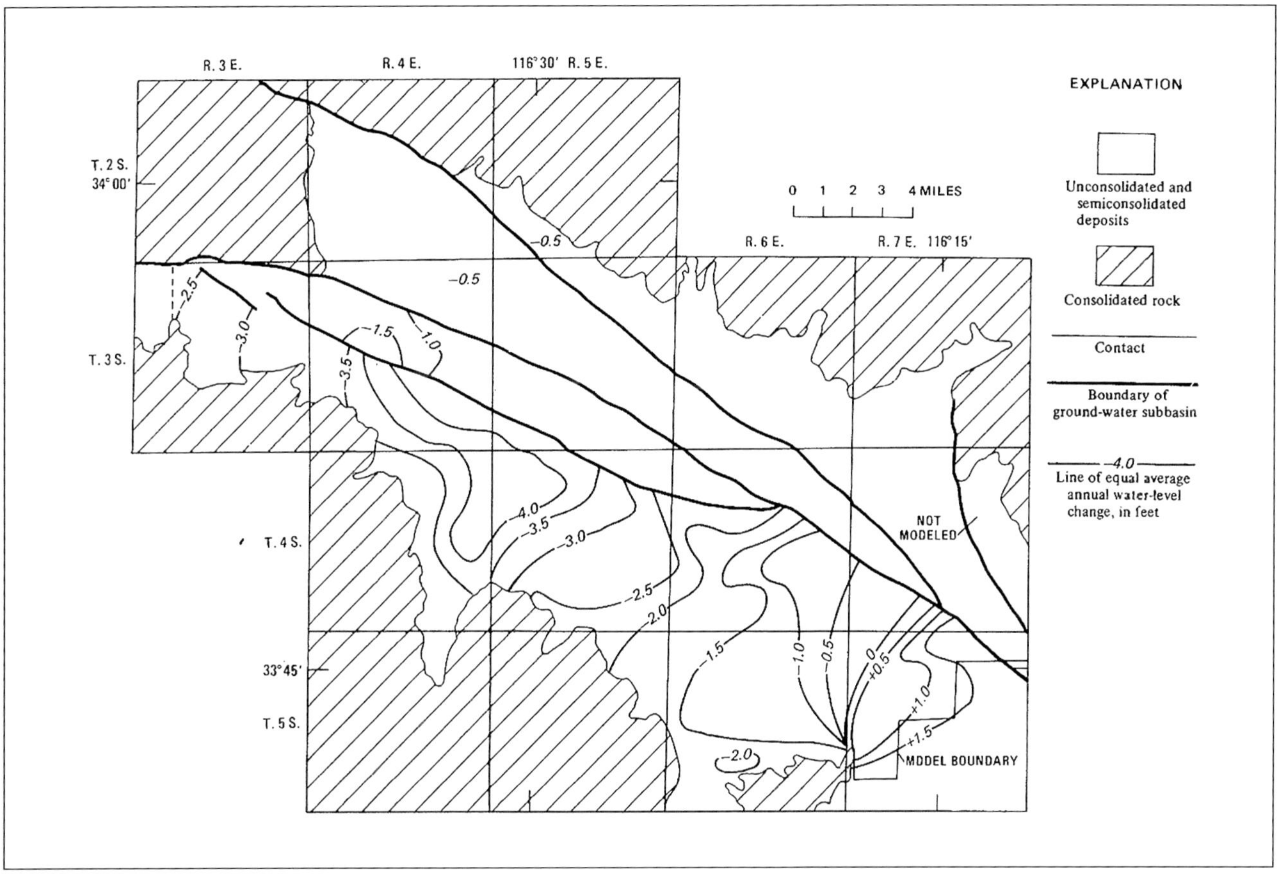

Figure 11. Average annual change in water-level elevation, 1953–67 (Tyley, 1974, Figure 17)

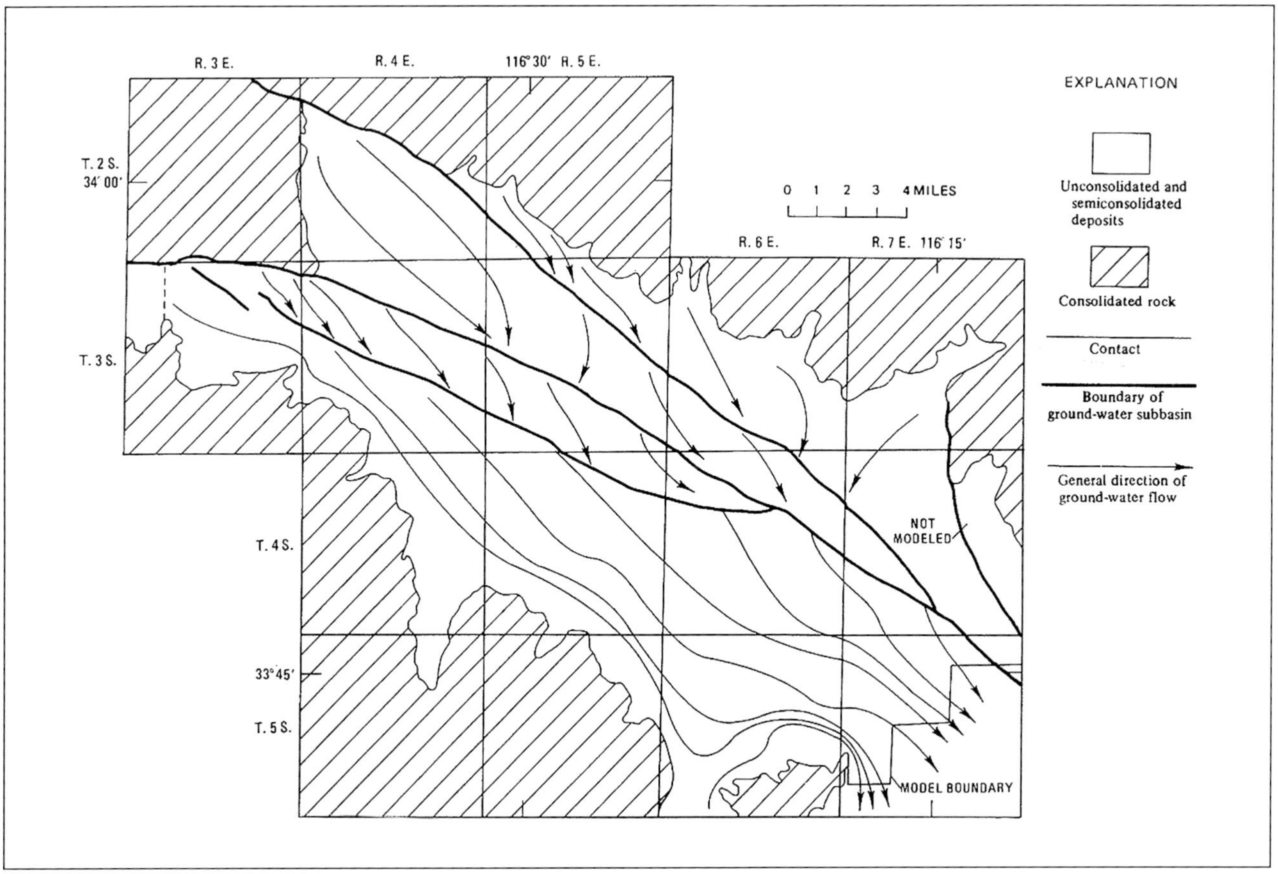

Figure 12. Generalized ground-water flow lines, 1937 (Tyley, 1974, Figure 16)

The Whitewater River drains into the northern end of the Coachella Valley. Prior to settlement in the area, the Whitewater River was a braided river that drained the mountain slopes and discharged into the Coachella Valley. Most times of the year, the river likely did not reach the Salton Sea (or Salton Sink), but rather the surface water most often would infiltrate into the alluvial sediments of the valley. During times of high flow, the river would certainly have discharged into the Salton Sink, or at some earlier time, ancient Lake Cahuilla, or even a more ancient expression of the Gulf of California (Sea of Cortez).

The other primary source of water in the valley is imported water from the Coachella Canal and Colorado Aqueduct. According to the CVWD, the 122-mile Coachella Canal diverts roughly 330,000 acre-feet of water per year for agricultural use. The remaining agricultural demand (25% to 30%) is satisfied by groundwater pumping by farmers using privately owned wells.

Water imported from the Colorado River Aqueduct (the water swapped from MWD through the SWP entitlement), is used to recharge the aquifer in the northern part of the valley by way of the Whitewater River infiltration gallery. Since 1973, imported water has been used to recharge the aquifer in the upper part of the valley. Records indicate that initial amounts of water recharged at the Whitewater infiltration gallery were on the order of 16,000 af per year. This amount increased dramatically over the period 1979 to 1986, from 16,000 af to 298,000 af. Since that time, recharge quantities have varied between 14 and 178,000 AF. As a result of groundwater infiltration at Whitewater, groundwater levels have risen over the past few decades in the northern part of the valley, while water levels have continued to decline in the southern part of the valley, which is unaffected by the Whitewater infiltration gallery. As a result, the CVWD has been evaluating options for the creation of infiltration galleries in the southern part of the valley at Dike 4, which is at the mouth of Martinez Canyon (Trip 1, Stop 9).

Water for domestic use is supplied by wells that penetrate the alluvial aquifer system of the valley. The CVWD produces about 125,000 af annually from groundwater for use by more than 100,000 homes and businesses, which increases monthly.

Another water source, for domestic irrigation and irrigation of golf courses and other outdoor landscaping, is recycled waste water, or gray water. The CVWD treats nearly 15,000 acre-feet of water annually. However, even with this conservation method, the predominant source of irrigation for golf courses is still groundwater. In the southern portion of the valley, gray water is increasingly used for irrigation on golf courses. CVWD has projects underway to deliver more gray water to the central part of the valley, reducing demand on groundwater.

OTHER FEATURES OF THE COACHELLA VALLEY

The **Palm Springs Aerial Tramway** takes passengers nearly vertically up to the top of Mount San Jacinto through Chino Canyon, from elevation 2,643 feet 2.5 miles up to the Mountain Station at an elevation of 8,516 feet. The face of the canyon is nearly 6,000 vertical feet. It is currently the largest rotating tramcar in the world. The other two rotating trams are in Engelberg, Switzerland, and Cape Town, South Africa. The concept of this tram was developed in 1935 by Francis Crocker, design began in 1950, and the tram opened on September 14, 1963. Crocker died in 1992, four years before the bonds issued to construct the tram were paid off. (Palm Springs Aerial Tramway, 2013)

In 2001 Mount San Jacinto was one of three final potential locations considered for a national **Deep Underground Scientific and Engineering Laboratory** (UCI, 2013). Other locations were the Homestake Gold Mine in the Black Hills near Lead, South Dakota, the Soudan Mine near Tower, Minnesota. Homestake Gold Mine was selected.

SUBSIDENCE DUE TO GROUNDWATER WITHDRAWAL

From Press Release (USGS, 2007)

A study by the U.S. Geological Survey confirmed CVWD concerns that land subsidence is occurring in areas of substantial groundwater use throughout the Coachella Valley.

The two agencies initiated the study in 1996 when it was first believed subsidence was occurring in the Coachella Valley. Scientists with the USGS California Water Science Center used Global Positioning System (GPS) surveying and a satellite mapping process known as interferometric synthetic aperture radar (InSAR) to document the drops in elevation between 1996 and 2005. GPS measurements are taken on the ground at specific locations, or benchmarks, while InSAR provides an areawide elevation-change map or snapshot.

At all of the GPS benchmarks, some subsidence occurred between 1996 and 2005. At three benchmarks, the drop was less than an inch, while at three others the subsidence was about a foot. At one benchmark, near the intersection of 54th Avenue and Jackson Street near Coachella, the one-foot drop in land-surface elevation happened from 2000 to 2005.

"The subsidence rates in many areas have more than doubled since 2000," said Michelle Sneed, USGS scientist and lead author of the study. "All the subsiding areas are near sites where ground-water levels declined between 1996 and 2005, and some water levels in 2005 were at the lowest levels in their recorded histories."

The research, which has cost about $790,000 since 1995, has been funded primarily by the USGS and the Coachella Valley Water District, with the city of Palm Desert contributing $17,000.

Since the 1920s, groundwater has been a major source of agricultural, municipal, and domestic supply in the Coachella Valley, resulting in significant groundwater pumping that has contributed to water-level declines of as much as 100 feet. The heavy groundwater use, in turn, has led to subsidence, which can disrupt surface drainage; reduce aquifer storage; cause earth fissures; and damage wells, buildings, roads and utility infrastructure.

In 2001, CVWD addressed groundwater overdraft by adopting a comprehensive Water Management Plan (WMP) to act as a blueprint for water reliability. The WMP takes a three-tiered approach to groundwater management: an increased imported water supply; promotion and assistance with conservation, and providing existing groundwater users an alternative source of water.

CVWD planned and prioritized nearly 50 programs and projects for the WMP. Current WMP efforts aimed at keeping groundwater levels stable include construction of the $70 million Mid-Valley Pipeline, which will enable up to 50 golf courses to utilize a blend of recycled water and Colorado River water in lieu of groundwater, and the $40 million groundwater recharge facility south of Lake Cahuilla.

"Addressing declining water levels has always been a priority for the water district, which is why the Coachella Canal was built in the 1940s and the Whitewater recharge ponds constructed in the 1970s," Robbins said. "As the Coachella Valley's popularity and prosperity continue, it's crucial that CVWD and other local water agencies continue making significant strides to ensure a reliable water supply."

The U.S. Geological Survey report, "Detection and Measurement of Land Subsidence Using Global Positioning System Surveying and Interferometric Synthetic Aperture Radar, Coachella Valley, California, 1996-2005," by Michelle Sneed, et al., can be found on the Internet at the USGS California Water Science Center's Web site at http://ca.water.usgs.gov/.

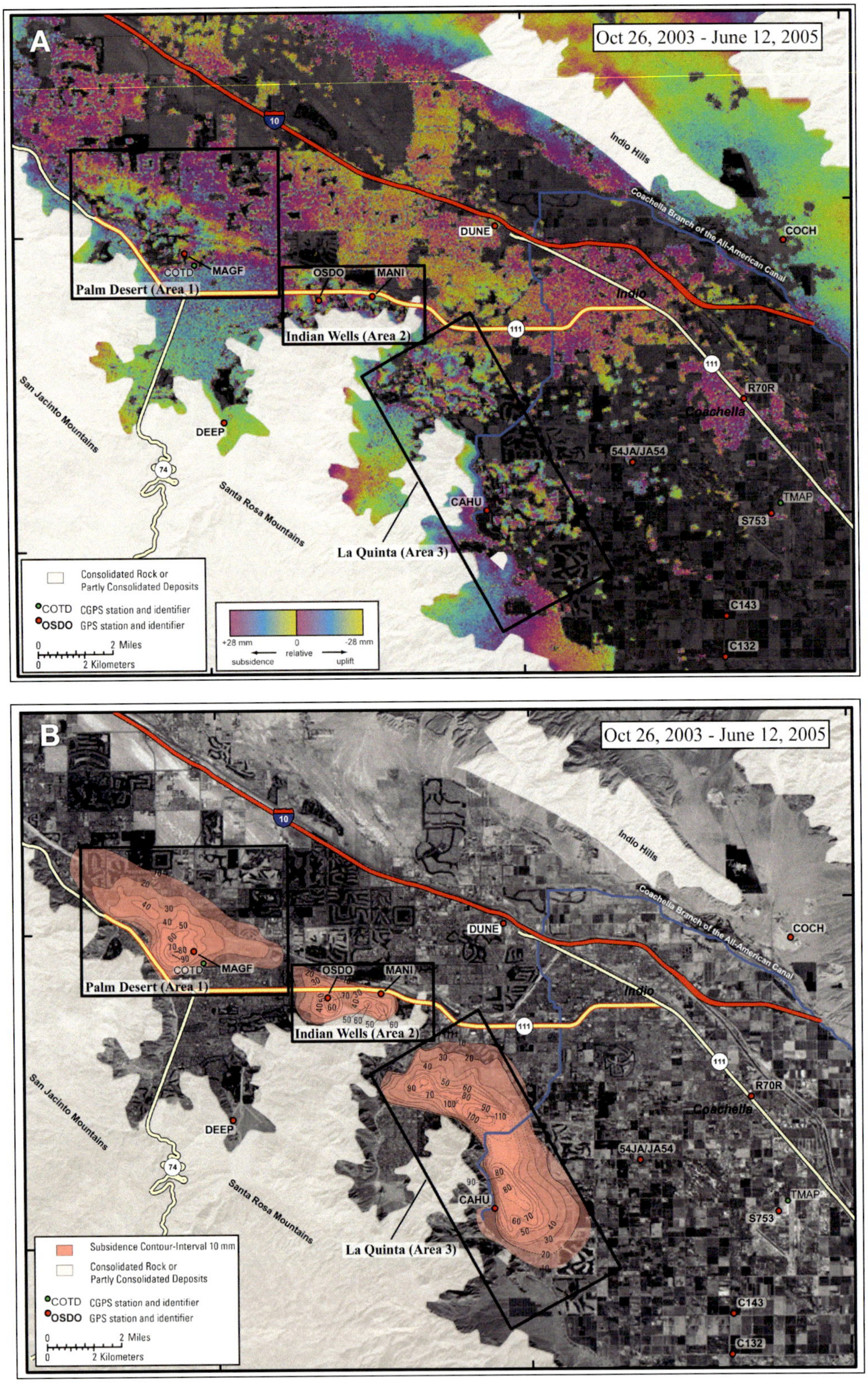

Figure 13A and 13B. Global Positioning System and Interferometric Synthetic Aperture Radar study shows continued land subsidence from groundwater extraction in the western Coachella Valley (USGS, 2013).

Maximum Measured Subsidence. GPS benchmark DUNE maximum measured subsidence is 13.11 inches for time period 1996–2005. InSAR maximum measured subsidence PALM DESERT is 12.96 inches for time period 1996–2000, and Spring 2003–Fall 2005 combined. GPS benchmark JA54 maximum measured subsidence is 12.28 inches for time period 2000–2005.

REFERENCES

Bigley, Steve, Coachella Valley Water District, 2000, Telephone conversation with D.A. Gamon, Southern District, California Department of Water Resources. October 31.

California Department of Water Resources (DWR), 1964, Coachella Valley Investigation. Bulletin 108. 145 p. 13 plates.

California Department of Water Resources, 1964, Coachella Valley Investigation: California Department of Water Resources, Bulletin 108, 145 p.

California Department of Water Resources, 1979, Coachella Valley Area Well Standards Investigation. Southern District Memorandum Report. 40 p.

California Department of Water Resources, 2003, California's Groundwater. Bulletin 118 October 2003 update. Accessed October 27, 2013, from http://www.water.ca.gov/groundwater/bulletin118/update2003.cfm

Coachella Valley Water District (CVWD), 2000a, Engineer's Report on Water Supply and Replenishment Assessment 2000/2001. Coachella, California. 46 p.

Coachella Valley Water District, 2000b, Coachella Valley Water Management Plan (Draft). Coachella Valley Water District. 272 p.

Coachella Valley Water District, 2011, Urban Water Management Plan, Final Report. July. 249 p.

Dawson, B.J.M., and Belitz, Kenneth, 2012, Groundwater quality in Coachella Valley, California: U.S. Geological Survey Fact Sheet 2012–3098, 4 p.

Dibblee, T. W., Jr., 1954, Geology of the Imperial Valley region, California; in Jahns, R. H., ed., Geology of southern California: California Division of Mines Bulletin 170, ch. 2, pt. 2, p. 21–28 and Plate 2 (Generalized geologic map of the Imperial Valley region, map scale 1 inch = 6 miles).

Levy, Tom, Coachella Valley Water District, 2002, Letter to Doug Osugi, Division of Planning and Local Assistance, California Department of Water Resources. April 5, 2002.

Matti, J.C., Morton, D.M. and Cox, B.F., 1992, The San Andreas fault system in the vicinity of the central Transverse Ranges province, southern California: U.S. Geological Survey Open-File Report 92-354, 40 p., scale 1:250,000.

Metropolitan Water District (MWD), 2007, Groundwater Assessment Study: A Status Report on the Use of Groundwater in the Service Area of the Metropolitan Water District of Southern California. Report Number 1308. September. Accessed October 27, 2013 from http://www.mwdh2o.com/mwdh2o/pages/yourwater/supply/groundwater/gwas.html

Mission Springs Water District (MSWD), 2000, Mission Springs Water District Urban Water Management Plan 2000.

Mission Springs Water District, 2000, Mission Springs Water District Urban Water Management Plan 2000.

Palm Springs Aerial Tramway, 2013, Palm Springs Aerial Tramway History. Accessed October 27, 2013 from http://www.pstramway.com/history.html

Planert, Michael, and John S. Williams, 1995, Ground Water Atlas of the United States – Segment 1 California Nevada. U. S. Geological Survey Hydrologic Investigations Atlas 730-B.

Reichard, E.G., and Meadows, J.K., 1992, Evaluation of a ground-water flow and transport model of the upper Coachella Valley, California: U.S. Geological Survey Water-Resources Investigations Report 91-4142, 101 p.

Slade, Richard and Associates LLC, 2000, Final Hydrogeologic Evaluation, Well Siting, and Recharge Potential Feasibility Study: Mission Creek Groundwater Subbasin, Riverside County, California. 59 p. Report prepared for ASL Consulting Engineers Pasadena, California and Mission Creek Water District, Desert Hot Springs, California.

Slade, Richard, 1981, Hydrogeologic Conditions in the Mission Creek Subbasin Upper Coachella Valley, California. In Geology of the San Jacinto Mountains. Annual Field Trip Guidebook No. 9. South Coast Geological Society. p. 151–163.

Sneed, Michelle and Brandt, Justin T., 2013, Detection and Measurement of Land Subsidence Using Global Positioning System Surveying and Interferometric Synthetic Aperture Radar, Coachella Valley, California, 1996–2005. U.S. Geological Survey Scientific Investigations Report 2007–5251. Version 2.0, June 2013. Accessed October 27, 2013, from http://pubs.usgs.gov/sir/2007/5251/

Tyley, S.J., 1974, Analog Model Study of the Ground-Water Basin of the Upper Coachella Valley, California, U.S. Geological Survey Water Supply Paper 2027. 77 p. Accessed October 27, 2013, from http://pubs.usgs.gov/wsp/2027/report.pdf

University of California, Irvine, 2013, Deep Underground Science & Engineering Lab at Mt. San Jacinto. Accessed October 27, 2013, from http://www.ps.uci.edu/~SJNUSL/

U.S. Geological Survey, 2003, Major Faults of Southern California, Inland Empire Region, Neotectonic Framework of the South-Central Transverse Ranges and Vicinity, Southern California Aerial Mapping Project (SCAMP). Last updated September 3. Accessed October 26, 2013 from http://geomaps.wr.usgs.gov/archive/scamp/html/scg_ie_neo.html

U.S. Geological Survey, 2006, Geology of the Coachella Valley and Salton Trough Region. May 26. Accessed October

27, 2013 at http://geomaps.wr.usgs.gov/archive/socal/geology/coachella_valley/index.html

U.S. Geological Survey, 2007, Study finds subsidence continuing in the Coachella Valley. Press Release dated December 17. Accessed October 27, 2013, from http://ca.water.usgs.gov/news/2007/release071217.html

U.S. Bureau of Land Management, 2012, Santa Rosa San Jacinto Mountains National Monument. Accessed September 12, 2012, from http://www.blm.gov/ca/palmsprings/points_of_interest.html. Checked October 25, 2013, from https://web.archive.org/web/20121018104518/http://www.blm.gov/ca/st/en/fo/palmsprings/santarosa/points_of_interest.html

Indian Canyons, 2008

Factors Determining the Occurrence of the Desert Fan Palm, *Washingtonia filifera*

James W. Cornett
JWC Ecological Consultants
P.O. Box 846
Palm Springs, CA 92253

ABSTRACT

The desert fan palm, Washingtonia filifera, *ranges throughout the Colorado Desert of southeastern California where springs, seeps or streams supply a permanent supply of water. The occurrence of this species is a result of three factors: a permanent moisture supply, suitable climate and dispersal opportunity.*

INTRODUCTION

The desert fan palm, *Washingtonia filifera*, is generally the largest plant occurring at permanent seeps, springs and streams in California's Colorado Desert. This fact, together with a rich green crown of fronds, make the desert fan palm a striking addition to the flora of the approximately 168 desert oases where it occurs (Cornett, 2010). Its visual dominance is so great that any oasis is popularly identified as a "palm oasis" when the desert fan palm is present (Henderson, 1961; Kirk, 1973).

Such visual distinctiveness is made all the more striking when contrasted with the surrounding terrain. The vast majority of palm oases occur in the Colorado Desert, the lowest, hottest and driest subdivision of the North American Desert (Jaeger, 1957). In this region vegetation is scant and what plants do exist are typically small and generally dormant. Desert fan palms stand in dramatic contrast to such surroundings. It is small wonder that they have been the subject of public curiosity and popular writing for decades.

Figure 1. Pushawalla Palms on the San Andreas Fault. James W. Cornett.

SOURCES OF MOISTURE

The desert fan palm requires a permanent or near permanent source of moisture in order for seeds to germinate and seedlings to grow to maturity (Cornett, 2010). Without a permanent

Adapted from Santa Bernadino County Museum Association, Special Publication 93-1, pages 37–38, with permission.

supply of moisture a tree will decline, as revealed by a reduction in leaf size and lack of fruit production. The decline may take place over a period of several years, and a palm can be revitalized should a permanent moisture source be reestablished.

Over 90 percent of wild desert fan palms are found either along the eastern base of the Peninsular Range in California and Baja California Norte, or the San Andreas fault in Riverside County, California (Cornett and others, 1986). Numerous locations in both of these regions provide a permanent source of moisture.

The Peninsular Range runs in a northwest direction beginning at the tip of the Baja Peninsula and extending northward approximately 900 miles to the San Gorgonio Pass in Riverside County, California (Oakeshott, 1971). The range reaches its greatest elevation, 10,834 feet above sea level, at its northern terminus.

The Peninsular Range intercepts eastward-traveling storms, resulting in appreciable amounts of precipitation falling on the mountain slopes. Year-round runoff in the form of streams or seeps occurs in many of the canyons draining the slopes. The reliability and quantity of this source of moisture support over 75 percent of the wild fan palm population. All of these trees occur in canyons draining the eastern or leeward side of the Peninsular Range, and the majority occur in the northeastern portion of the range where elevations are higher and watersheds greater. In addition, the northerly portions of the range receive more precipitation in winter since the region lies closer to Southern California's winter storm track (Bailey, 1966).

THE SAN ANDREAS FAULT

Approximately 15 percent of all wild desert fan palms occur along the San Andreas fault in the Colorado Desert of southeastern California (Figure 1). At no less than 15 locations along the fault, groundwater reaches the surface and is available to desert fan palm.

The finely ground soil along the fault prevents groundwater from passing through it as the moisture flows downslope from north to south (Iacopi, 1964). With the continual flow of water, the water table rises at the fault, eventually reaching the surface. Water does not seep upward over the entire length of the fault; it only reaches the surface at a dozen or so locations where underground conditions promote the rise of groundwater. Although it is generally believed that earth movements can rapidly affect the amount and location of seepage, this has not yet been documented along the Colorado Desert portion of the San Andreas fault.

DESERT SPRINGS

There are about a dozen additional palm oases that do not occur at the base of the Peninsular Range or the San Andreas Fault. These oases occur at isolated desert springs where faults allow water to rise to the surface, a condition similar to those along the San Andreas fault.

THE INFLUENCE OF CLIMATE ON THE OCCURRENCE OF WILD DESERT FAN PALMS

It appears there is no locality in North America where high, summertime temperatures directly prohibit the occurrence of the desert fan palm. The species occurs throughout the Colorado Desert, the hottest subdivision of the North American Desert, and Death Valley, where summer temperatures average the hottest in the world (Jaeger, 1957).

Low temperatures, however, do seem to prohibit the occurrence of the desert fan palm. Desert fan palms have never been found at elevations over 4,000 feet, although there are Colorado Desert locations at such altitudes where a permanent supply of moisture is available. Nor have they been found north of latitude 38°N (Cornett, 1988). Again, it is assumed that this reflects an intolerance to cold conditions. It should be noted that at high latitudes and elevations, growth of desert fan palms is considerably slower than in warmer climates.

It is less obvious why desert fan palms do not occur in the coastal climates of southern California. Winter temperatures are relatively mild and locations where permanent moisture can be obtained are numerous. It is the author's belief that relatively cool summer temperatures and/or frequent fires are responsible for the absence of the palms. Cool summer temperatures may slow the growth of desert fan palms so that they cannot compete effectively with other arborescent species around springs. A more frequent fire regime might either eliminate young palms or reduce the competitiveness of palms. Although the desert fan palm is quite resistant to fire, it may not be able to maintain the viability of its populations when subjected to the kind of fire regime experienced by coastal plant communities of southern California.

LACK OF DISPERSAL OPPORTUNITY AS AN INFLUENCE IN PALM DISTRIBUTION

The present distribution of desert fan palm represents the locations where the desert fan palm has occurred since its appearance as a distinct species, or the locations to which the seeds have been effectively transported by birds, humans or other mammals. It is likely that the desert fan palm occurs at certain locations because of successful dispersal by animals other than humans.

There are numerous locations where the conditions seem suitable for the occurrence of the desert fan palm, but none are present. Their absence could be attributed to a lack of dispersal opportunity. There are numerous springs in western Arizona, southern Nevada and the Mojave Desert where the climate seems suitable, but the palms are absent. The most reasonable explanation for their absence is that there has been no bird or mammal that has yet transported the seeds to the sites. At some point in the future seed transport may occur, and the desert fan palm may become established.

ACKNOWLEDGMENTS

The author would like to thank the Richard King Mellon Foundation of Pittsburgh, Pennsylvania, for the financial support that made research on the ecology of desert fan palms oases possible. Appreciation is also extended to Barbara McAlpine who reviewed and edited the manuscript.

REFERENCES CITED

Bailey, H.P., 1966. Weather of southern California. University of California, Berkeley.

Cornett, J. W., T. Glenn and J. M. Stewart. 1986. The Largest Desert Fan Palm Oases. Principes 30(2):82–84.

_______, 1988. Naturalized populations of the desert fan palm *Washingtonia filifera*, in Death Valley National Monument, in Plant biology of eastern California. C.A. Hall Jr. and V. Doyle-Jones, eds. Los Angeles, University of California Press.

_______, 2010. Desert Palm Oasis: A Comprehensive Guide. Nature Trails Press, Palm Springs, California.

Henderson, R., 1961.On desert trails. Los Angeles, Westernlore Press.

Iacopi, R., 1974. Earthquake country. Menlo Park, Lane Books.

Jaeger, E.C.,1957. The North American deserts. Stanford University Press.

Kirk. R., 1973. Desert the American Southwest. Boston, Houghton-Mifflin Company.

Oakeshott, G.B., 1971. California's changing landscapes. New York, McGraw-Hill Book Company.

Exploration of the Hydrogeology of the Mission Creek Fault: Faulting as a Barrier to Groundwater Flow

Jan Gillespie
State University California, Bakersfield,
Bakersfield, California

Alex Mayer
Michigan Technological University,
Houghton, Michigan

BACKGROUND

Stop 1 of Trip 2, 2012 South Coast Geological Society Road Log, is located on the fault scarp of the Mission Creek Fault, a strand of the San Andreas Fault in the northern Coachella Valley. The San Andreas Fault is a large right-lateral strike slip fault zone that extends from near Mendocino in northern California to the Gulf of California in Mexico—a distance of approximately 800 miles. In the northern Coachella Valley, the San Andreas splits into two strands. The Mission Creek Fault at Desert Hot Springs, California is the north-eastern strand of the San Andreas and the Banning Fault is the south-western strand (Figure 1). A recent study by Catchings et al. (2009) presents seismological and seismic data that suggest that these faults merge to form a single strand at depths of about 10–12 kilometers.

At Trip 2, Stop 1, a poorly consolidated outcrop of Pleistocene alluvium (probably the Ocotillo Conglomerate) occurs between the Mission Creek Fault on the south end of the outcrop and the Miracle Hill Fault on the north side (Figure 2a). The Mission Creek Fault is marked by a conspicuous northwest-trending scarp which is easily seen in the photo in Figure 2b. The scarp is marked by numerous springs and seeps.

Slip on the fault is right lateral oblique and the fault is nearly vertical. The northeastern block of the fault is the up thrown block. Seismic data interpreted by Catchings et al. (2009) suggest that

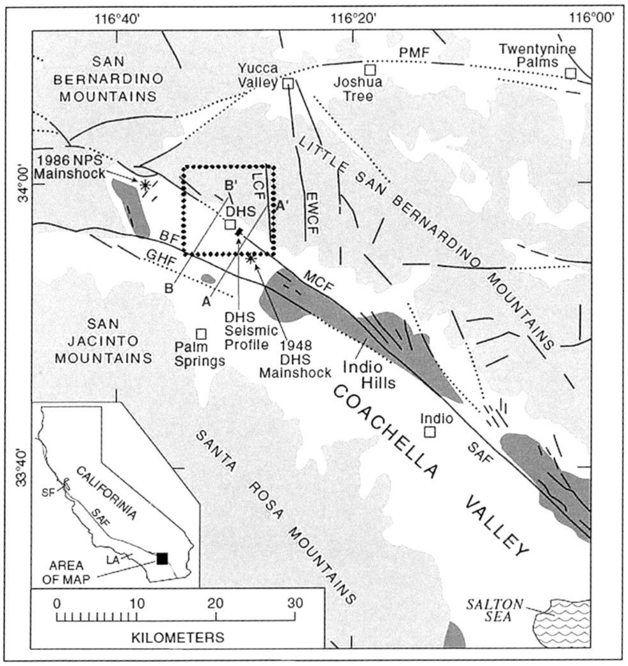

Figure 1. Location map of the northern Coachella Valley area (from Catchings et al., 2009) with major faults and cities. Abbreviations are as follows: Banning Fault (BF), Desert Hot Springs (DHS), East Wide Canyon Fault (EWCF), Garnet Hill Fault (GHF), Long Canyon Fault (LCF), Mission Creek Fault (MCF), Pinto Mountain Fault (PMF), and San Andreas Fault (SAF). Asterisks depict the epicenters of the 1948 DHS and the 1986 North Palm Springs (NPS) earthquakes (Nicholson, 1986). Quaternary alluvium shown in white, Pliocene and Pleistocene stratified rock shown in dark gray, and pre-Cenozoic crystalline rocks, shown in light gray. The bold line labeled DHS Seismic Profile marks the location of the outcrop visited in the field trip (from Catchings et al., 2009).

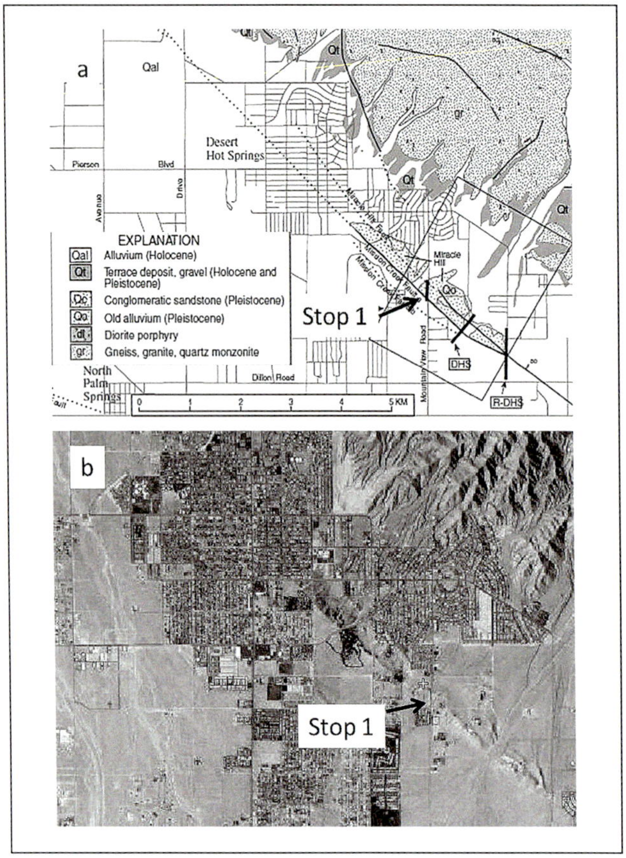

Figure 2a. Geologic map (from Proctor, 1968) of the Desert Hot Springs area showing the location of Trip 2, Stop 1 (modified from Catchings et al. (2009).

Figure 2b. Aerial photo of the location of Trip 2, Stop 1. The Mission Creek Fault is marked by a pronounced northwest-trending scarp and the presence of springs.

reflectors representing more consolidated rocks are offset 150 meters across the Mission Creek Fault.

The outcrop at Trip 2, Stop 1 lies on the up thrown block of the Mission Creek Fault and is composed of north-dipping layers of sandstone and conglomerate. The outcrop is cut by numerous predominantly vertical to south-dipping faults (Figure 3). Many of the faults are filled with white, calcareous precipitates which are more resistant than the surrounding sandy matrix. These are easily seen on the north end of the outcrop both to the right and left of the power pole in Figure 3.

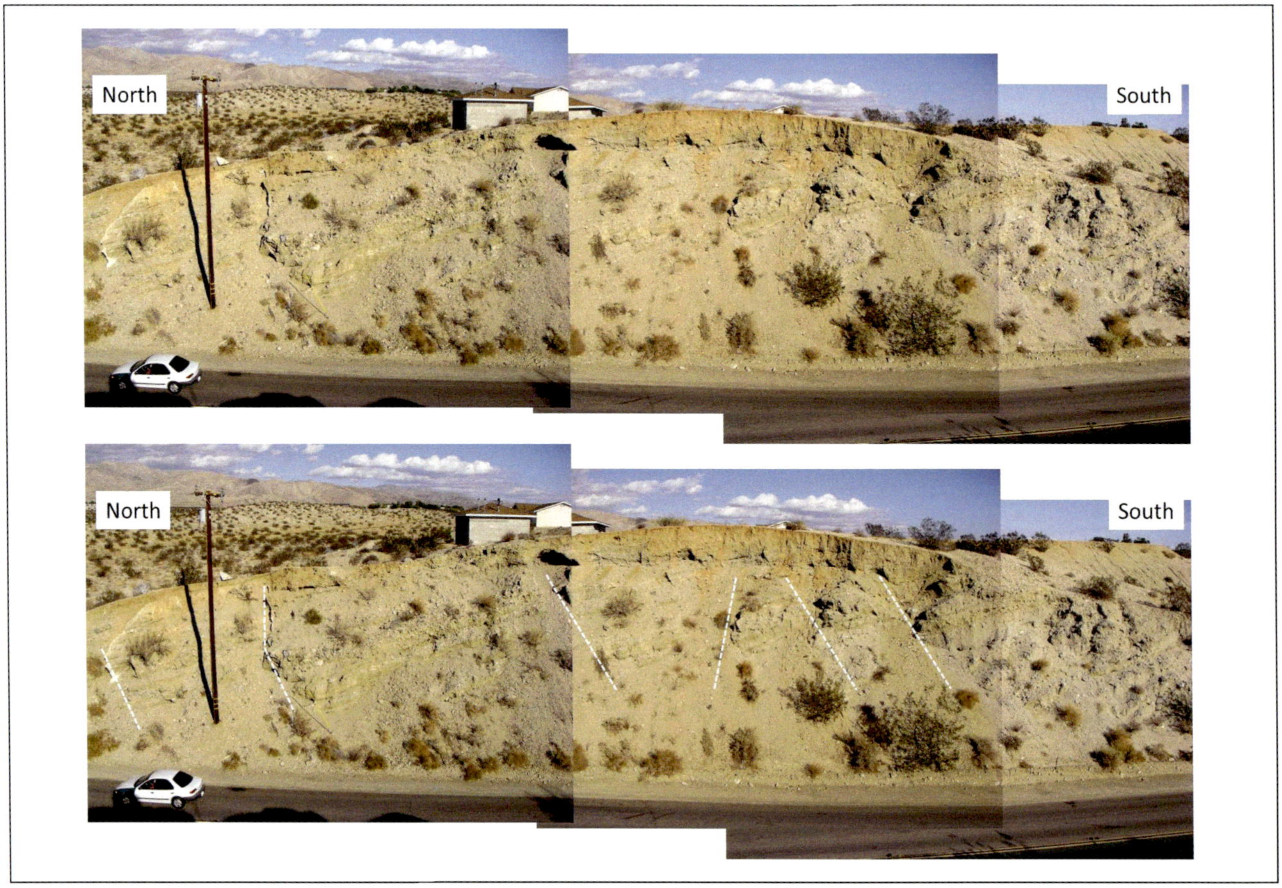

Figure 3. Roadcut at Stop 1 along Mountain View Road. View is to the east. Poorly consolidated sandstones and conglomerates are cut by vertical to south-dipping faults. These faults are shown by dashed lines in the lower photo.

ESTIMATION OF THE HYDRAULIC CONDUCTIVITY OF THE MISSION CREEK FAULT

Both the Mission Creek Fault and the Banning Fault are effective barriers to groundwater movement, as evidenced by offset water levels (on the order of 200 to 250 feet), fault springs, and changes in vegetation (Coachella Valley Water District, 2012). Groundwater TDS values are high on the north side of the Mission Creek Fault and ranged from 800 to 1000 mg/liter. Hot water wells, by the city of Desert Hot Springs, in the sub-basin along the Mission Creek Fault, have groundwater temperatures on average of 118°F (California Department of Water Resources, 1964). On the south side of the Mission Creek Fault, temperatures are in the normal range for groundwater and TDS values are typically less than 500 mg/liter.

We undertook field data and groundwater modeling studies to determine the hydraulic conductivity of the Mission Creek Fault in 2004-2007. The field study focused on the fault scarp at Trip 2, Stop 1, whereas the modeling study covered a 15-kilometers section of the fault that intersected the location of the fault scarp at Stop 1.

FIELD DATA COLLECTION

A lightweight syringe-based air mini-permeameter (Davis et al., 1994) was used to obtain in situ values for permeability on the outcrop. The device consists of three components: a glass syringe which applies pressure to the outcrop, a timing device which measures the rate at which the syringe piston displaces a known volume and a tip seal through which air is injected into the outcrop. The device is most accurate in weakly lithified exposures such as those studied at this outcrop. Three lithosomes were sampled: debris flow matrix, pebbly sand and pebbly sand within 2 inches of a fault trace. The values are shown in Table 1 below.

The data suggest that the hydraulic conductivity of the pebbly sand near the fault zone is approximately half the hydraulic conductivity of the normal pebbly sand. The hydraulic conductivity of the calcareous fill was too low to be accurately measured by the mini-permeameter.

Lithosomes	Mean hydraulic conductivity (m/s)	Standard deviation (m/s)
debris flow matrix	7.6 ´ 10^{-7}	3.7 ´ 10^{-7}
pebbly sand	9.5 ´ 10^{-7}	2.0 ´ 10^{-7}
pebbly sand near fault	5.8 ´ 10^{-7}	9.1 ´ 10^{-7}

Table 1. Hydraulic conductivity values for materials in the outcrop at Trip 2, Stop 1.

MODELING STUDY

A numerical model of the two hydrologic subbasins separated by the Mission Creek Fault was developed using the USGS MODFLOW numerical model for simulating groundwater flow (Harbaugh et al., 2000). Details on the modeling study can be found in Mayer et al. (2007). The MODFLOW horizontal flow barrier (HFB) package (Hsieh and Freckleton, 1993) was used to simulate the effect of the Mission Creek fault on groundwater flow. The HFB package models the fault through a term that simulates the leakance between two horizontally adjacent finite-difference cells, based on the assumption that the fault is vertically oriented. The leakance is controlled by hydraulic conductance, $K^* = K/x$, where K and x are the hydraulic conductivity and lateral thickness of the fault zone.

The model is based on a single layer, 170-column by 98-row, rectangular grid oriented along the strike of the Mission Creek Fault (see Figure 4). Each grid cell is 125 meters by 125 meters. Subsurface inflows from rivers along the northwest and northeast are simulated with constant flux terms at the relevant grid cells, as are the outflows along the Banning Fault and the eastern flux boundary. All other boundary grid cells are no-flow cells. Pumping at 21 wells is represented in the model grid. The pumping rates are based on historical pumping records. The model was calibrated against over 40 locations, including historical groundwater elevations taken from observation wells and elevations inferred from electrical resistivity surveys taken near the Mission Creek Fault in a 2004 field campaign.

The conductance along the Mission Creek Fault is estimated as changing linearly with distance along the fault from the northwest to the southeast, yielding variations in the fault hydraulic conductance of about an order of magnitude along the fault length: $2 \times 10^{-11}\ s^{-1}$ to $4 \times 10^{-10}\ s^{-1}$. A sensitivity analysis indicates that assumptions made in the conceptual model do not significantly impact estimated fault hydraulic conductances. Assuming reasonable values of fault zone thickness of 35 m, fault zone hydraulic conductivities estimated from the model are 7×10^{-10} m/s to 1×10^{-8} m/s, which are from three to five orders of magnitude lower than hydraulic conductivities for the surrounding materials.

SYNTHESIS

Comparison between the LSAMP and modeling results are confounded by the fact that the LSAMP measurements are point-wise, whereas the calibrated vales from the model are averaged across the fault zone. However, assuming a value of fault zone thickness of 35 meters, hydraulic conductivities estimated from the model are at least two orders of magnitude lower than those estimated with the LSAMP methodology. The discrepancy between the two methodologies may be due to increased hydraulic conductivity in the outcrop associated with weathering processes or overestimation of hydraulic conductivity in the outcrop because of difficulties in achieving a sufficient seal with the LSAMP. A coordinated campaign of pointwise hydraulic conductivity measurements and pumping-test measurements in the vicinity of the fault zone could yield improved comparisons.

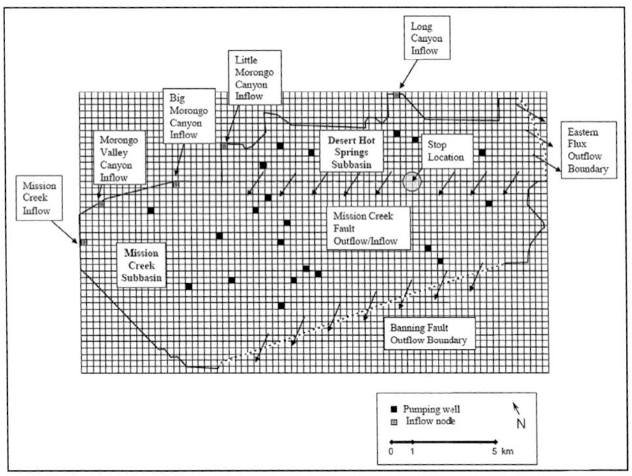

Figure 4. Model grid and location of boundaries and pumping wells (after Mayer et al., 2007).

REFERENCES CITED

California Department of Water Resources (DWR). 1964. Coachella Valley Investigation. Bulletin 108. 145 p. 13 plates

Catchings, R.D., Rymer, M.J., Goldman, M.R., Gandhok, G. (2009) San Andreas Fault geometry at Desert Hot Springs, California and its effects on earthquake hazards and groundwater: Bull. Of the Seismological Soc. Of America v. 99, n. 4, p. 2190–2207.

Davis, J.M., Wilson, J.L. and Phillips, F.M., 1994, A portable air-minipermeameter for rapid in situ field measurements: Ground Water v. 32, p. 258–266.

Harbaugh, A. W., Banta, E. R., Hill, M. C., & McDonald, M. G. (2000). MODFLOW-2000, the US Geological Survey modular ground-water model: User guide to modularization concepts and the ground-water flow process (p. 121). US Department of the Interior, US Geological Survey.

Mayer, A.S., May, W., Lukkarila, C. and Diehl, J. 2007. Estimation of fault zone conductance by calibration of a regional groundwater flow model – Desert Hot Springs, California, *Hydrogeology Journal*, DOI 10.1007/s10040-007-0158-0.

Nicholson, C., 1996, Seismic behavior of the southern San Andreas Fault zone in the northern Coachella Valley, California: Comparison of the 1948 and 1986 earthquake sequences: Bulletin of the Seismological Soc. Of America v. 86, p. 1331–1349.

Proctor, R.J., 1968, Geology of the Desert Hot Springs-Upper Coachella Valley area: Calif. Div. of Mines Spec. Report v. 94, 50 p.

The Indio Hills "Desert Badlands"

Robert M. Norris
University of California, Santa Barbara

Photo 1. Row of *Washingtonia* Palms along the Banning Fault Trace, South Face of the Indio Hills near Pushawalla Canyon.

Motorists traveling east on Interstate 10 between Thousand Palms and Indio may give no more than a passing glance at the pale tan, extensively gullied, low hills to the north. However, more discerning observers will notice the sharply contrasting rows of dark trees at the base of the hills facing Coachella Valley (Photo 1; Figure 1). These interesting hills are certainly worth a closer look because they have a wealth of fault related landforms of remarkable clarity, an array of interesting erosional features, and some striking examples of the influence of ground water on desert vegetation.

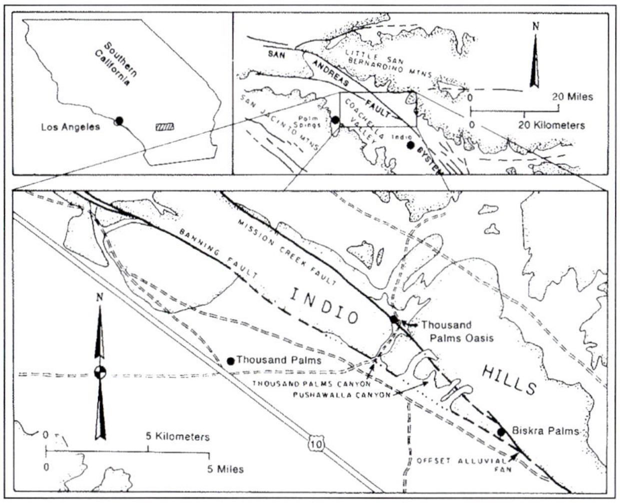

Figure 1. Map of Indio Hills area. *Modified from Norris and others, 1979.*

Topographically, the Indio Hills represent a fine example of desert badlands. The hills are dissected by a closely-spaced network of usually dry stream channels that characteristically develop on soft, nearly impervious rocks in an arid climate and feed a number of alluvial fans on the Coachella Valley side. Aridity results in sparse vegetation, leaving much of the ground surface unprotected and the soft, exposed rocks erode very rapidly when the rains come. Furthermore, an appreciable portion of the rain occurs as monsoonal thunderstorms in summer. Because these storms may be very intense though of short duration, much of the erosion that affects the Indio Hills occurs during these erratically spaced storms. The average annual rainfall of 2.5 to 4 inches (7 to 10 centimeters) is somewhat misleading because much of it falls as light winter rains that do little more than dampen the surface.

During the 1980s, several severe thunderstorms raked the Coachella Valley, causing widespread flooding and damage. In the Indio Hills, considerable erosion resulted and many mature palms were uprooted, swept downstream, and left stranded here and there on the fan surface some distance from the hills. Remains of these uprooted trees can still be seen.

Although in a given decade only two or three such events may occur, they are relatively common over the long term, a century or more. Many of the shaggy palm trees appear quite old, but probably very few have reached 100 years. Palms grow rapidly, and young trees continually replace those destroyed by fire or flood (Photo 2).

Paper adapted from California Geology, September/October 1994, pages 127–133. Used with permission from California Department of Conservation, California Geological Survey. Photos by author unless otherwise noted.

Despite the fact that most of the stream channels that cut the Indio Hills are dry nearly all of the time, there are some places where surface water is almost always present or where the ground is damp a short distance below the surface. These are the areas that support the palms and other water-loving vegetation. The dark lines of trees seen from the highway are rows of California's only native palm, *Washingtonia filifera*. These palms are a sure sign that there is surface water or near-surface ground water. In some of the canyons, such as Thousand Palms and Pushawalla, pools of cool standing water surrounded by dense groves of palms are always present (Photo 3).

As might be expected in such an arid region, the trees, shade, and water attract a large range of birds and other animals. Fortunately, some of the best of such oases are on reserves owned by Riverside County and The Nature Conservancy.

The lines of palms mark fault traces where clayey fault gouge and other impermeable material block movement of ground water toward the valley and force it toward the surface. Ground water, very likely, moves downhill slowly to the southwest from the Little San Bernardino Mountains toward the Indio Hills and Coachella Valley. Although motorists traveling on Interstate 10 can see only a single line of palms marking the Banning Fault at the base of the hills, a closer look shows there are several, more or less parallel lines of trees. Some are in the central part of the hills, others on the northern side, but all closely follow the traces of the main faults that cut the area.

The faults responsible for these oases belong to the San Andreas Fault System, consisting mainly of two converging strands, the Mission Creek Fault on the north, and the Banning Fault on the south. These two faults merge near the southeastern end of the hills (Figure 1).

Where strike-slip faults converge, compression subjects the block between the faults to uplift. This activity pushed the Indio Hills above the Coachella Valley and gave them a generally anticlinal form. On a much larger scale, the San Jacinto Mountains rose between the converging San Andreas and San Jacinto faults.

Photo 2. Deformed Pleistocene Ocotillo Conglomerate in the Walls of Pushawalla Canyon Downstream from the Oasis.

Photo 3. *Washingtonia* Palms at Thousand Palms Oasis.

Strike-slip faults like the San Andreas usually consist of discrete strands or segments that die out along strike and are over lapped and replaced by other strands or segments either to the right or left. Where a right-slip fault steps left, as the Banning Fault does at Pushawalla Canyon, the area between the strands is squeezed and uplifted, and rocks are deformed. This is evident in the walls of Pushawalla Canyon and can also be seen in Figure 2 near the letter "s" in the word "HILLS," where a high ridge has formed. On the other hand, where the fault steps to the right, as it does at several places between Pushawalla Canyon and Biskra Palms to the southeast, there is a pull-apart or sag, forming small undrained depressions.

Another consequence of squeezing due to a left step in the fault can be seen at Pushawalla Canyon. The normal profile of a stream is concave upward, but at Pushawalla Canyon between the fault strands, the profile has a subtle upward bulge, demonstrating that continued offset on the faults is recent enough that Pushawalla Stream has been unable to maintain its normal concave-upward profile.

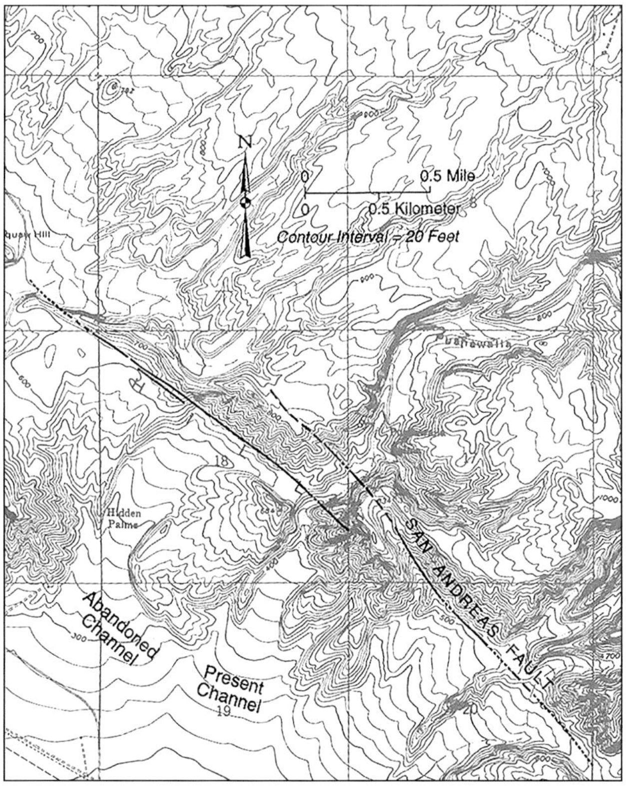

Figure 2. Topographic Map Showing the Left Step in the San Andreas Fault (line dashed where inferred, dotted where concealed) and Present and Abandoned Channels of Pushawalla Canyon. *Base map from U.S. Geological Survey 7.S-minute Myoma quadrangle. From Norris and others, 1979.*

Because the major part of the slip on the San Andreas Fault and its individual strands is right-lateral, streams crossing the fault often show deflections. We do not know the magnitude of recent slip because no major earthquakes have occurred on this part of the San Andreas in historic time. However, observations elsewhere on the San Andreas show that streams may be deflected as much as 30 feet (10 meters) during a single large earthquake. The June 28, 1992, M7.5 earthquake on the Emerson Lake Fault near Landers was not on the San Andreas Fault System, but produced an offset of about 20 feet (6 meters).

Of course, a single displacement on the fault may fail to sever a stream course and instead produce only a slight kink. Subsequent displacements may enlarge the kink or possibly bring an older beheaded stream channel into line with a headwaters drainage, establishing a straight course across the fault. Because of many variations in stream spacing, channel width, and magnitude of offset, and a host of other reasons, it is no surprise that in the Indio Hills some streams show considerable deflection while others show little or none.

A particularly fine example of the continued deflection of a stream and its ultimate straightening is illustrated at Pushawalla Canyon (Figure 2, Photos 4 and 5). At this location the old channel of Pushawalla Canyon is deflected to the right by about half a mile (1 kilometer), undoubtedly the result of many earthquakes and some final event that allowed Pushawalla Stream to abandon its deflected course and occupy a new channel directly across the fault. The nature and magnitude of this critical event is not known. Stream capture may have resulted when an old, beheaded channel was moved into line with upper Pushawalla Canyon, or when a flood overtopped the point of deflection. Whatever the story, the half-mile (l kilometer) deflection probably took 16,000 to 20,000 years to develop, assuming the annual slip rate along the fault in this area is between 1 and 1.4 inches (2.5 and 3.5 centimeters). The deflection could easily represent several hundred separate events. Capture, however, almost certainly occurred within the last few thousand years because there has not been sufficient time for desert varnish or pavement to have developed on the floor of the offset channel.

The abundance of well-defined, very young fault-formed features in the Indio Hills demonstrates that tectonic activity is still very much in progress despite the fact that the only historic earthquake of note on this part of the San Andreas Fault System was the M6.5 event at Desert Hot Springs in 1948. The April 1992 Desert Hot Springs Earthquake, with a magnitude of 6.1, was not on the San Andreas Fault. Since we do not have much of an earthquake history on this part of the San Andreas Fault System, it is difficult to guess when the next major earthquake might occur (trench records indicate the last one was 200-400 years ago). The probability of this part of the fault system experiencing a strong earthquake in the next 20 years is about 40 percent.

Preliminary soils studies by E.A. Keller (University of California, Santa Barbara) and his students provide interesting independent estimates of the slip rate on the San Andreas Fault in the Indio Hills. Their estimates are consistent with values obtained from other evidence.

Much of the evidence comes from the southeast end of the hills, near Biskra Palms, where the students studied an old pediment-fan complex, now cut off from its headwaters by movement on the San Andreas Fault. The headwater channel that once fed the fan-pediment now lies about 765 yards (700 meters) southeast of the fanhead. The age of the soils on the fan suggests that the offset developed over the last 20,000 years or so.

In several ways, this interesting fan pediment contrasts sharply with nearby active, more recent fans. First, its surface is distinctly darker owing to the gradual development of desert varnish on rocks long exposed on the fan surface (Photos 6 and 7). Second, the old stream channels on the beheaded fan are

Photo 4. Abandoned channel of Pushawalla Canyon Along Trace of the Banning Fault. View to the west.

Photo 5. Upstream End of Abandoned Bend of Pushawalla Canyon. View to the west.

Photo 6. Aerial View of Offset Pediment-Fan Complex. Note contrast between dark color of old pediment-fan complex and modern stream deposits in lower left corner. *Photo by John C. Crowell.*

Photo 7. Ground View of Old Pediment-Fan Complex. Note heavily varnished rocks and degraded drainage channels.

much more degraded than those on active fans; the channels are less distinct, channel banks are gently sloping, and varnish is developing on rocks in the channel. On active fans, channels are sharp and clear, and typically have steep or vertical banks and unvarnished pebbles. Third, soils recorded a complex history of offsets as the fan was alternately disconnected from its upland feeder streams and connected to other drainages, and the extensively gullied Indio Hills moved southeast along the fault. During periods when the fan drainage was largely cut off, soils began to develop, and when drainage was reestablished, an influx of sediment buried parts of these older soils and started the soil forming process anew.

In a large gravel quarry close to the eastern margin of the fan, a series of buried soils has been exposed, each unit representing a period of soil development when the drainage was largely cut off. Should dateable materials be found in these soils, which is not very likely in such an arid environment, a more detailed chronology of offset along the fault might be worked out and dates of major earthquakes established. Even without dateable material, soils yield a useful, though rough, chronology because they go through a series of well-defined changes over time.

In places where major climatic shifts or other important disruptions have not occurred, soil development can yield a chronology, and soil ages in years can be estimated within a 25- to 50-percent accuracy. Among the more important changes observed in soils are the gradual accumulation of clays in the subsoil (B-horizon) and their depletion in the top soil (A-horizon), and accumulations of calcium carbonate in arid and semiarid climates. Calcium carbonate deposition often results in nearly impervious hardpan near the base of the subsoil (C-horizon). Soil chronologies in the Indio Hills agree generally with other evidence regarding the long-term slip rate on the San Andreas Fault derived from such things as offset bedrock bodies and structural features. It is therefore possible to determine the history of a particular drainage from soils studies. At Pushawalla Canyon, for example, there are four or five elevated terraces, each with a distinctive soil profile. These terraces permit the reconstruction of the drainage history, helping us understand how and over what period the Indio Hills evolved.

REFERENCES

Crowell, J.C., 1975, The San Andreas Fault in southern California, *in* Crowell, J.G., editor, San Andreas Fault in southern California: California Division of Mines and Geology, Special Report 118, p. 7–27.

Dibblee, TW., 1954, Geology of the Imperial Valley region, California, *in* Jahns, R.H., editor, Geology of southern California: California Division of Mines and Geology, Bulletin 170, p. 21–18.

Nettleton, W.O., Witty, J.E., Nelson, RE., and Hawley, J.W., 1975, Genesis of argillic horizons in soils in desert areas of the southwestern United States: Soil Science Society of America Proceedings, v. 39, no. 5, p. 919–926.

Norris, A.M., Keller, E.A., and Meyer, G.L., 1979, Geomorphology of the Salton Basin: selected observations, *in* Abbott, P,L., editor, Geological excursions in the southern California area: Department of Geological Sciences, San Diego State University, p. 19–46.

Sylvester, A.G. and Smith, RR, 1976, Tectonic transpression and basement- controlled deformation in San Andreas Fault Zone, Salton Trough, California: American Association of Petroleum Geologists Bulletin, v. 60, p.2081–2102.

Robert Norris (1921–2012), *emeritus professor of geology, University of California, Santa Barbara, where he co-founded the Department of Geology, taught courses in general geology, geomorphology, and marine geology. His research interests involved desert erosional processes, sand dunes, and nearshore processes.*

Valerie Jean Date Shop, Thermal. Photo from the Collection of Steve Lech.

Ancient Lake Cahuilla and Indian Fish Traps

Harry M. Quinn

INTRODUCTION

Many ancient lakes have occupied the Salton Sink during the past two to three million years. During the late Miocene (about 5.4 million years ago) through early Pliocene (about 4.4 million years ago), the Salton Sink was a northern extension of the Gulf of California. By about the late Pliocene, the Colorado River developed a delta at its mouth large enough to cut off the Salton Sink from the gulf. The Salton Sink has apparently remained separated from the gulf ever since.

The Salton Sink has been the location of a number of large fresh water lakes since being cut off from the gulf. Most of these lakes have resulted from the shifting of the Colorado River. When the river flowed south to the delta directly into the gulf, the Salton Sink contained only small playas and salt marshes that developed when runoff from the surrounding mountains settled in low areas. However, when the river's course shifted northerly off of the delta, directly into the sink, a large fresh water lake would form. If the northern flow lasted long enough, the lake would fill to a level that would be determined by the spill point into the Gulf of California. This level would then be retained until the river shifted to the south again. If the level remained at a set elevation long enough, a beach line would develop.

When the river shifted away from the sink area, the large fresh water lake would begin to dry up because of the high evaporation rate common to this desert region. This would result in the formation of a single or group of small saline lakes or salt marshes. In early historical times, small saline lakes and surrounding salt marshes occupied the lower portion of the Salton Sink until 1905 when the Colorado River broke through to form the present day Salton Sea. The Salton Sea is able to maintain its presence, and even rise at times, because of the excess water brought in from the Colorado River for irrigation.

Adapted from CVAS Occasional Paper No. 1, January, 1997, pages 20–28.

ANCIENT LAKE CAHUILLA

What has been commonly called Lake Cahuilla was the last of the large fresh water lakes to occupy the Salton Sink (de Stanley, 1966; Balch and Balch, 1974). This lake appears to have had a high water level of around 42 feet above mean sea level. The lake is believed to have remained at a full level until the river shifted around 500 years ago (Wilke and Lawton, 1975; Brown, 1985). Littlefield (1966) indicates that the lake was present until about 300 years ago. Tom Rockwell (1995) places the last high stand at between 1660 and 1680 AD, based on carbon-14 dating of peat found along the old shoreline. The high water level was maintained long enough to develop beach ridges in the alluvial fans and to cut a small notched shoreline into the bedrock areas along the northwestern shore. Much of this old beach line can still be seen today.

Ancient Lake Cahuilla was probably not large enough to have been affected by large tides, like those of oceanic areas. This lake, however, was no doubt affected by winds in much the same way that the Salton Sea is today. Some paleo-wind directions can be postulated from 1) the distribution of pumice along the old shorelines, 2) the build up of local sand bar beach areas, and 3) the development of local tufa-covered bedrock areas along portions of the old shoreline.

1. Pumice. Large amounts of pumice were present along the eastern shorelines north of the town of Niland, California (Figure 1). During the early and middle 1950s, my family and I gathered pumice from this area when we camped at the old hot mineral well near Frank Spring. This pumice had a source at the volcanic fields near the southern end of the present day Salton Sea (Figure 1). Its distribution would indicate a southwest-to-northeast prevailing wind direction or could result from a lake current pattern that would have been south along the west shore and north along the east shore (Figure 1). Such a lake current pattern could have been developed by a paleo-wind direction of northwest-to-southeast, very similar to that found in the Salton Sea area today.

2. Sandy Beaches. Two large sandy beaches, or sand bars, are present along the northwestern shoreline area (Figure 2). Both of these sandy beaches developed in an east-west direction at a nearly right angle due to change in the shoreline. In both cases, they occupy the southern leg of the right angle bend (Figure 2). The development of these sandy areas could have resulted from a near-shore southerly flowing current.

3. Travertine. Travertine, a variety of calcium carbonate, is known to be deposited in areas where waters experience a loss of carbon dioxide, coupled with high evaporation rates and increased saturation levels (Chesterman and Kleinhampl, 1991). Tufa, or porous travertine, is a variety of calcium carbonate commonly deposited along lake shorelines, or by hot springs. Tufa forms in crusty layers and is known to form rather rapidly in some hot spring areas (Chesterman and Kleinhampl, 1991). The tufa deposits present at rocky points along the northwestern shoreline could have resulted from strong winds blowing from northwest-to-southeast. If the winds were strong enough to push the lake waters to the south, thus lowering the lake levels in the north, these rocky point areas would have been exposed and subject to wave action. The wave action at these point areas would have created a spray above the wave level, and this spray would experience a loss of carbon dioxide, coupled with a high evaporation rate, algal growth and an increased saturation level. When the wind subsided and the lake level returned to normal, these tufa deposits would be covered by water and protected from weathering. These tufa deposits would only have been exposed during the time when more tufa was being deposited. Areas above the lake level would have been exposed to weathering all of the time. Tufa was probably deposited above the normal lake level, but the constant weathering in this area removed it faster than it could be maintained.

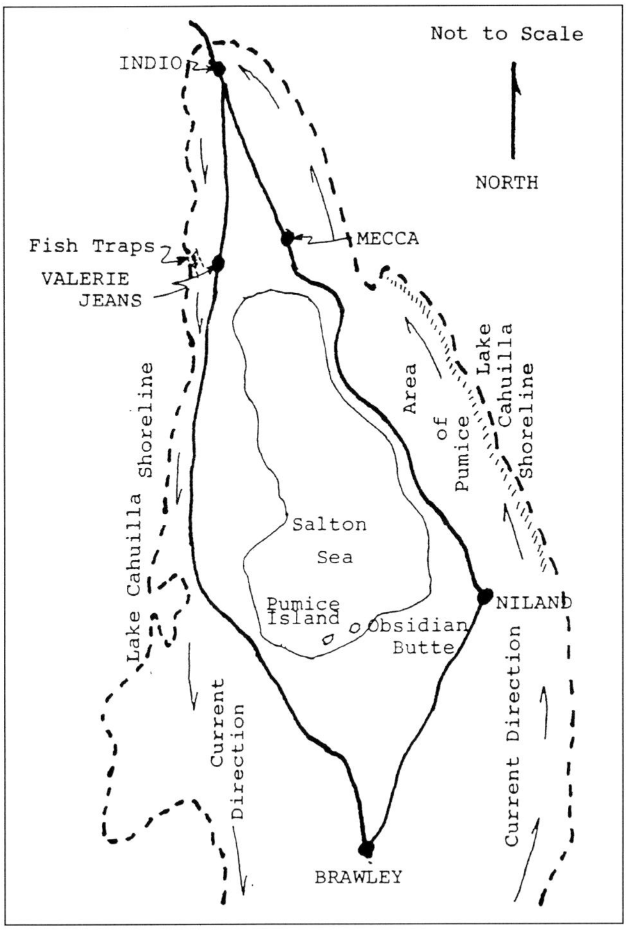

Figure 1. Site Location Map with Current Directions and Pumice

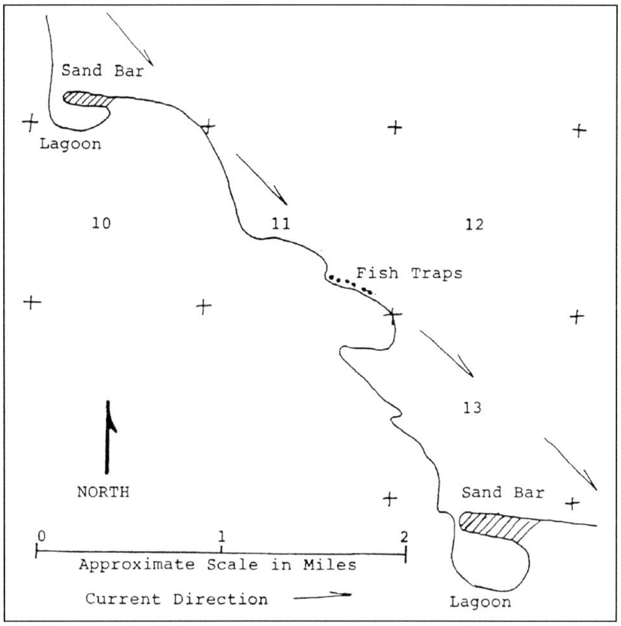

Figure 2. Sheltered Areas Containing Sand Bars and Fish Traps

THE FISH TRAPS

The fish traps are located about eight miles south of the city of Indio and approximately two miles west of the town of Valerie Jean (Figure 1). These fish traps consist of a series of rectangular stone weirs constructed along the old lakeshores (Balch and Balch, 1974). The stone weirs are constructed in a continuous line and share common sidewalls. Like the sand bar beaches to the northwest and southeast, the fish traps occupy the southern limb of a nearly right angle bend in the shoreline (Figure 2). This location would have been protected during storms rising out of the northwest when open portions of the lake would have been rather choppy. The calm areas, or coves, would have been logical places for fish to congregate during such storms.

During the early stages of a strong wind out of the northwest, water would potentially have been piled up along the north-facing shores in the protected areas. As the wind continued to blow, the lake waters would be pushed to the southern end of the lake causing a drop in water levels at the northern end of the lake. If fish were seeking calm waters in the area of the fish rraps during a storm, they could easily enter the stone weirs when the waters rose and then be trapped when the water level fell.

There are a series of four main levels of nearly parallel stone weirs, with each level having a 6 to 8 foot difference in elevation. There is the possibility of some isolated stone weirs above the four main levels and some hollowed out areas below them. The evaporation rate for the present Salton Sea is estimated to be about 6 to 8 feet a year (de Stanley, 1966). Littlefield (1966) shows the average annual evaporation rate between 1948 and 1962 to have been around 5.8 feet. If one assumes the same evaporation rate for Lake Cahuilla, then these fish traps, or stone weirs, may have served their purpose only once per tier. The weirs appear to have been constructed by removing stones to produce a basin and then using their moved stones to construct the walls around the basin. By this presumed construction method, the exposed row of tiers would not have to be destroyed to build the new ones. The upper levels of Fish Traps do not appear to have been destroyed as the lake level dropped, supporting this presumed construction method.

The uppermost main tier of stone weirs is approximately 80 feet below the high water shoreline of Lake Cahuilla. The lower tier is some 20 feet below that. This means that a substantial amount of water had evaporated from the lake prior to the construction and use of the fish traps. The water in the lake may have even been brackish by the time it had dropped to the level at the fish traps.

There is an old aerial photo of the fish traps hanging on the wall of the Valerie Jean Date Shop (Figure 3). As indicated on the photo, it was taken by Mr. Dewey Moore from an altitude of about 2,200 feet, but no date is included. The photo shows three distinct sets of stone lined pits. The upper series are larger, fewer in number and more squared in shape than the lower two. The second series consists of more, but smaller pits than the upper ones. The lower series has the most pits, but they are the smallest of the group. There are 14 well-defined shorelines above the upper tier of fish traps and 5 to 7 poorly-defined shorelines above that. If each of the shorelines represents one year's evaporation, then the first set of fish traps was constructed about 19 to 21 years after the lake began to subside. I would like to thank the owners of the Valarie Jean Date Shop for allowing me to use the information from their photo.

Figure 3. Aerial photograph of the fish traps, early 1900s. Courtesy of Southwest Museum of the American Indian Collection

One fish that is known to be able to tolerate brackish to salt water is the mullet. The mullet is a saltwater fish that migrated from the Gulf of California into the Salton Sea by way of the Colorado River when the river broke through in 1905-1906 (de Stanley, 1966).

The area below the lower set of main stone weirs begins to flatten out and contains few rocks. There are a few cusp-shaped depressions scooped out of the soil below the lower main tier of stone weirs and just above the flat area. These depressions might have been associated with a last attempt to trap fish in this area. The flat area below the cusps is presently being fanned and would have probably been a muddy "Tidal Flat" when the lake level dropped. The position of the fish traps would indicate that the area was not suitable for trapping fish when the lake was full and fresh and was not usable after the water level dropped below the rocky shoreline. This drop in water level may have also made the water too saline for the type of fish being trapped.

Based on some of the ethnographic information available on the Desert Cahuilla Indians (Bean, 1975; Bean, Vane and Young, 1991; Hooper, 1920; Kroeber, 1908; Strong, 1929), the Indians wintered on the desert floor and collected food crops in the mountains during the summer. If the same patterns of life were practiced long ago when the lake was there, the fish traps should have been constructed during the winter months. This would have been when the evaporation was at its lowest and the local mountain runoff was at its peak, which would have permitted the lake level to have remained rather constant. When the summer season came and the people were away gathering food, the high evaporation would cause the lake level to drop. This drop in water level would have required a new set of traps to be constructed for the next winter season of fishing. The present Salton Sea usually reaches its high point in April and its low point in October (de Stanley, 1966).

A different time for using the fish traps can be postulated from information available on the present Salton Sea. In the Salton Sea today, biologists report that during August and September the bottom waters become deficient in dissolved oxygen. This dissolved oxygen deficiency produces a condition that forces the Corvina to seek depths of less than 20 feet (de Stanley, 1966). During this same period, there has been a reported heavy die-off of Gulf Croaker (de Stanley, 1966). If a similar bottom water condition existed in Lake Cahuilla after it dropped down to the fish trap level and not before such a drop, then fish would have been driven into shallow water during this time. This condition would have only lasted a couple of months, presumably August and September. This short duration for fish in shallow water could explain why the fish traps were not moved down as the water dropped.

Eugene Shepard (1965) examined the stone features and concluded that they were the remains of a large Pueblo, which he called "Fish Trap Pueblo." He described the apartment row as having six levels at the northern end and dwindling down to two rows at the southern end. He surmised that the rock walls were once coated with poles and mud, and that cottonwood beams had been used to support the roofs. He further correlated the nearby petroglyphs with pueblo clan symbols found elsewhere in the west.

B. E. McCown (1956 and 1964) examined the stone weirs and concluded that there was a good chance that they were "fish traps." He advocated a theory that the fish were drawn into the traps by food (bait) being placed in the pits. After being trapped in the pits, he also thought that the Indians might have used astringent plants to force the fish to the surface.

Regardless of which fishing scenario is considered, the fish traps appear to have been only used for a short duration, probably about four to six years. A short-term use of the fish traps is further indicated by the lack of any well-worn trails between the traps and known habitation sites. No indication of any habitation site in the immediate area of the fish traps has been observed by the author or found in any reports. A number of well-defined trails can still be found that connected habitation sites in the desert with food gathering sites in the mountains. These trails were probably used on an annual basis for a century or more. If the indicated short-term use of the Fish Traps is correct, then this time of a "fishing culture" had only a limited impact on the life style of the local Indians who used them.

REFERENCES

Anonymous, 1986, Map of the Salton Sea (with soundings): Triumph Press, Inc., Los Angeles.

Balch, A. R and Balch J. W., 1974, Ancient Lake Cahuilla's Fish Trappers: Published by the Balchs, 92 p.

Bean, L. J., ed., 1975, The Cahuilla Indians of the Colorado Desert: Ethnohistory and Prehistory: Ballena Press, 73 p.

Bean, L. J., Vane, S. B., and Young, J., 1991, The Cahuilla Landscape: The Santa Rosa and San Jacinto Mountains: Ballena Press, 116 p.

Brown, James T.,1985, Harvest of the Sun: An Illustrated History of Riverside County: Windsor Publications, p. 20–22.

Chesterman, C. W. and Kleinhampl, F. J., 1991, Travertine Hot Springs, Mono County, California: California Geology, v. 44, no. 8, p. 171–182.

de Stanley, M., 1966, The Salton Sea Yesterday and Today: Triumph Press, Inc., Los Angeles, 128 p.

Hooper, L., 1920, The Cahuilla Indians: University of California Publications in American Archaeology and Ethnology, v. 16, no. 6, p. 315–380; Reprinted 1978 by Malki Museum Press.

Kroeber, A. L., 1908, Ethnography of the Cahuilla Indians: University of California Publications in American Archaeology and Ethnology, v. 8, no. 2, p. 29–68; Reprinted 1978 by Malki Museum Press.

Littlefield, W. M., 1966, Hydrology and Physiography of the Salton Sea, California: U.S. Geological Survey Hydrologic Investigations Atlas, HA-222, 1 plate.

McCown, Benjamin E., 1956, The Indio 'Fish Traps' Re-Examined: The Masterkey, p. 133.

McCown, Benjamin E., 1964, The Indio 'Fish-Traps' Re-Examined: in McCown, B. E., 1964, Collected Papers of Benjamin Ernest McCowan: Archaeological Survey Association of Southern California Paper Number Six, p. 97–98.

Rockwell, Tom, 1995, Notes from lecture given at the Coachella Valley Archaeological Society (CVAS) meeting on March 17, 1995, on file.

Shepard, Eugene, 1965, Fish Trap Pueblo: Pacific Coast Archaeological Society Quarterly, v. 1, no. 1, January, p. 15–16.

Strong, W. D., 1929, Aboriginal Society in Southern California: University of California Publications in American Archaeology and Ethnology, v. 26, p. 36–86; Reprinted 1972 by Malki Museum Press.

Valerie Jean Date Shop, 1992, Photo of Fish Traps by Mr. Dewey Moore.

Wilke, Philip J. and Lawton, Harry W., 1975, Early Observations on the Cultural Geography of Coachella Valley: in Bean, Lowell John, ed., 1975, The Cahuilla Indians of the Colorado Desert -Part1: Ethnohistory and Prehistory: Ballena Press Anthropological Paper No. 3, p. 9–12.

Hurkey Creek Campground, San Jacinto Mountains.
Late 1930s. Photo from the Collection of Steve Lech.

Mid to Late Pleistocene Dynamic Landsliding in the Coachella Fan Fault Zone, between the Mecca and Indio Hills, Northeast of the San Andreas Fault, California

Miles D. Kenney
Kenney GeoScience
KenneyGeoscience@gmail.com

Fault trenching and detailed mapping of Pliocene to present age sediments within over 5400 acres in the Coachella Fan between the Mecca Hills and the Indio Hills, and northeast of the San Andreas Fault provides constraints on relative timing of Quaternary deformation, strike-slip distribution on regional faults, and northern Salton Trough tectonics. Figure 1 shows the general Coachella Fan area. The oldest sediments in the area consist of isolated exposures of faulted and steeply folded Palm Spring Lower Formation (Qpl) and an underlying member of the Canebrake Formation that are likely Pliocene to possibly early Pleistocene (see Figure 5 for regional stratigraphic section). The deformational structures identified in these units demonstrate that early Pleistocene age transpressive style deformation occurred in the area between similar transpressive areas of the Mecca and Indio hills. After folding and tilting of the Palm Spring Lower Formation and the underlying Canebrake member, the region was eroded to a relief in close approximation with the Coachella Valley Floor with the exception of a few low lying erosional remnants. The fluvial and lacustrine Palm Spring Upper Formation (Qpu, ~750,000 years ago) was deposited upon this new erosion surface in the northeast of the San Andreas Fault and within the Coachella Valley. The relatively fine grained Palm Spring Upper Formation lacks syndeformational structures indicating that the region was likely not deformed during deposition.

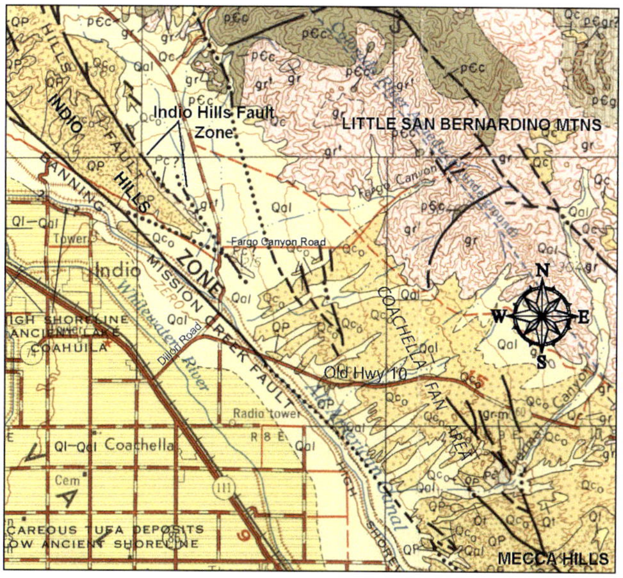

Figure 1. Regional geologic map, southern San Bernadino Mountains, Indio Hills, and Coachella Fan.

Palm Spring Upper Formation deposition ceased in the Coachella Fan Region roughly between 400,000 to 700,000 years ago as a tectonic driven northwest striking antiform and/or southwestward tilting developed within the Coachella Fan causing relatively higher relief in this area compared to the Coachella Valley. The uplift of the Coachella Fan Region coincided with localized erosion of the Palm Spring Upper Formation and subsequent deposition of late Pleistocene coarse grained Ocotillo Conglomerate (Qo). Sediment material in the Ocotillo Conglomerate was primarily derived from the Little San Bernardino Mountains to the north. In the Coachella Fan Region, the Ocotillo Conglomerate/Palm Spring Upper Formation contacts exhibit both structurally conformable (but possibly disconformities) and angular unconformities indicating that the magnitude of uplift, deformation, and erosion into the Palm Spring Upper Formation was varied across the region.

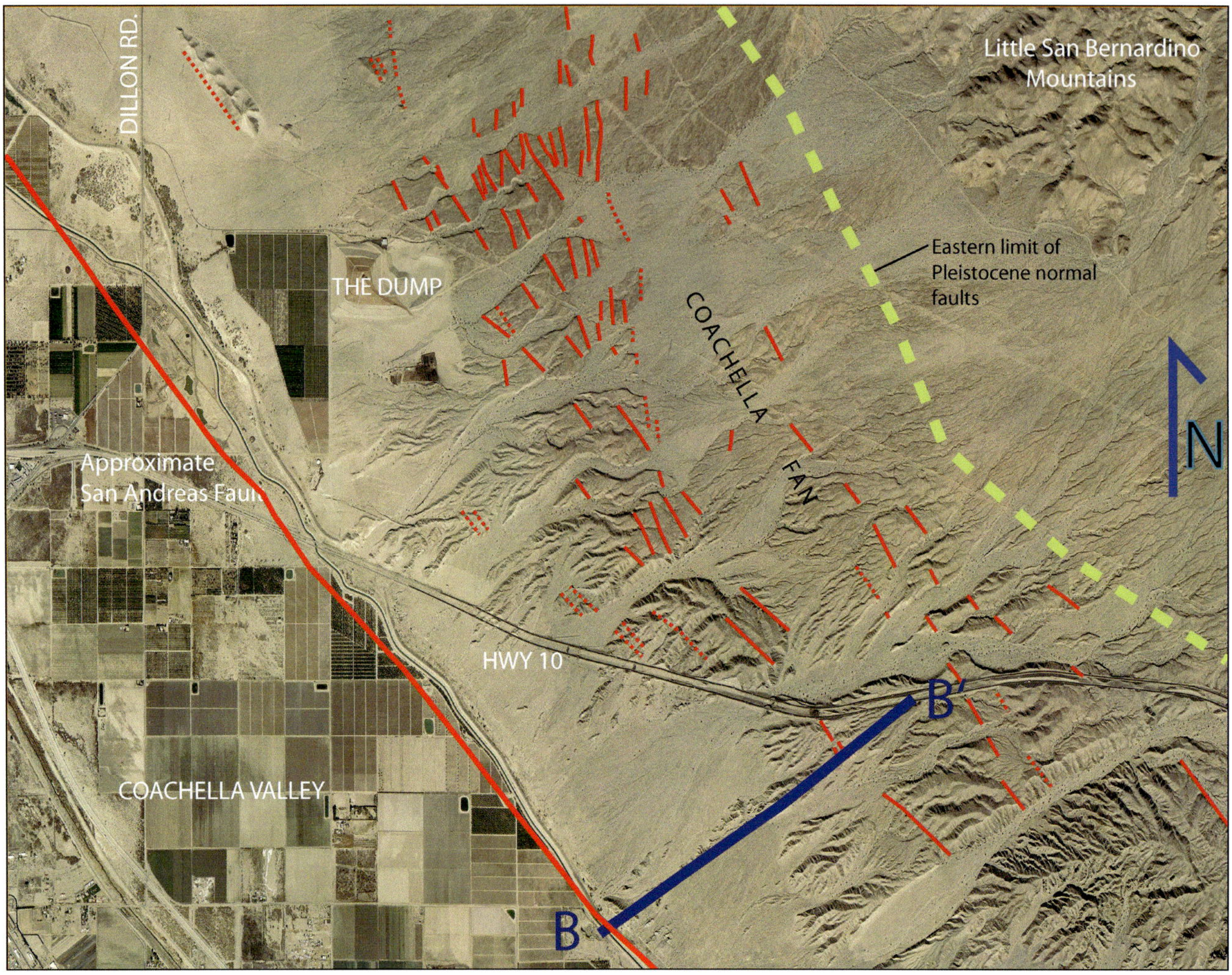

Figure 2. Aerial Photograph of the Coachella Fan Fault Zone.

Within the Ocotillo Conglomerate, some late Pleistocene inter-unit angular unconformities and deformed fan surfaces provide structural markers indicating ongoing warping (uplift) during and after deposition of the Ocotillo Conglomerate. This deformation was likely due to tectonic forces. During internal fan uplift (warping), abundant N to NW striking normal listric "faults" with parallel to bedding basal shear surfaces formed on the uplifted crest. The vast majority of the listric faults strike parallel to existing fan contours and there is a SW dip direction allowing for downslope movement toward the SW (Figures 2 and 3). Compressional

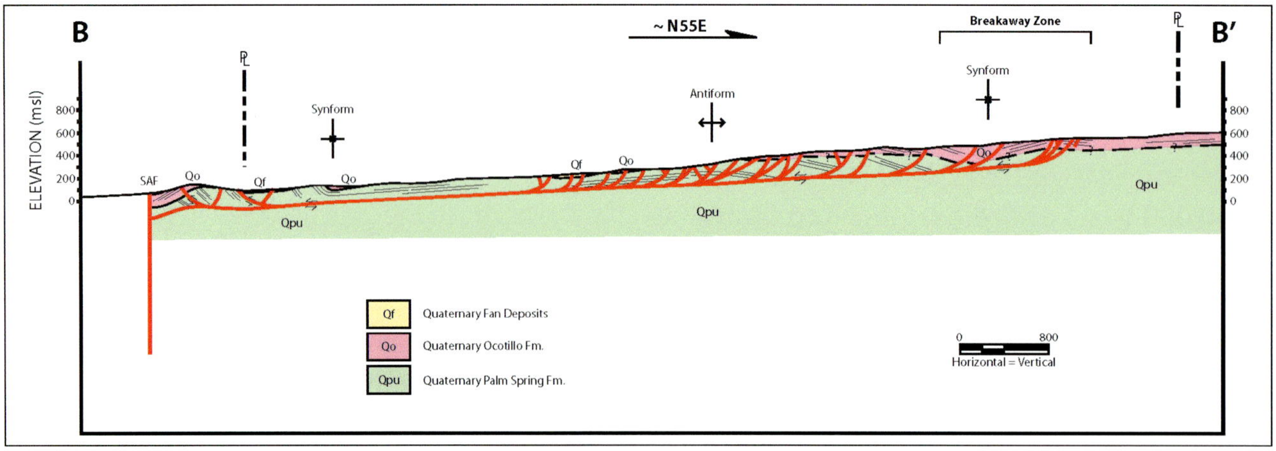

Figure 3: Cross Section of a typical section of the Coachella Fan

deformation occurred within the Palm Spring Upper Formation and Ocotillo Conglomerate in a zone within 2000-feet northeast of the San Andreas Fault at the same relative time as the normal listric faults within the Coachella Fan Region (Figure 4). The compression may be due to the San Andreas Fault acting as a sliding buttress (opposing fault normal forces) and/or a newly recognized local restraining bend in the San Andreas Fault. The normal listric faults may be attributed to landsliding occurring during major earthquakes on the San Andreas Fault during wetter conditions of the Pleistocene. The landsliding occurred on steeply SW dipping listric normal "faults" that connected with basal shear surfaces primarily within fine grained members of Palm Spring Upper Formation at depth. The basal shear surfaces were stacked with some deeper than others. A number of relatively shallow parallel to bedding shear surfaces were identified in 12-foot deep fault trenches exposing the Palm Spring Upper Formation. In a few localities where a rare fine grained member of Ocotillo Conglomerate was identified, southwest transport basal shear surfaces with accompanying listric normal faults were also exhibited. A few NW trending, steeply dipping faults within the Coachella Fan Fault Zone also exhibit right-lateral strike slip motion that may be due to landsliding or tectonic right lateral shear in the region as well similar to that observed in the Mecca Hills.

Structural relationships indicate that the spacing of the listric normal slide surfaces became progressively wider toward the end of the Pleistocene. For example, the normal shear surface spacing within unit Qpu are commonly on the order of just a few feet, some of these faults do not penetrate overlying unit Qo, and the spacing of the normal shear surfaces within unit Qo are wider than within unit Qpu. In addition, the head-scarp normal displacement structures within the preserved Qo fan surfaces that range in age from 70 kya to over 300 kya are typically separated by 100's of feet. The increase in normal shear surface

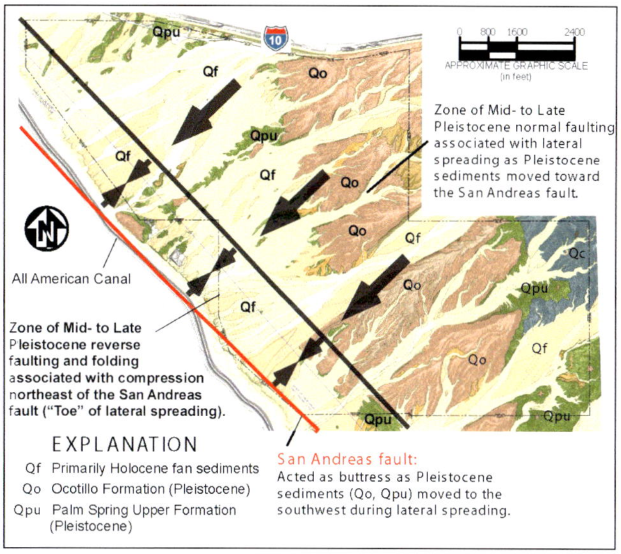

Figure 4. Generalized geologic map for lateral spreading model.

spacing toward the end of the Pleistocene is interpreted to be associated with deepening of the utilized basal shear surfaces along weak bedding planes within unit Qpu as groundwater levels lowered during this time. Very few of the normal listric slide surfaces have moved since the latest Pleistocene. This suggests that as the region dried out and the water table lowered, that the dynamic landsliding movement was greatly reduced.

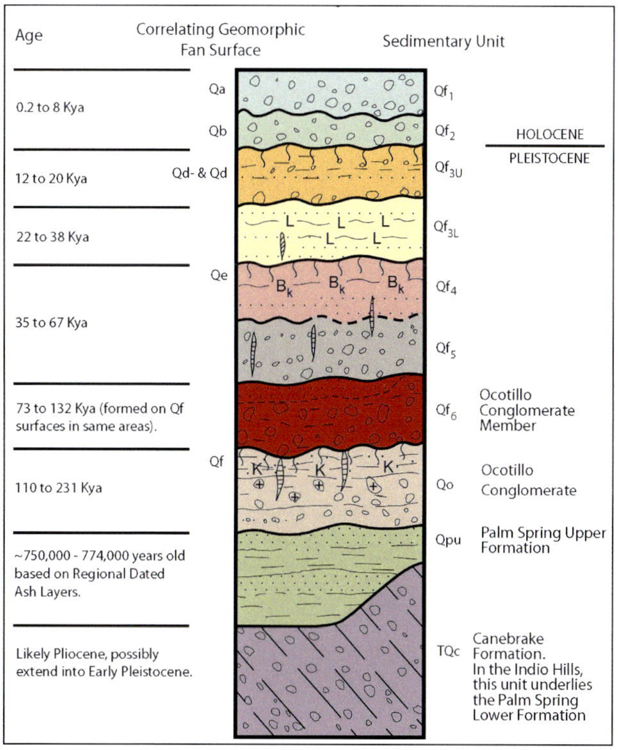

Figure 5. Generalized stratigraphic column.

The mode of landsliding appears to have involved lateral spreading (shearing along liquefaction layers) at lower elevations and classic wet condition landsliding upslope. A dynamic slide analysis performed by S.V. (Jag) Jagannath, PhD with Petra Geotechnical along Cross Section B-B′ (Figure 3) indicates that the slide mass moved 8-inches during a 7.4 magnitude earthquake on the local San Andreas fault with a relatively shallow water table.

In summary, based on our studies, the majority of the published late Pleistocene faults in the CF area (Clark, 1984) are likely associated with gravitational (creep) and/or ground shaking forces, and to a lesser degree to tectonic far-field forces. The initial uplift (warping) in the CF region that created the local relief leading up to the landsliding was likely due to transpressional tectonic forces associated with the San Andreas fault and local block kinematics. No significant through going strike-slip faults are recognized in the CF region to provide a good candidate for more than 0.5 mm/yr transfer of slip into the Eastern California Shear Zone. Evidence provided by lack of mid-Holocene dip-slip offset across the SAF, and lack of deep incision in the CF region since the latest Pleistocene indicates that the Coachella Valley portion of the Salton Trough has not experienced significant tectonic down-dropping during that time.

REFERENCE

Clark, M.C., 1984, Map showing recently active breaks along the San Andreas Fault and associated faults between Salton Sea and Whitewater River-Mission Creek, California: U.S. Geological Survey Miscellaneous Investigation Series Map I-1483, scale 1:24,000.

Miles Kenney discusses results of fault trench study on the Coachella Fan, 2008.

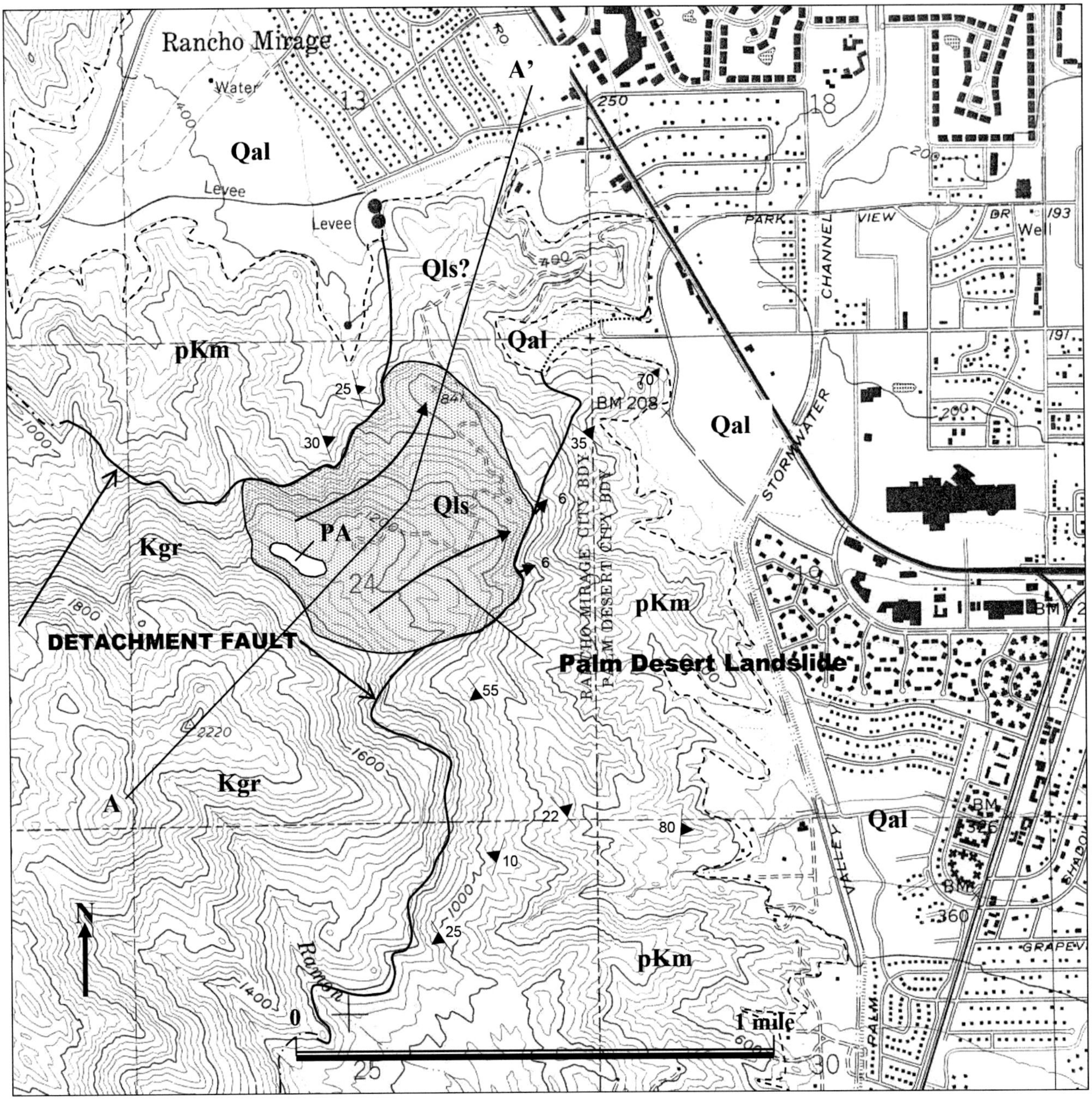

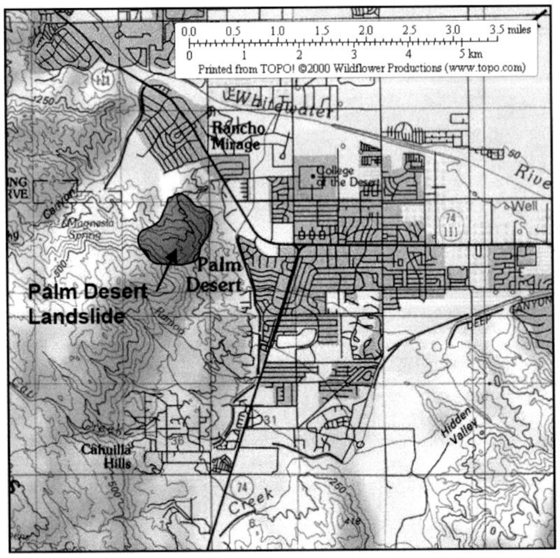

Figure 1 (above). The Palm Desert Landslide, Palm Desert, California; PA: ponded alluvium; Qls: Landslide; Kgr: granitic rocks; pKm: pre-batholithic metamorphic rocks, includes thinly foliated schist, gneiss, and marble.

Figure 2 (left). Location Map, the Palm Desert Landslide.

Photo 1 (opposite). Mike Hart on detachment fault surface.

Quaternary Landsliding along a Miocene Detachment Fault, Palm Desert, California

Michael W. Hart, Consultant
P.O. Box 261227
San Diego, California, 92196

INTRODUCTION

Perched above the town of Palm Desert on the eastern slopes of the Santa Rosa Mountains is a large landslide in granitic and metamorphic rocks that can be easily viewed from Highway 111 near its intersection with the Pines to Palms Highway (Highway 74). The late afternoon sun reveals a prominent east-facing scarp and broad landslide bench developed in granitic rocks of the Peninsular Ranges Batholith. The sharp contact with these coarsely crystalline rocks and pre-batholitic metamorphic rocks can also be seen from the vantage of the highway (Figures 1, 2, and 3).

Previous studies have shown that landslides are, or have been, an important geomorphic process in the Santa Rosa Mountains as well as other mountain ranges in the Anza-Borrego Desert (Bock, 1977, Baldwin, 1986, Hart, 1991, and 2008; and Wagner, 1996). The dry climate seems at odds with the occurrence of large bedrock landslides that in other areas of California are related primarily to the presence of weak clay-rich rocks and high precipitation. The occurrence of such large landslides in generally strong rocks is an apparent incongruity. Granitic rocks, at least at outcrop scale, appear to be highly competent and extremely resistant to landsliding. Research into the causes and geologic environment leading to landsliding has revealed that most of the landslides in crystalline terrane are intimately related to a combination of intense seismic activity, the presence of fault planes that serve as rupture surfaces, and the presence of highly fractured or jointed rocks (Hart, 1991 and 2008).

Figure 3. The Palm Desert Landslide looking west from Highway 111. (solid line: detachment fault; Qls: landslide; pKm: pre-batholithic metamorphic rocks; Kgr: granitic rocks.

GENERAL GEOLOGY

The Palm Desert Landslide occurs in granitic rocks consisting of moderately to highly foliated tonalite and granodiorite. Metamorphic rocks consisting of biotite schist, strongly foliated gneiss, marble, and amphibolite underlie the landslide and adjacent terrain to the north and south. The contact between the granitic and metamorphic rocks in the vicinity of the landslide is formed by a generally east-dipping Miocene detachment fault.

Detachment faults in the extreme easterly part of the Peninsular Ranges of California occurred as a result of east-west extension and separate a lower core of gneissic Late Cretaceous granodiorite and Jurassic metasedimentary rocks from an upper plate consisting in many areas of granitic rocks (Frost and others, 1996). In some areas there is a thick zone of unconsolidated, unmetamorphosed megabreccia overlying the fault surface. The presence of the low-angle fault surface and highly fractured and jointed rocks associated with the fault zone is conducive to landsliding particularly where the fault surface dips at moderate angles in an out-of-slope direction.

An excellent example of the unconsolidated breccia that often overlies the fault occurs in road cuts along Highway 78 opposite the Tamarisk Grove Campground in the Anza-Borrego Desert State Park. Eric Frost (personal communication) describes these rocks as remnants of massive Miocene gravity landslides that occurred as a result of nearby detachment faulting. There are excellent exposures of similar megabreccia along the Pines to Palms Highway approximately 4.2 miles south of Highway 111 in Palm Desert. This megabreccia is composed of angular boulder to cobble-sized clasts set in a matrix of light gray sand to silt-sized rock particles. The overall appearance of the material is identical to breccias created by Quaternary landsliding. The similarity can result in the incorrect identification of landslides. Adding to the confusion is the possibility that the Miocene aged detachment breccia has been remobilized by Quaternary landsliding.

LANDSLIDE GEOMETRY AND MORPHOLOGY

The landslide is approximately 3000 feet in length and has an irregular, crudely oval shape that is highly modified by erosion. The slide extends from an elevation of 800 feet at the toe to an elevation of 1400 feet attained at the base of a nearly 800 feet high headscarp (Figure 1). If the feature mapped as a possible landslide (Figures 4 and 5) is included within the area of landsliding then the overall length of the landslide would be approximately one mile. Because of a lack of diagnostic geomorphic criteria to accompany the structural evidence, a landslide origin for the megabreccia making up the possible landslide can not be confirmed and the breccia may be solely a result of detachment faulting.

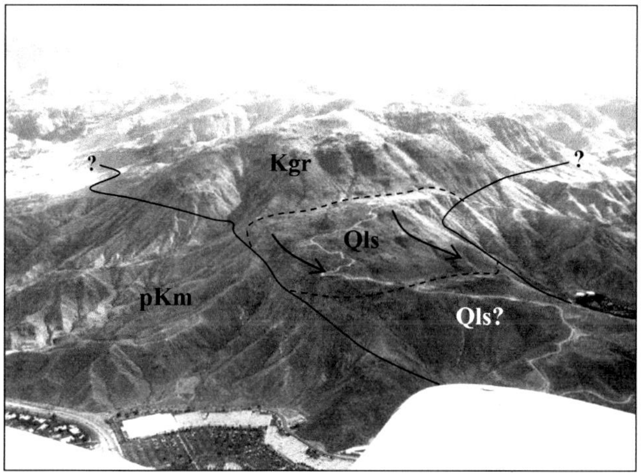

Figure 4. View to west of the Palm Desert Landslide (Qls). Qls?: possible landslide; pKm: pre-batholithic metamorphic rocks; Kgr: granitic rocks; solid line: detachment fault.

The topographic bench at the head of the landslide has been highly modified by erosion that has also dissected a 20 feet thick alluvial deposit at the base of the scarp (Figure 1). The alluvium consists of very coarse carbonate-rich sand formed in the pull-apart or "graben" zone at the head of the landslide. The presence of this alluvium, the broad topographic bench at the head of the slide, and the prominent headscarp are strong evidence that the feature was created by Quaternary landsliding as opposed to being an erosional outlier of Miocene-age detachment fault breccia.

In the terrain adjacent to the northern and southern flanks of the landslide the fault forms the contact between the lower plate pre-batholithic metamorphic rocks and the upper plate granitic rocks. The detachment fault plane is somewhat undulating north of the landslide and is nearly horizontal up to the point where it merges with the basal rupture surface at the landslide's southern flank. Here the dip of the fault plane changes abruptly and maintains an approximately 10 degree northeasterly dip until it disappears beneath the alluvium of Coachella Valley (Figure 5).

LANDSLIDE AGE

The Late Pleistocene age of the landslide is suggested by several lines of evidence. First, the landslide is highly modified by erosion and is dissected by a deep gully that must have formed after the final sliding event. Secondly, the steep slopes that lie adjacent to the northern and southern flanks of the slide and directly below the well-exposed basal rupture surface are free of thick deposits of talus or landslide debris that would be present if the landslide were only a few thousand years old or had undergone recent reactivation. Lastly, the twenty feet thick deposit of alluvium at the head of the landslide is highly dissected and is only a remnant of what was a more extensive deposit. The presence of well-developed interstitial carbonate and carbonate veins in the alluvial sands also suggests a Late Pleistocene age.

FAILURE MECHANISM AND POTENTIAL FOR REACTIVATION

The causes of landslides can be differentiated into three categories: preparatory, triggering, and

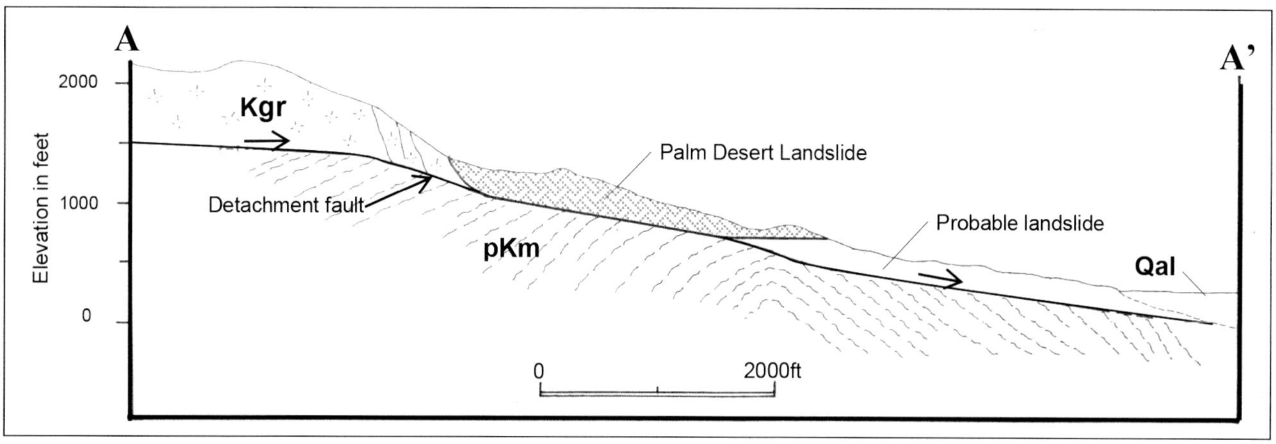

Figure 5. Geologic section A-A' through the Palm Desert Landslide.

sustaining (Dikau and others, 1996). Preparatory factors are those that prepare the slope for movement and include such factors as rock strength, the presence of faults and fractures, and presence of groundwater. The most common triggering mechanisms are seismic shaking, an increase in groundwater levels, and the removal of support by erosion or grading. Sustaining causes are those that tend to keep the slope in motion.

The most significant preparatory factor leading to massive landsliding on the slopes above Palm Desert is the presence of a detachment fault (the Salton Sea Detachment; Frost and others, 1996) that can be traced over 50 miles southward from the vicinity of San Gorgonio Pass to the southern end of the Santa Rosa Mountains. This fault provides the necessary low-strength basal rupture surface that makes landsliding possible in otherwise strong rock. An additional critical factor in landsliding at this location is the abrupt change in dip of the fault surface from near horizontal to an easterly dip of 10 degrees. Without a relatively steep valleyward dip it is unlikely the landslide would have occurred.

The triggering mechanism can only be surmised based on the apparent age of the landslide and its tectonic environment. In this area of high seismic activity 8 miles west of the San Andreas fault, the slopes above Palm Desert have been subject to numerous episodes of intense ground shaking that when combined with the much wetter conditions during the Late Pleistocene would have resulted in conditions conducive to landsliding.

Probably the best evidence there is a very low potential for landslide reactivation is that the landslide does not appear to have been active in the last several thousand years. If the landslide had undergone even moderate movements in the last few thousand years, the slopes adjacent to the northern and southern flanks of the slide would be covered with talus or younger secondary slide debris.

CONCLUSION

The hills are shadows, and they flow
From form to form, and nothing stands;
They melt like mist, the solid lands,
Like clouds they shape themselves and go.
—Tennyson

REFERENCES

Baldwin, J., 1986, Martinez Mountain rock avalanche, in Guptil, P., Gath, E., and Ruff, R., eds., *Geology of the Imperial Valley, California,* Annual Field Trip Guidebook #14, South Coast Geological Society, Inc., pp. 37–48.

Bock, C.G., 1977, Martinez Mountain rock avalanche, in, Coates, D.R., ed., *Reviews in Engineering Geology,Vol. III, Landslides,* Geological Soc. America, pp. 155–168.

Dikau, R., Brunsden, D., Schrott, L., and Ibsen, M., 1996, Introduction, in Dikau, R., Brunsden, D., Schrott, L., and Ibsen, M., eds., *Landslide Recogition, Identification, Movement, and Causes,* Report No. 1, of the European Commission Environment Programme Contract No. EV5V-CT94-0454, pp. 1–12.

Frost, E., Suitt, S., and Fattahipor, M., 1996, Emerging perspectives of the Salton Trough region with an emphasis on extensional faulting and its implications for later San Andreas deformation, In, Abbott, P. and Seymour, D., eds., *Sturzstroms and Detachment Faults Anza-Borrego State Park, California,* pp. 81–121.

Hart, M.W., 1991, Landslides in the Peninsular Ranges, southern California, In, Walawender, M.J., and Hanan, B.B., eds. *Geological Excursions in southern California and Mexico,* Geologic Society of America guidebook for the 1991 annual meeting, pp. 349–365.

Hart, M., 2008, Structural and geomorphic characteristics of landslides at Coyote Mountain, Anza-Borrego Desert State Park, California, *Environmental and Engineering Geoscience*, Vol. XIV, No. 2, pp. 81–96.

Wagner, D., 1996, Geologic Map of the Tubb Canyon 7.5 Minute Quadrangle, San Diego County, California, California Division of Mines and Geology Open File Rept. 96-06.

Pinyon Flat Landslide, San Jacinto/Santa Rosa Mountains, Riverside County, California

With Notes on Deep Geologic Space and Time, and some of the Essentials of Surficial Gravity Tectonics

William J. Elliott
Consulting Engineering Geologist
P. O. Box 541
Solana Beach, California 92075

Raymond J. Scaramella
Consulting Engineering Geologist
1951 Falmouth Drive
El Cajon, California 92020

ABSTRACT

The approximately 15 mile-long by 27 mile-wide composite bedrock block-glide Pinyon Flat landslide occupies about 310 square miles of real estate along the west side of the Coachella Valley. It is a mega-slide that extends southeastward from Palm Springs to Martinez Canyon. Palm Canyon and the steep northeastern flank of the Santa Rosa Mountains generally define the back- and side-scarp. We estimate that the toe may extend an additional 3 to 6 miles northeasterly, beyond the mountain front, under Coachella Valley. Top-to-buried-toe relief may be on the order of 2 miles.

Evidence for the landslide includes: a generally arcuate back-scarp, springs, prominent mid-slope benches (including Pinyon Flat), hummocky topography, disrupted and atypical drainages (such as Deep Canyon), deck-of-cards *slickenlines and smoothed striated surfaces in the back-scarp area, opened-and-subsequently-filled fractures (also in the back scarp area), and subsidiary slides within the main slide mass (for example, the Martinez Mountain rock avalanche).*

The Pinyon Flat landslide scarp occurs primarily within the Santa Rosa Shear Zone. This old zone of weakness, coincident with shearing and brecciation associated with mid-Tertiary extension likely represents pre-existing low-strength shear planes which were utilized as landslide slip-plane surfaces.

Mid-Tertiary extensional basins, including the proto-Gulf of California provided extreme topographic relief required for large-scale landslides. In this area of high relief, even typically strong, resistant rock masses sometimes cannot resist gravity, and fail at zones of weakness. Such is the massive complex of late Mesozoic igneous and metamorphic rock that slid into the Coachella Valley.

The exact age of sliding is unknown, but it is bracketed by the opening of the proto-Gulf of California about 6 million years ago and post-landslide Coachella Valley fill.

INTRODUCTION

"Landslides are defined as the bulk transfer of soil [and/] or rock debris downslope under the direct influence of gravity." (West, 2007, p. 117).

The Pinyon Flat landslide can probably best be described as a large, coalescing, composite translational mass of broken hard-rock debris, or alternatively, a disrupted bedrock block-glide (Varnes, 1978, p. 13–14).

Recent landslides, with fresh scarps, well defined mid-slope benches, hummocky topography, disrupted drainages, and a clearly defined pile of rubble and debris at the toe are hard to miss — even for the uninitiated. Older landslides that have been subjected to the ravages of weathering, erosion, and perhaps even further mass wasting can be a little harder to identify. With rounded and softened features, these old timers can be elusive, even for the most ardent of landslide enthusiasts (Keaton and DeGraff, 1996, p. 188–189).

And then there are those old grand-daddy mega-slides that are so easy to walk or drive over without even knowing you are trudging on unstable ground. Often times, it takes peering out the window of an airplane at 30,000 feet (à la Elliott, 2003), or critically looking at small-scale satellite imagery, before — the a-ha comes — with a gasp! — and a sudden realization that an out-of-place looking chunk of dirt has all the earmarks of a mega-slide.

Such is the case described herein.

The Pinyon Flat landslide is so big, that, unless one looks at it from a distance, it is difficult to corral the whole image into one plausible landslide hypothesis. Ray and I were chatting one day about a project I had in the Hemet area, not too far west of the Pinyon Flat landslide back scarp. At his urging, after finishing my chores, I drove east over to State Route 74 (Palms to Pines Highway) and started to look for telltale signs of landsliding.

Along this oftentimes nearly shoulder-less stretch of narrow and winding two-lane pavement, it was a heart-stopping adventure to get a good look at road-cuts as speeding traffic whizzed past my backside — by drivers negotiating blind, decreasing-radius-curves at break-neck speed with nary a care for unexpected obstacles, like a geologist trying to drive a rock pick into a chunk of crumpled earth…

With abundant caution and a modicum of good luck, I was able to get tolerable views of several tight inside-curve road-cuts and determine that there appears to be more evidence for landsliding than evidence against (Photos 1 and 2). Additional evidence for sliding was found along an-unnamed jeep trail on the east side of the Santa Rosa Mountains (Photos 3 and 4).

Back at the office, I pulled out my old yellowing copy of a plastic raised-relief topographic map of the Santa Ana 1°x2° degree sheet (Defense Mapping Agency, 1979). And, not surprisingly, topo-geomorphic evidence for the presence of an ancient mega landslide seemed to just pop off the map (Figure 1).

Upon receipt of a stack of black and white aerial photographs to provide stereo coverage of the landslide area, I was rewarded with clear photo-geomorphic evidence for landsliding (United State Department of Agriculture, 1984).

Large- and small-scale landslides and rockfalls in granitic and metamorphic terrane in Central and Southern California are not without precedent. For example, see: Copenhaver, 2006, p. 83–86; Elliott, 2003, p. 111–116; Evans, J. R., ed., 1988, 87p.; Hart, 2003, p. 193–209; Shreve, R. L., 1987, p. 109–114; and Wieczorek, G. F., 2002, p. 165–190.

PREVIOUS INTERPRETATIONS OF THE PROPOSED MEGA-SLIDE BACK SCARP

Older maps show thrust faults, or no faults at all, along the trace of what we believe is the back-scarp of the Pinyon Flat landslide. (Fraser, 1931; Rogers, 1966; Jennings, 1975.)

As to the Palm Canyon segment of the proposed Pinyon Flat mega-slide back-scarp, Ruff and others (1981) noted the following on page 204 of their road log:

"… Palm Canyon … is structurally controlled by the Palm Canyon Fault. This fault was mapped by Fraser for nearly 6 miles along the bottom of the canyon. Evidence for faulting was based on wide zones of crushed and intensely weathered rocks, and numerous springs along the canyon bottom (Fraser, 1931). Dibblee and Brown [however] could not find evidence for this [Palm Canyon] fault (Brown, pers. com.). Palm Canyon is eroded generally along the strike of the east-dipping contact of mylonitic quartz diorite and gneissic metasedimenatry rocks…" (Dibblee, 1981, 1982; Brown, 1981).

Photo 1 (above). Head scarp fracture fillings in State Route 74 road-cut (photo by W. J. Elliott, 4-27-2004).

Photo 2 (above right). Close-up view of Photo 1 (photo by W. J. Elliott, 4-27-2004).

Photo 3 (right). Slickensides, mullions, striations and chatter marks at side of fire road (photo by W. J. Elliott, 4-27-2004).

Photo 4 (bottom right). Slickensides, mullions, striations and chatter marks in crown of fire road (photo by W. J. Elliott, 4-27-2004).

So, whether the curvilinear Palm Canyon is an artifact of one or more of the following: tectonic faulting, detachment faulting, a conventional landslide back scarp, or simply the result of differential weathering and erosion, it will be up to you to decide. Any thoughts?

Other segments of the proposed mega-slide back scarp, south of Palm Canyon, are equally unresolved. Since much of this portion of the slide is in areas of limited access, field work will require youngsters with the stamina to hike in and out, carrying everything needed to sustain life during the field reconnaissance. Are you up to the challenge?

GEOLOGIC HISTORY

Some of this analysis may get a little "fuzzy," as not all aspects of Southern California geologic history are as well understood as we would like. But, let's give it a try, and start in late early-Jurassic.

Stick with me, this all relates and will lead us logically to the ingredients necessary for the location and timing of the Pinyon Flat Landslide.

Farallon Plate. At ~180 Ma (million years before present) the Farallon Plate occupied most of the eastern Pacific basin. Drifting easterly toward certain death at a convergent plate margin, it was being subducted under the western margin of the North American Plate. This process continues today, with small remnants of the Farallon Plate (known today as the Explorer, Juan de Fuca, and Gorda plates) subducting off the coast of Northern California, Oregon, Washington, and British Columbia — from Cape Mendocino to the northern end of Vancouver Island. (Englebretson, and others, 1985; Debiche, and others, 1987).

Weathering And Erosion. All the while, weathering and erosional processes continued wearing-down and wearing-away exposed regions of the North American Plate. This debris was likely carried by gravity and water in many directions, including westerly into the Pacific Ocean. As these aggrading clastic wedges became thicker and thicker, material at and near the bottom was compressed, squeezed and warmed, leading to regional (dynamo-thermal) metamorphism. Typical metamorphic products resulting from this kind of heat and pressure include: gneiss, schist, phyllite, slate, quartzite and marble. (Longwell and others, 1932, p. 409–437 and 346–361).

Hang in there, this is all relates….

Plutonism. Returning to the northern Peninsular Ranges of Southern California, several plutonic episodes took place during mid-Cretaceous time in response to Farallon Plate/North American Plate subduction, friction, and intense heating. (The jury is apparently still out on the debate surrounding exactly why this event took place in mid-Cretaceous time and not before or after [Engebretson and others, 1985, p. 2].)

In the San Jacinto/Santa Rosa mountains these plutonic bodies intruded into previously altered (regionally metamorphosed) country rock, such as noted in the paragraphs above. And, to add insult to injury, they created additional thermal (contact) metamorphism and metasomatism in the immediately adjacent host rock (Parcel, 1981, p. 147).

Individual plutonic events occurred during the approximate intervals of: 135–129, 124–116, 113–103, 100–92, and 84–83 Ma. From west to east, across the Peninsular Ranges, individual plutons generally decrease in age, and compositions typically change from more basic to more acidic (Kistler, and others, 2003).

How do we know all of this? Uplift, weathering and erosion over the eons have brought the results of these once deeply buried events to the present-day ground surface (Longwell and others, 1932, p. 409–437). Geologic mapping by the likes of the late Thomas W. Dibblee Jr. (for example, 1954, 1981, 1997) and uncounted others over the years have brought us (through two-dimensional maps and best-guesstimate structure sections) three dimensional models of what likely happened long, long ago.

Santa Rosa Shear Zone. In the San Jacinto/Santa Rosa mountains a thin belt of generally northwesterly trending older metamorphic rocks occurs sandwiched between younger plutonic rocks. This belt, known as the Santa Rosa Shear Zone, is flanked on the west by typically early late-Cretaceous (~100–92.5 Ma) granodiorites, and on the east by late mid late-Cretaceous (~84–83 Ma) granodiorites of varying names and compositions. (Parcel, 1981; Kistler and others, 2003.)

Long before all of this geology was uplifted and exposed, while still buried at depths estimated to be on the order of about 7 to 14 miles, these old metamorphic rocks, as well as adjacent book-end plutonics were sheared by tectonic transport in a right lateral direction (Parcel, 1981). (Sounds to me like an early version of the present-day San Andreas shear couple.)

Interestingly, however, linear striations, mullions, and scratch marks observed on exposed surfaces today dip down moderately in an easterly direction (Parcel, 1981), toward the Coachella Valley — counter intuitive to an ancient lateral transport/shear system. Do you suppose we might be looking at an artifact of geologically recent landsliding?

The age of this shearing event has been estimated to be post batholith cooling/crystallization (post ~84–83 Ma) and prior to deposition of a non-conformably overlying Eocene (~34–56 Ma) conglomerate (Sharp, 1966, referenced in Parcel, 1981, p. 147; Kistler and others, 2003). More recent work by George and Dokka (1994) suggests that, "This zone of ductile deformation appears to have formed in late Mesozoic time [Campanian] as indicated from fission-track uplift ages that record a time of rapid cooling ending at about 72 Ma…" (Frost and others, 1997, p. 65).

Laramide Orogeny, Un-Roofing, And Sedimentation. Moving along, beyond the metamorphic and plutonic events noted above, sometime after about early late-Cretaceous (~85 [or so] Ma) the Farallon Plate subduction angle shallowed under the Peninsular Ranges and Western North America (Abbott, 1999, p. 36). This resulted in the Laramide Orogeny in the Western States — from Eastern California to the Rocky Mountains (Bunge and Grand, 2000; Earle, 2000). This event lasted from ~80-40 Ma (Engebretson and others, 1985, p. 2).

During this time, and continuing to the present, uplift, weathering and erosion was un-roofing these afore-mentioned deep-seated metamorphic/plutonic terranes. All the while, sedimentary debris was being spread far and wide (Longwell and others, 1932, p. 409–437) — much of it being carried in a westerly direction toward the Pacific Ocean. (See also, for example, Kimbrough and others, 2001, p. 491–494.)

Evidence of this un-roofing can be found today in the sequence of sedimentary deposits between the crest of the Peninsular Ranges and the western off-shore margin of the North American Plate. Between San Diego and the Los Angeles basin, named sedimentary formations of just about every age, from late Cretaceous to Recent, can be found. To name a few that you might recognize, from youngest to oldest these are:

Terrace deposits	upper Pleistocene
La Habra Formation	lower Pleistocene
Pico Formation	upper Pliocene
Repetto Formation	lower Pliocene
Puente Formation	upper Miocene
Topanga Formation	middle Miocene
Vaqueros Formation	lower Miocene
Sespe Formation	Oligocene
Santiago Formation	middle Eocene
Silverado Formation	Paleocene
Williams Formation	Upper Cretaceous
Ladd Formation	Upper Cretaceous
Trabuco Formation	Upper Cretaceous

This stratigraphic column for the northern Santa Ana Mountains (located about 50 miles westerly from Pinyon Flat) has been pirated from Oakeshott and others (1954). Some names and ages have been changed and tweaked over the years, but I think this makes the point of nearly continuous sedimentation from late Cretaceous to present.

Pacific and North American Plates Meet. Now, at ~30 Ma, the Pacific Plate, which had been following behind the Farallon Plate across the Pacific basin finally reached the west coast of North America. With a new dance card in hand, the Farallon Plate detached from, and then left the Pacific Plate behind as it continued its journey under the North American Plate. (See Atwater, 1970; Engebretson and others, 1985; Irwin, 1990, p. 75.).

Two things happened here.

First: extension and detachment faulting.

As the Farallon Plate continued to slide easterly under the North American Plate, it began to break apart and fracture into individual segments (Menard, 1978) – apparently due to lack of cohesive bonding and confining forces. In a sense, it began to look like the cupped and curled-up mud cracks in a dry mud-puddle. This likely resulted in islands of warm and cool slab material — an artifact of heat escaping upward from the asthenosphere through breaks in the slab. In the process, this imbalance could have resulted in localized crustal thinning, tweaking, and isostatic instability. And, this may have led to areas of uplift and down-warp that might have looked something like a corrugated ridge/trough sub-scape.

With the cold Farallon Plate disappearing to the east, the North American Plate was now beginning to rest more and more on the underlying warmer asthenosphere. The stage was now set for crustal extension in the Basin and Range provinces of eastern California, Nevada, Arizona, New Mexico and western Colorado (Lister and Davis, 1989; Pridmore and Frost, 1992; Frost and others, 1997; Sylvester, 1999; Dokka, 1993; Earle, 2000; Christie-Blick and others, 2007; Haines and van der Pluijm, 2010). Thought to be perhaps continuing today, the heyday of extension, metamorphic core complexes and detachment faulting and extensive volcanism occurred principally in later Oligocene (post ~30 Ma) and throughout most of the Miocene (~23-5.3 Ma) (Pridmore and Frost, 1992, p. 7; Dokka, 1993, p. 711; Christie-Blick and others, 2007, p. 426; Haines and van der Pluijm, 2010, p. TC4028 - 1 of 20).

Beyond the early Cenozoic stress/strain couple that created the Santa Rosa Shear Zone, it is thought that mid-Tertiary extension and detachment-faulting may have added additional zones of weakness to the area now occupied by the Pinyon Flat landslide, for example, the Haystack Mountain detachment fault (Pagnillo, 2003). See also, Frost and others, 1997, p. 68, Figure 11; Pridmore and Frost (1992, p. 4).

Second: San Andreas Fault and extensional rifting.

In the following paragraphs, principal literature consulted includes: Atwater, 1970; Englebretson and others, 1985; Lonsdale, 1989; Irwin, 1990, Sylvester, 1999.

The new Pacific Plate/North American Plate boundary also drew a new dance card and changed from a collisional/compressional plate boundary to a right lateral transform plate boundary. From ~30 to 6 Ma this right lateral transform plate boundary resided just off shore from the western edge of the sub-aerial portion of the North America. With time, as the Pacific Plate made its way northwesterly toward Alaska's Aleutian Islands, the fault grew in length to extend from the Rivera triple junction off the tip of Baja California, north to the Mendocino triple junction located about 25 miles south-southwest from Eureka, California.

Then at ~6 Ma, with extensional weaknesses and crustal thinning (Frost and others, 1997, p. 74) in the western North American crust, this right lateral plate boundary fault jumped from its long-occupied watery home at the eastern edge of the Pacific Ocean to its present location where we know it today as the San Andreas fault. With this leap, a relatively thin slice of western North America was sheared off and joined to the eastern edge of the Pacific Plate. Currently this slice travels northwesterly at about 2 inches per year. (Don't hold your breath, because at this rate it will take ~14 Ma before the daily commute from San Diego to San Francisco will be short enough to be practical in an all-electric car.)

In addition to this lateral movement, there is a tensional component, which is pulling the Pacific Plate away from the North American Plate between the tip of Baja California and White Water, located about 10 miles northwest of Palm Springs, California. From there to well north of San Francisco, the two plates are in direct contact — wherein plate motion is characterized by both a-seismic creep (Hollister, California) and major earthquake events (1857 and 1906) (Wallace, 1990).

The Gulf of California and Salton Trough (Imperial and Coachella valleys) were created by this oblique tensional plate boundary strain. Early on in the opening process, sea water was able to intrude as far north as White Water, California, where late Miocene marine Imperial Formation strata can be found today as evidence of that invasion (Dibblee, 1997, p. 43, Figure 4 — a thin north-south outcrop about 1 mile east of Whitewater Canyon). It did not take long, however, for the newly reorganized Colorado River to create a large delta and close off the rift, or ditch, to sea water entry north of about the present day USA/Mexico boundary. Following this event, the Salton Trough sedimentary environment changed from marine to continental and lacustrine.

The scene was now set for mass wasting to fill the northern end of the ditch — also known as the Salton Trough (Coachella-Imperial valleys).

Because the Salton Trough did not open until ~6 Ma, the age of the Pinyon Flat Landslide cannot be older than ~6 Ma.

Now that the scene has been set, by reviewing geologic history and events that preceded and ultimately led up to the Pinyon Flat Landslide, let's take a closer look at landsliding itself.

INGREDIENTS FOR LANDSLIDING

What are the ingredients for mega-scale mass wasting — such as a mega-scale bedrock landslide. Really, they are not much different from landsliding and slope instability at any scale. Here are some of the basic ingredients that typically result in gravity-tectonics-driven landsliding.

Excellent references detailing ingredients leading to landsliding can be found in: Terzaghi (1950, p. 88–115), Leighton (1966, p. 161–169), Varnes (1978, p. 26–28), Cruden and Varnes (1996, p. 26–28), and West, 2007, p. 117–118). If you get the chance to delve into one or more of these pertinent references, pay close attention to external versus internal conditions that relate to various ingredients that ultimately lead to gravity tectonics – that is landsliding (including most forms of mass wasting) and all of its variations thereof.

For landslides to occur there must first be a set of conditions that lead to instability and failure. Some of these are listed below.

1. Open and Unsupported Free Face — that is, a place into which a mass of earth can move. First and foremost, for a landslide to occur, there needs to be an interface between a mass of earth (rock and/or soil) and a fluid, such as air and/or water, that is unable to support or resist downward and outward lateral movement of the mass. That is to say, any place where there is a sufficient elevation difference such that a mass of earth at the higher elevation is rendered unstable and outward/downward movement will bring the mass back into relative equilibrium with a given set of driving forces (gravity) and resisting forces (Φ and c). (See Terzaghi and Peck, 1948, p. 97; Johnson and DeGraff, 1988, p. 29, for the Mohr-Coulomb equation and definitions of Φ [phi] and c.). (The Mohr-Coulomb formula is: t = s tan(Φ) + c [where t = shear strength, s = normal stress, tan = tangent, and c = cohesion]).

The Coachella Valley is a good example. When the San Andreas transform fault system opened the rift, or ditch, which today we call the Salton Trough, also known as the Imperial and Coachella valleys, (Irwin, 1990; Sylvester, 1999, p. 5–6), an approximately 15,000 to 20,000 feet high free face was created between the top of the San Jacinto/Santa Rosa mountains and the bottom of the Coachella Valley. Today, Mount San Jacinto stands at 10,804' and Toro Peak stands at 8,716' (United States Geological Survey, 1959–69). Below the Coachella Valley, approximately 8,000 to 10,000 feet of late Miocene through Recent sediments must be penetrated to reach basement (Biehler, S., 1964; Smith, 1964, sheet 2 of 2; Cromling, J., 1973; Downey and Clinkenbeard, 2005, p. 63; Lovely and others, 2006, p. 1887; Dorsey, R., 2012, Figure 1). Prior to the Coachella Valley being filled to its present elevation of a few tens of feet above sea level (United States Geological Survey, 1959–69), a high and steep free face must have existed between the San Jacinto/Santa Rosa mountains and the bottom of the emerging Salton Trough.

It is into this trough that the Pinyon Flat landslide moved to restore the balance of forces and gain eventual equilibrium.

2. Planes of Weakness. A solid mass of isotropic earth with high Φ and c is unlikely to fail, even given a substantial near-vertical free-face. Good examples of this hard-rock condition can be found in Yosemite Valley, California. Here, technical rock climbers are delighted with the straight-up challenges of Half Dome and El Capitan. In both cases, these 2,000 to 3,000 feet high, near-vertical to overhanging early Cretaceous granodiorite and late Cretaceous granite (respectively) edifices have stood since the end of the last major glacial advance (approximately 10,000 years ago) without failure (Huber, Bateman, and Wahrhaftig, 1989) — save small and occasional rock falls.

In contrast, inherently and relatively weak anisotropic soft-rock earth masses are subject to, and more prone to failure. Good examples along the Southern California coastline include classic landslides such as:

Middle-late Miocene Monterey Formation strata underlying the Palos Verdes Peninsula (Haydon, 2007). And of particular interest, the well-known composite Portuguese Bend group of coalescing slides (Brown, 2007 p. 15).

Upper Eocene-lower Miocene Vaqueros/Sespe formations and the middle Miocene Topanga Formation at Malibu's Big Rock Mesa (Evans, 1984; Robertson, 1984).

Middle to upper Miocene Monterey Formation and Pliocene Pico Formation at La Conchita, near the southwestern corner of Ventura County (Jibson, 2005).

Internal loci of weakness include, but are not limited to: bedding planes, partings, joints, fractures, faults, cleavage, and foliation.

For those who have worked on landslides in coastal southern California over the years, almost any kind of out-of-slope planar structures are well-known, classic red-flag bad actors.

Can you add to this list? Write them in the margin for future reference.

3. Driving Forces Overcoming Resisting Forces. This one is pretty simple. When a pushing force increases to the point where it exceeds and over powers a resisting force, something has to give. And, in the opposite sense, when a resisting force decreases sufficiently, again, something has to give. In engineering terms, when the apparent factor of safety falls below 1.0, failure either occurs or is imminent.

For example, when a speeding car hits a parked car, the driving force of the moving car far exceeds the resisting force of the stationary car. This results in both cars sliding for some distance before skidding friction brings them to a stop – where equilibrium has been reestablished!

A laundry list of existing landslides, say in southern California, is an identical list of failures where the apparent factor of safety has dropped below 1.0. For developers, this is a perpetual nightmare, but for engineering geologists and geotechnical engineers it is job security! Will they ever learn?

4. Changes that Increase Driving Forces and/or Reduce Resisting Forces.

a. Let's start with driving forces. Gravity, the power that never sleeps, is the ultimate driving force that urges everything to seek equilibrium at a lower elevation. When masses of earth are out of equilibrium, the force of gravity operates to restore balance by moving the unstable part or parts to a lower, more stable position.

Driving forces can be increased by adding mass. However this is done, either naturally or by human activity, the result is more weight and thus more driving force. Adding artificial fill at the top of a potentially unstable land mass, as well as a rapid addition of water can either permanently or temporarily increase driving forces. It should be noted, however, that ground water at and in the vicinity of the slip surface has an opposite, or buoying effect, reducing shearing resistance (Terzaghi, 1950, p. 91).

b. Now to resisting forces. These internal strength parameters are a measure of the mass of earth and/or discontinuities within the mass, to resist breaking and/or shearing (sliding one piece past the other). Soil strength is often measured and reported as the angle of internal friction, Φ, and cohesion, c. Rock strength is often measured and reported as unconfined compressive strength in pounds per square inch (psi), or megapascals (Johnson and DeGraff, 1988, p. 128-134).

Several conditions can act on an earth mass to reduce its internal shearing strength. (Equally, a reversal of some of these same conditions can act to increase internal strength.)

But, for sake of this discussion, let's focus on shear strength reduction. Here are a few examples from Terzaghi (1950, p. 88-110), Cruden and Varnes (1996, p. 36-75), and Wieczorek (1996, p. 76-90), along with a few thoughts of my own.

i. Increase in pore water pressure, for example, resulting from infiltration of transient precipitation, elevating ground water levels (artesian pressures), and changes in groundwater regimens.

ii. Re-orientation of platy minerals, such as clays and micas, to facilitate a less resistive, parallel grouping of plates which can more easily slide one past the other. Imagine a deck of slippery playing cards or shingles on a roof.

iii. Weathering processes over time that can lead to changed and/or new mineral compositions. These seemingly subtle rearrangements at the atomic-level can then lead to new and different internal structures, configurations and/or orientations that have inherently less internal strength. For example, the slow process of chemically weathering high internal strength feldspars and their cousins (with relatively high internal strength parameters) to clay minerals (with relatively low internal strength parameters).

iv. Progressive structural changes in earth masses, such as isostatic adjustments. This may manifest as crustal compression (including transpression) leading to mountain building, and/or crustal extension (including transtension) leading to subsidence. Structural weaknesses are created in the process and existing structural weaknesses are exacerbated. In the case of extension, dilation can create pathways and reservoirs for transmission and collection of water.

v. Fault movements that grind, shear, rearrange and change minerals along the shear plane to weaker, similarly oriented platelets.

vi. Folding, which can, like catastrophic fault movements, rearrange mineral constituents and create zones and planes of weakness. For example, bedding plane shears/faults, so common — especially in tightly folded rocks.

vii. Mylonitization provides an extreme example of tectonic forces altering rocks caught in an active shear couple between two opposing rock masses. Such is the case of the Santa Rosa Shear Zone, the suspect zone of weakness for the Pinyon Flat Landslide.

That is quite a list, not entirely complete, and with a few of my own twists, but a good start. Now it is your turn, see how many you can add here in the margin.

5. Triggering Events. Most folks are familiar with events that can trigger a landslide. But let's lay a few of them out here for the uninitiated. Most of what follows can be found in Terzaghi (1950, p. 88-110), Cruden and Varnes (1996, p. 36-75), and Wieczorek (1996, p. 76-90), along with a few tidbits of my own thrown in just for good measure.

a. Earthquakes. A triggering event, such as earthquake ground shaking can be the equivalent of the straw that broke the camel's back. Rapid and intense ground vibrations can increase pore-water pressure in a potential failure zone resulting in decreased shearing stress, leading to failure. In this case, for just long enough, the higher value coefficient of static friction is reduced to the lower value of kinetic friction — and movement is initiated. Such a scenario is a prime suspect in the case of the Pinyon Flat landslide, as the adjacent Pacific Plate/North American Plate boundary is a well-known locus of large magnitude earthquakes.

b. Rainfall. A transient hydrologic event, such as an intense or prolonged rain event may temporarily add weight (driving force), by filling pore spaces and natural surface depressions in an existing slide masses — for just long enough to drive the apparent factor of safety below 1.0.

c. Chemical Weathering. Over a sufficient period of geologic time, chemical weathering of minerals along zones of potential shearing and failure can change once higher shear strength minerals (such as feldspars and their cousins) to lower shear strength minerals (such as clays and their cousins).

d. Physical Reorientation of Minerals at the Molecular Level in a Potential Shear Zone. As above, over a sufficiently long period of geologic time, and with sufficient shearing stress, strain (deformation) at the molecular level can reorient minerals in a potential slip zone. That is to say, a physically random arrangement of minerals, when subjected to sufficient stress, can rearrange orientations to reduce strain and thus increase the likelihood of reduced shearing resistance.

e. Other. The above-noted triggers are focused on naturally occurring situations, such as might have been the case for the Pinyon Flat landslide.

Doubtless, most of you are wondering about typical ways in which human activities lead to landsliding, such as loading the top and/or cutting away the toe of a slope. These are all well-documented in many text books and professional publications, but in this case artifacts of the Anthropocene are not applicable to the Pinyon Flat Landslide.

So, now it is your turn. What additional potential natural and/or anthropogenic triggering mechanisms can you add to the list? Go ahead, you have probably been thinking about some, just waiting for me to add them to the list … now that's your job.

GEOLOGIC AND GEOMORPHIC EVIDENCE FOR THE PINYON FLAT MEGA LANDSLIDE

Geologic and geomorphic evidence for landsliding has been detailed by numerous authors, including, but not limited to: Stout, 1991, 1992, 1993; West, 2007). Much like searching for clues to an unsolved murder mystery, there are certain give-away clues that a trained observer looks for when on the hunt for the presence or absence of landslides. In the case of the Pinyon Flat landslide, some of the more obvious and useful clues are noted below.

Arcuate Back Scarp. An irregularly-shaped, arcuate back scarp is outlined in Figure 1. Starting at Palm Springs, from north to south, the back scarp travels up Palm Canyon, and then follows Omstott Creek across State Route 74 (near Pinyon Flat Camp) and then heads southeasterly along Horsethief Creek, at the base of the Santa Rosa Mountains escarpment. At the headwaters of Horsethief Creek, the back scarp travels southeasterly down Martinez Canyon and then finally turns easterly (as a sided-scarp) as it approaches the mouth of Martinez Canyon (located about 3 miles southwesterly from the intersection of State Routes 86 and 195). From there it goes easterly and northeasterly, to where it finally disappears under Coachella Valley alluvium (United States Geological Survey, 1956, 1959a, 1959b, 1959–69).

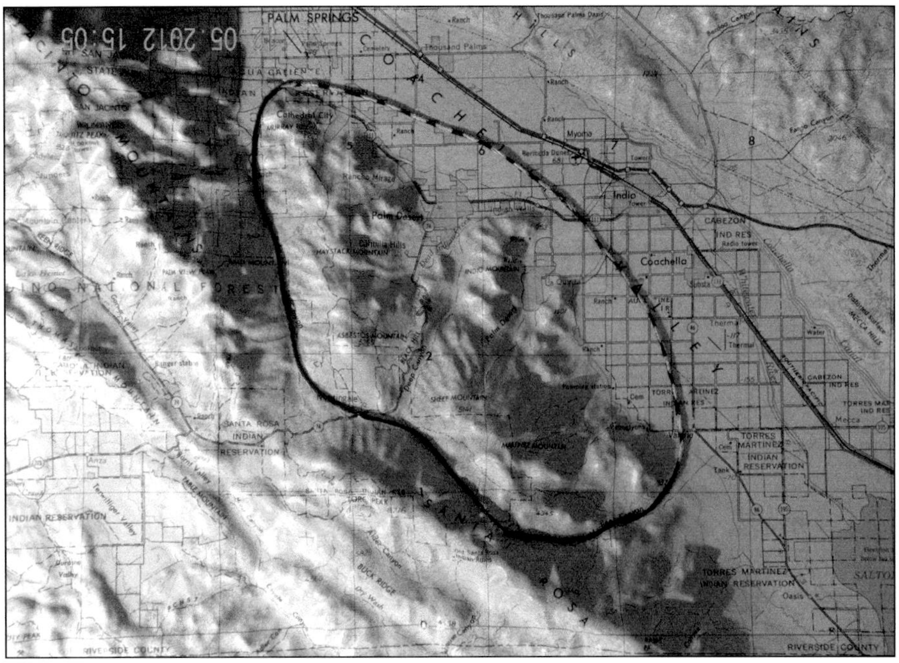

Figure 1. Location Map. Santa Ana, California, 1° x 2°, raised-relief topographic map sheet (modified after Defense Mapping Agency, 1979; map width is approximately 41 miles).

Benches. A distinctive mid-slope bench, in the form of Pinyon Flat, is a telltale giveaway for the presence of landsliding (Figures 1, 2, and 3); United States Geological Survey, 1959b, 1996b).

Additionally, smaller mid-slope benches occur southward from Pinyon Flat, namely, Little Pinyon Flat and Pinyon Alta Flat (United States Geological Survey, 1959b). To the north, broad flat bench-like areas occur at Potrero Spring and at Rock Canyon (United States Geological Survey, 1959b).

Benches and ponds are not long-term natural physiographic components of mature mountain-side landscapes (Longwell, and others, 1932, p. 409–437; Longwell and Flint, 1962, p. 177–195; McAlester and Hay, 1975, p. 264–297; Sharp, 1982). They are ephemeral.

An excellent illustration of the natural physiographic maturing/aging process following a landslide event can be found in Keaton and DeGraff (1996, p. 188–189). Here, in Figure 9-7, eight black and white line drawings illustrate how fresh, sharp, youthful topographic features melt gradually over the years of weathering and erosion to a state of being almost incognito — especially to the untrained observer (Figures 4 and 5).

Bulging Toe. Although the actual toe of the Pinyon Flat landslide is interpreted to be most likely buried under Coachella Valley fill (Figures 1, 2 and 3), a sub-aerial sense of a toe can be seen in Figure 1, where toe-of-mountain-front land forms protrude northeasterly into the Coachella Valley at Indian Wells and at La Quinta (United States Geological Survey, 1959b, 1959–69).

Structural Fabric. To start with, the Pinyon Flat Landslide head scarp coincides with the Santa Rosa Mountain Shear Zone (Figures 2 and 3). The convenience of this seeming coincidence is that these mylonitized metamorphic and granitic rocks are weakened to start with, and display down-to-the basin (Coachella Valley) slickenlines, mullions and linear striations (Dibblee, 1981, in particular Figure 9, page 20; Parcel, 1981), all strongly suggestive of landsliding. Good photogenic examples of this consistent linear fabric were found in unimproved truck-trail road-cuts, one mile south of State Route 74 in the N½, N½, NW¼, Section 14, T7S, R5E, SBBM; United States Geological Survey, 1996b). See Photos 3 and 4.

Boots-on-the-ground field reconnaissance forays led to the impression that back-scarp sliding has likely occurred across a broad zone of weak rock, much like the shuffling of a deck of playing cards, as opposed to being confined to a single shear surface. For example, see Frazer (1931, p. 521).

Mid-slide break-away shears within the overall landslide mass need to be investigated with respect to location, style and composition – a task beyond the scope of this progress report.

Tension Fractures. Another clue to the presence of break-away lateral movement and tension in the head-scarp area is the occurrence of once-open, but now filled tension fractures. A fracture filled with red-orange-brown gravelly clayey sand was found in a road cut along the south side of State Route 74, one mile west of Palm Canyon Drive (E½, E½, SE¼, Section 8, T7S, R5E, SBBM; United States Geological Survey, 1996b). See Photos 1 and 2.

Although, like may lines of evidence which may be attributed to other origins, these do not detract from the landslide hypothesis. In fact, they reinforce the idea.

Immature Geomorphic Features. In a mature physio-topographic setting, most if not all of the following would be expected to be absent, or at least subdued. In the case of the Pinyon Flat landslide, all are present, some more pronounced than others. At the end of this example list, see if you can add a few more.

Hummocky Topography. Figure 1, the low sun-angle plastic raised relief map shows, in a broad way, hummocky topography enclosed within the Pinyon Flat Landslide area. Compare this to surrounding topography and look for subtle differences in topographic forms. For this exercise, you may want to go back and review your class notes from either geomorphology (or physiography if you took this class in the geography department).

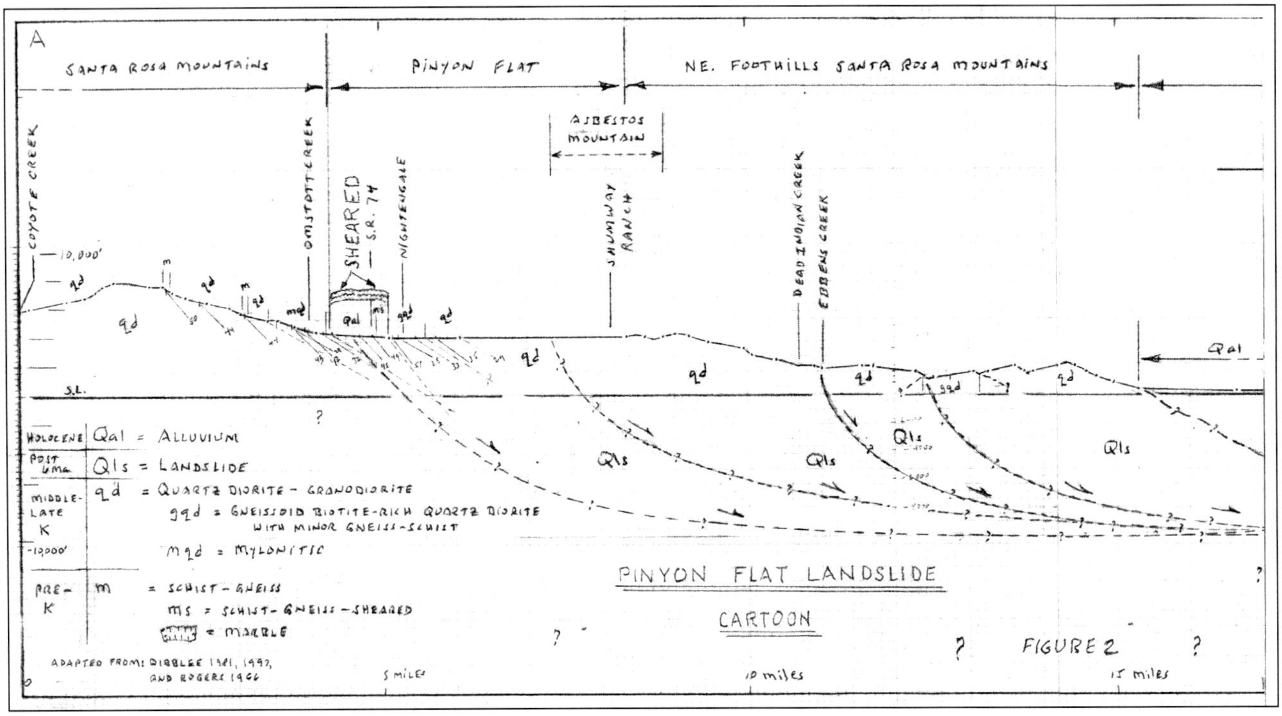

Figure 2A (above) and 2B (opposite). Structure Section A-A'.

Disrupted, Displaced, and/or Out-of-Balance Drainage. Again, we are back to the basics of geomorphology. Most of these features are difficult to define at the scale of Figure 1. One in particular, however, stands out. That one is Deep Canyon on the southern side of State Route 74. Its deep and narrow dimensions seem incongruous with its relatively small precipitation catchment basin. That is, it seems out of place in this otherwise desert environment. (See notes on maturation of topographic forms above, under the previous heading, "*Benches.*")

The question then is: where did the energy and erosive power come from to carve so deeply into hard crystalline rock? One answer that seems to make internal sense is that the relatively small drainage took advantage of a preexisting fracture pattern that can trace its origins back to the Pinyon Flat landslide. With a deep fracture system already in place, most of the depth was likely already there and all the little stream had to do was clean it out and let mass wasting do a little widening. At its most impressive depth and width, in the vicinity of Hidden Palm Canyon, distinctly "V"-shaped Deep Canyon is about 800 to 1,000 feet deep and 2,000 feet wide (United States Geological Survey, 1959b, 1996b).

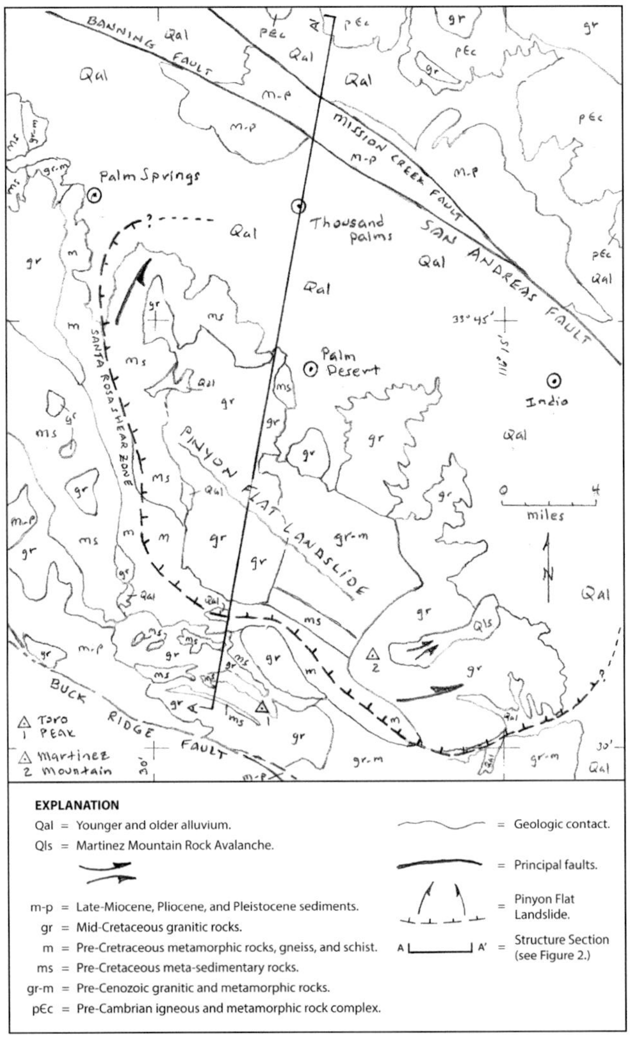

Figure 3. Geologic Map (modified and simplifed after Rogers, 1966).

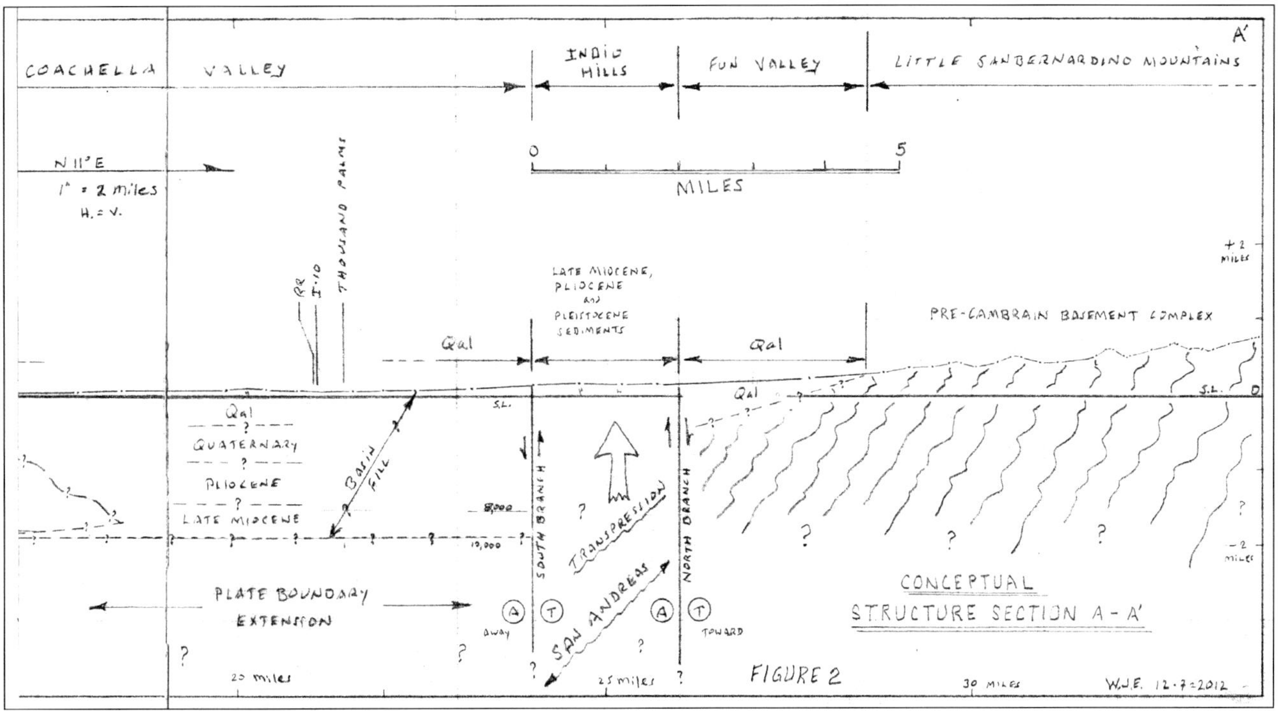

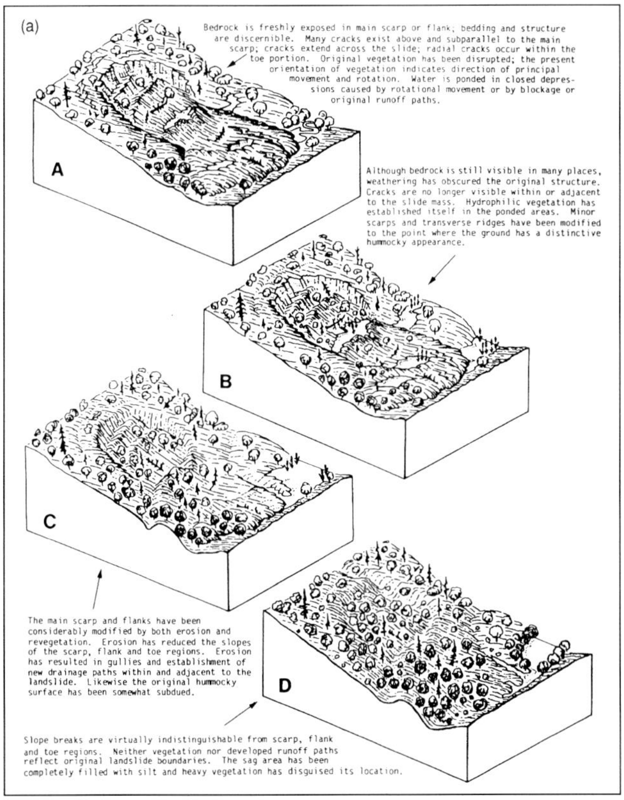

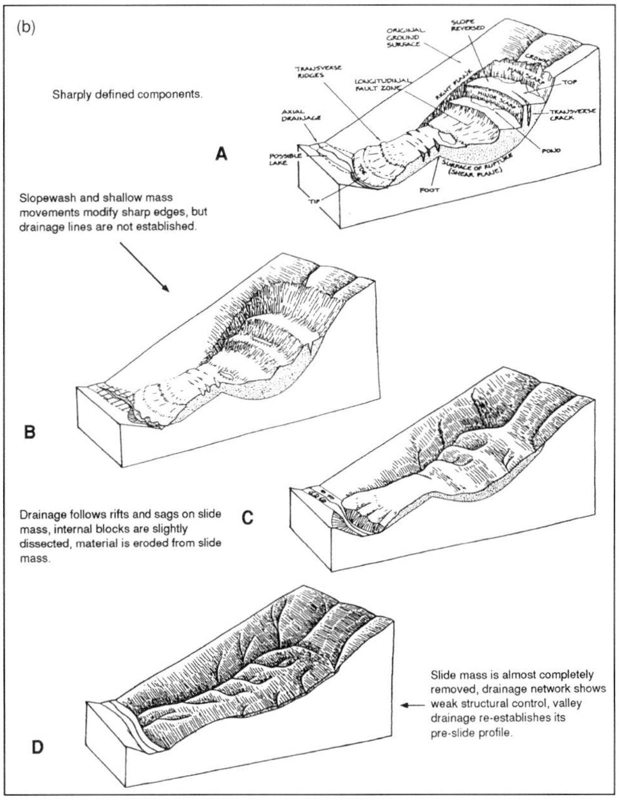

Figure 4. Block diagrams of morphologic changes with time of idealized landslide in a humid climate (modified from Keaton and DeGraff, 1996, p. 188).

Figure 5. Block diagrams of morphologic changes with time of idealized landslide in a semiarid climate (modified from Keaton and DeGraff, 1996, p. 189).

Look carefully at the shape and orientation of other drainages within the Pinyon Flat Landslide. None are as impressive as Deep Canyon.

Now, surely there must be a number of alternative explanations for the presence of Deep Canyon. This is where you and T. C. Chamberlin (1890) come in. Using The Method of Multiple Working Hypotheses, make a list of other explanations for the presence of Deep Canyon — come on now, they are just sitting there on the tip of your tongue.

Springs. Springs are a common occurrence in landslide terrane. This seems to be, at least in part, an artifact of the presence of broken and fractured rock. Some of the named springs noted on United States Geological Survey topographic maps (1959a,1959 b; 1996b) include: Agua Bonita, Potrero, Magnesia, Asbestos, Cactus, and Agua Alta springs. See also, Fraser (1931, p. 521).

Subsidiary Landslides. Within the main mass of weakened landslide debris, subsidiary landslides often times occur.

One such case is the Martinez Mountain rock avalanche. It is located within the southern portion of the Pinyon Flat landslide, on the east flank of Martinez Mountain, between Guadalupe Creek and Toro Canyon (United States Geological Survey, 1959b; United States Department of Agriculture, 1984, Photo Nos. 162–163). This event occurred prior to Cahuilla Indian presence, approximately 300 to 1600 years ago (Baldwin, 1986, p. 42), and is thought to have likely been triggered on the order of 15 to 20 thousand years ago by a moderate-sized local earthquake during a wetter climatic period of geologic time (Baldwin, 1986, p. 47).

See also, Quaternary Landsliding along a Miocene Detachment Fault, Palm Desert, California by Mike Hart in this volume.

Other. Now it is your turn again. How many other out-of-place topographic/physiographic features can you think of that are common to landslide terrane? Do any of these apply to the Pinyon Flat landslide?

DETACHMENT FAULTING

Although mid-Tertiary extension and detachment faulting may well have contributed planes of weakness and pre-existing slip surfaces to the overall Pinyon Flat landslide, this hypothesis is still in need of investigation. For example, additional studies, like that for the Haystack Mountain Detachment Fault (Pagnillo, 2003), located about 7 miles north of Pinyon Flat Camp (and State Route 74) in the SE¼, of Section 33, T5S, R5E, and the NE¼ of Section 4, T6S, R5E, SBBM (United States Geological Survey, 1959b), may shed additional light on mid-Tertiary extensional history in this corner of Southern California.

Curiously, in Frost and others (1997, p. 68, Figure 11), the mapped trace of the Salton Sea Detachment Fault appears to more or less coincide with portions of the southern end of the Pinyon Flat Landslide, between about Little Pinyon Flat and Pinyon Alta Flat. To the north, however, the Salton Sea Detachment Fault trace turns easterly and then wanders northwesterly along the base of the mountain front toward Palm Springs.

Perhaps you will be the lucky one to find, piece together, and unravel the many circuitous and obscure threads of this mystery.

OUTRAGEOUS HYPOTHESIS?

We think not! No, we don't have direct evidence of internal slide geometry, or even a drill core containing the basal slip surface. But the rest of the evidence noted in the paragraphs above is hard to cast aside with so many lines of compelling evidence that all seem to point to a common conclusion — that Pinyon Flat Landslide really is a **MEGA-SLIDE**.

Multiple working hypotheses, yes, we think Thomas Chrowder Chamberlin (1890) would be proud. Nothing has been ruled out. In fact, all good bets are still on the table. Our challenge to you is to come up with equally valid and convincing reasoning why the geologic/geomorphic evidence could result from one or more sets of alternative circumstances.

CONCLUSIONS

At this level of reconnaissance investigation, there appears to be abundant compelling direct and circumstantial evidence to suggest the presence of the Pinyon Flat mega-landslide.

Age of sliding is post ~6Ma and pre-Coachella Valley fill.

A lot of work still needs to be done to refine this geo-approximation of the Pinyon Flat landslide and to determine its apparent factor of safety against future failure events.

CHALLENGES FOR FUTURE INVESTIGATORS

To date, only the bare rudiments of evidence for the presence of the Pinyon Flat landslide have been illuminated. It is our hope that future investigators will be able to, with boots-on-the-ground, walk and map the essential ingredients that will either verify or scuttle our landslide hypothesis. This work will be especially important in the head scarp area, along Palm Canyon, Horsethief Creek, and Martinez Canyon. Additionally, a careful look into Deep Canyon will likely be a very useful adventure in an effort to understand structural and stratigraphic history.

Review of stereographic aerial photographs (for example, United States Department of Agriculture, 1984) of various ages, as well as topographic maps at various scales (for example, United States Geological Survey topographic maps listed in the references section below) will likely be helpful in directing future investigators to key places to visit on the ground.

A series of LiDAR (Light Detection and Ranging) images of the landslide and immediately surrounding areas may illuminate surface features not seen on conventional stereographic aerial photographs and/or by careful visual examination while walking on the ground.

A review of previous "black-box" geophysical studies (magnetic, gravity, seismic and resistivity), as well as new focused studies, will likely be useful in delineating three dimensional slide geometry.

A review of existing test boring logs, as well as drilling new holes will also lend valuable information. Old water, geothermal, and/or hydrocarbon (oil/gas) well logs may be available from private parties and/or public agencies (such as municipal water districts, and the California Division of Oil, Gas and Geothermal Resources).

Back calculations using estimated soil strength parameters (Φ and c), ground water levels, bulk density, and slide geometry at the time of failure(s) will likely be useful in shedding light on stress/strain fields active at the time(s) of failure.

And, as is almost always the case, one good field investigation leads to something not previously on the agenda. So, keep an open mind and let your inquisitive and curious mind be your guide…

ACKNOWLEDGEMENTS

Our thanks go to Dr. Monte Marshall for assisting us with current wisdom on regional tectonic and structural aspects of the Coachella segment of the Salton Trough and adjacent highlands. We are also indebted to Dr. Elaine Hanford for her critical review which greatly improved the content and flow. Thank yous also go to Carole Ziegler and David Bloom for their tireless help in threading this essay through the pre-publication process. And, as always, Wendy Elliott has again patiently awaited the return of our living room to an uncluttered space for relaxation, entertainment and family gatherings.

SELECTED REFERENCES

Abbott, P. A., 1999, The Rise and Fall of san Diego – 150 Million Years of History Recorded in Sedimentary Rocks: Sun Belt Publications, San Diego, California, 231p.

Atwater, T., 1970, Implications of Plate Tectonics for the Cenozoic Tectonic Evolution of Western North America: Geological Society of America Bulletin, v. 81, p. 3513–3536.

Baldwin, J. E., II, 1986, Martinez Mountain Rock Avalanche, in Guptil, P. D., Gath, E. M., Ruff, R. W., eds., Geology of the Imperial Valley, California: South Coast Geological Society, Inc., Annual Field Trip Guidebook No. 14, p. 37–48.

Biehler, S., 1964, Geophysical Study of the Salton Trough of Southern California [PhD Dissertation]: California Institute of Technology, on line at: http://resolver.caltech.edu/CaltechETD:etd-10142002-105706

Brown, A. R., 1891, Structural History of the Metamorphic Granitic and Cataclastic Rocks in the Southeastern San Jacinto Mountains, in Brown, A. R., and Ruff, R. W., eds., Geology of the San Jacinto Mountains: South Coast Geological Society, Annual Field Trip Guidebook No. 9, p. 100–138.

Brown, A. R., 2007, Geological Overview of the Palos Verdes Peninsula, in Brown, A. R., Shlemon, R. J., and Cooper, J. D., eds., Geology and Paleontology of Palos Verdes Hills,

California — A 60th Anniversary Revisit to Commemorate the 1946 Publication of the U. S. Geological Survey Professional Paper 207: Pacific Section of the Society of Economic Paleontologists and Mineralogists (Society for Sedimentary Geology), Fullerton, CA, Book 103, p. 5–22.

Bunge, H-P, and Grand, S., 2000, Mesozoic Plate-Motion History Below the Northeast Pacific Ocean from Seismic Images of the Subducted Farallon Slab: Nature, v. 405, p. 337–340 (May 2000).

Chamberlin, T.C., 1890, The Method of Multiple Working Hypotheses: Science, v. 15, p. 92–96.

Christie-Blick, N., Anders, M. H., Wills, S., Walker, C D., and Renik, B., 2007, Observations from the Basin and Range Province (Western United States) Pertinent to the Interpretation of Regional Detachment Faults, in Karner, G. D., Manatschal, G., and Pinheiro L., M., eds., Imaging, Mapping and Modeling Continental Lithosphere Extension and Breakup: Geological Society of London Special Publications, 282, p. 241–441 (DOI: 10.1144/SP282.17).

Copenhaver, G. C., Jr., 2006, Main Stem – Geology of the San Diego & Arizona Eastern Railway, San Diego and Imperial Counties, California, U. S. A., and Baja California, Mexico, in Farquharson, P. T., Bloom, D. M., and Ziegler, C. L., eds., Geology and History of Southeastern San Diego County, California: San Diego Association of Geologists annual field trip guidebook for 2005 and 2006, Sun Belt Publications, San Diego, California, p. 63–90.

Cromling, J., 1973, Geothermal Drilling in California: Journal of Petroleum Technology, p. 1033. On line at: http://www.onepetro.org/mslib/servlet/onepetropreview?id=00004177

Cruden, D. M., and Varnes, D. J., 1996, Chapter 3, Landslide Types and Processes, in Turner, A. K., Schuster, R. L., eds., Landslides – Investigation and Mitigation: Transportation Research Board, National Research Council Special Report 247, National Academy Press, Washington, D. C., p. 36–75.

Debiche, M. G., Cox, A., and Englebretson, D., 1987, The Motion of Allochthonous Terranes across the North Pacific Basin: Geological Society of America Special Paper 207, 49p.

Defense Mapping Agency, 1979, Raised Relief 1° x 2° Plastic Topographic Map, Santa Ana, California: prepared and for sale by Hubbard Scientific, Inc., Chippewa Falls, Wisconsin, 54729, scale 1:250,000, contour interval 200', vertical exaggeration 2X, map I.D. No. NI-11-8.

Deméré, T. A., 2006, The Imperial Sea — Marine Geology and Paleontology, in Jefferson, G. T., and Lindsay, L., eds., Fossil Treasures of the Anza-Borrego Desert: Colorado Desert District of the California State Parks, Anza-Borrego Desert State Park, The Anza-Borrego Foundation and Institute, and Sunbelt Publications, San Diego, California, p. 29–41.

Dibblee, T. W., Jr., 1954, Geology of the Imperial Valley Region, California: California Division of Mines Bulletin 170, Chapter II, Contribution 2, p. 21–28, and Generalized Geologic Map of Imperial Valley Region, California, Chapter II, Contribution 2, Plate 2, map scale 1"= 5.9 miles.

Dibblee, T. W., Jr., 1981, Geology of the San Jacinto Mountains and Adjacent Areas, in Brown, A. R., and Ruff, R. W., eds., Geology of the San Jacinto Mountains: South Coast Geological Society, Annual Field Trip Guidebook No. 9, p. 1–47.

Dibblee, T. W., Jr., compiler, 1982, Geologic Map of the Idyllwild (15 minute) Quadrangle, California (Map No. 4), in Brown, A. R., ed., Geologic Quadrangle Maps of the San Jacinto Mountains and Vicinity, California: South Coast Geological Society, Santa Ana, California, map scale 1" = 1 mile.

Dibblee, T. W., Jr., 1997, Geology of the Southeastern San Andreas Fault Zone in the Coachella Valley Area, Southeastern California, in Baldwin, J., Lewis, L., Payne, M., and Roquemore, G., eds., Southern San Andreas fault – Whitewater to Bombay Beach, Salton Trough, California: South Coast Geological Society, Annual Field Trip Guidebook No. 25, p. 35–56.

Dillon, J. T. and Ehlig, P. L., 1993, Displacement on the Southern San Andreas Fault, in Baldwin, J., Lewis, L., Payne, M., and Roquemore, G., eds., Southern San Andreas fault – Whitewater to Bombay Beach, Salton Trough, California: South Coast Geological Society, Annual Field Trip Guidebook No. 25, p. 188–205. (Reprinted from: Dillon, J. T. and Ehlig, P. L., 1993, Displacement on the Southern San Andreas Fault, in Powell, R. E., Weldon, R. J., II, and Matti, J. C., eds., The San Andreas Fault System – Displacement, Palinspastic Reconstruction, and Geologic Evolution: Geological Society of America Memoir 178.)

Dokka, R. K., 1993, Original Dip and Subsequent Modification of a Cordilleran Detachment Fault, Mojave Extensional Belt, California: Geology, v. 21, n. 8, p. 711–714.

Dorsey, R., 2006, Stratigraphy, Tectonics, and Basin evolution in the Anza-Borrego Desert Region, in Jefferson, G. T., and Lindsay, L., eds., Fossil Treasures of the Anza-Borrego Desert: Colorado Desert District of the California State Parks, Anza-Borrego Desert State Park, The Anza-Borrego Foundation and Institute, and Sunbelt Publications, San Diego, California, p. 89–104.

Dorsey, R., 2012, San Andreas Fault Zone in Coachella Valley, Southern California: see Becky Dorsey's home page (last updated, March 2012) at: http://pages.uoregon.edu/rdorsey/rjd.html

Downey, C., and Clinkenbeard, J., 2005, An Overview of Geologic Carbon Sequestration Potential in California: California

Geological Survey Special Report 183, Department of Energy Contract Number: DE-FC26-03NT41984, 64p.

Earle, S., 2000, The History of Subduction beneath Western North America: Earth Science News, on line at: http://records.viu.c...llon-may00 .htm (Society for Science and the Public, sciencenews.org).

Elders, W. A., Rex, R. W., Meidav, T., Robinson, P. T., and Biehler, S., 1972, Crustal Spreading in Southern California – The Imperial Valley and the Gulf of California Formed by the Rifting Apart of a Continental Plate: Science, October 6, 1972, v. 178, p. 15–24. (Reprinted from: Fischer, P., ed., Imperial Valley Regional Geology and Geothermal Exploration: American Association of Petroleum Geologists, 1973, Annual Meeting, Anaheim, California, Field Trip No. 2, p. 39–48.)

Elliott, W. J., 2003, Geomorphic Evidence for Ancient Bedrock Landsliding at Buckman Springs, San Diego County, California, in Murbach, M. L., and Hart, M. W, eds., Geology of the Elsinore Fault Zone, San Diego Region/Hot Springs and Tourmalines of Eastern San Diego County: a joint publication of the San Diego Association of Geologists and the South Coast Geological Society, distributed by Sunbelt Publications, San Diego, California, v. 31, p. 111–116.

Englebretson, D. C., Cox, A., and Gordon, R. G., 1985, Relative Motions Between Oceanic and Continental Plates in the Pacific Basin: Geological Society of America Special Paper 206, 59p.

Evans, D. A., 1984, Big Rock Mesa Landslide – Preliminary Conclusions, in Hollingsworth, R. A., Field Trip Chairman, and Robertson, H. S., Field Trip Co-Chairman, eds., Field Guide to Selected Engineering Geologic Features – Malibu, Los Angeles, County, California: Southern California Section of the Association of Engineering Geologists Annual Field Trip June 2, 1984, p. 8–10.

Evans, J. R., ed., 1988, Landslides in Crystalline Basement Terrain: San Diego Association of Geologists, annual field trip guidebook, 87p.

Fraser, D. M., 1931, Geology of San Jacinto Quadrangle South of San Gorgonio Pass, California: California Journal of Mines and Geology Bulletin, v. 27, n. 5, p. 494–540, map scale 1:125,000.

Frost, E. G., Suitt, S. C. and Fattahipour, M., 1997, Emerging Perspectives of the Salton Trough Region with an Emphasis on Extensional Faulting and its Implications for Later San Andreas Deformation: in Baldwin, J., Lewis, L., Payne, M., and Roquemore, G., eds., Southern San Andreas fault – Whitewater to Bombay Beach, Salton Trough, California: South Coast Geological Society, Annual Field Trip Guidebook No. 25, p. 57–97.

George, P. G., and Dokka, R. K., 1994, Major Late Cretaceous Cooling Events in the Eastern Peninsular Ranges, California and their Implications for Cordilleran Tectonics: Geological Society of America Bulletin, v. 106, p. 903–914.

Haines, S. H. and van der Pluijm, B. A., 2010, Dating the Detachment Fault System of the Ruby Mountains, Nevada – Significance for the Kinematics of Low-Angle Normal Faults: Tectonics, v. 29, TC4028, 20p. The American Geophysical Union (DOI:10.1029/2009TC002552)

Hart, M. W., 2003, Landslides in the Peninsular Ranges, in Murbach, M. L., and Hart, M. W, eds., Geology of the Elsinore Fault Zone, San Diego Region/Hot Springs and Tourmalines of Eastern San Diego County: a joint publication of the San Diego Association of Geologists and the South Coast Geological Society, distributed by Sunbelt Publications, San Diego, California, v. 31, p. 193–209. (Reprinted from Geological Excursions in Southern California and Mexico, Guidebook for the 1991 Annual Meeting, Geological Society of America, San Diego, California.)

Haydon, W. D., 2007, Overview of Landslides of the Palos Verdes Peninsula, in Brown, A. R., Shlemon, R. J., and Cooper, J. D., eds., Geology and Paleontology of Palos Verdes Hills, California — A 60th Anniversary Revisit to Commemorate the 1946 Publication of the U. S. Geological Survey Professional Paper 207: Pacific Section of the Society of Economic Paleontologists and Mineralogists (Society for Sedimentary Geology), Fullerton, CA, Book 103 p. 259–270.

Huber, N. K., Bateman, P. C., and Wahrhaftig, C., compilers, 1989, Geologic Map of Yosemite National Park and Vicinity, California: United States Geological Survey Map I-1874, map scale 1:125,000, contour interval 200'.

Irwin, W. P., 1990, Chapter 3 — Geology and Plate-Tectonic Development, in Wallace, R. E., ed., The San Andreas Fault System, California: United States Geological Survey Professional Paper 1515, p. 61-80. (Also, reprinted in Baldwin, J., Lewis, L., Payne, M., and Roquemore, G., eds., Southern San Andreas fault – Whitewater to Bombay Beach, Salton Trough, California: South Coast Geological Society, Annual Field Trip Guidebook No. 25, p. 99–118.)

Jennings, C. W., compiler, 1975, Fault Map of California with locations of Volcanoes, Thermal Springs, and Thermal Wells: California Division of Mines and Geology — Geologic Data Map No. 1, map scale 1:750,000.

Jibson, R. W., 2005, Landslide Hazards a La Conchita, California: United States Geological Survey Open-File Report 2005-1067, 12p.

Johnson, R. B., and DeGraff, J. V., 1988, Principals of Engineering Geology: John Wiley & Sons, New York, 497p.

Keaton, J. R., and DeGraff, J. V., 1996, Subsurface Observation and Geologic Mapping (Chapter 9), in Turner, A. K., and Schuster, R. L., eds., Landslides – Investigation and Mitigation, Special Report 247: Transportation Board, National Research Council, National Academy Press, Washington, D. C., p. 178–230.

Kimbrough, D. L., Smith, D. P., Mahoney, J. B., Grove, M., Gastil, R. G., Ortega-Rivera, A., and Fanning, C. M., 2001, Forearc-Basin Sedimentary Response to Rapid Late Cretaceous Batholith Emplacement in the Peninsular Ranges of Southern and Baja California: Geology, June 2001, v. 29, n. 6., p. 491–494.

Kistler, R. W., Wooden, J. L., and Morton, D. M., 2003, Isotopes and Ages in the Northern Peninsular Ranges Batholith, Southern California: United States Geological Survey Open-File Report 03-489, 45p.

Leighton, F. B., 1966, Landslides and Hillside Development, in Lung, R., and Proctor, R., eds., Engineering Geology in Southern California: Association of Engineering Geologists Special Publication, p. 149–207.

Lister, G. S., and Davis, G. A., 1989, The Origin of Metamorphic Core Complexes and Detachment Faults Formed During Tertiary Continental Extension in the Northern Colorado River Region, U.S.A.: Journal of Structural Geology, v. 11, n. 1/2, p. 65–94.

Longwell, C. R., and Flint, R. F., 1962, Introduction to Physical Geology, second edition: John Wiley & Sons, Inc., Ney York, 504p.

Longwell, C. R., Knopf, A., and Flint, R. F., 1932, A Textbook of Geology – Part I, Physical Geology: John Wiley & Sons, Inc., New York, 514p.

Lonsdale, P., 1989, Geology and Tectonic History of the Gulf of California, in Winterer, E. L., Hussong, D. M., and Decker, R. W., eds., The Eastern Pacific Ocean and Hawaii: Geological Society of America, The Geology of North America, v. "N", p. 499–521.

Lovely, P., Shaw, J. H., liu, Q., and Tromp, J., 2006, A Structural Vp Model of the Salton Trough, California and Its Implications for Seismic Hazard: Bulletin of the Seismological Society of America, v. 96, n. 5, p. 1882–1896.

McAlester, A. L., and Hay, E. A., 1975, Physical Geology – Principals and Perspectives: Prentice-Hall, Inc., Englewood Cliffs, New Jersey, 439p.

McCalpin, J., 1984, Preliminary Age Classification of Landslides for Inventory Mapping, in Proceedings 21st Engineering Geology and Soils Engineering Symposium, University of Idaho, Moscow, p. 99–120.

Menard, H. W., 1978, Fragmentation of the Farallon Plate by Pivoting Subduction: Journal of Geology (University of Chicago), v.86. p. 99–110.

Oakeshott, G. B., Jennings, C. W., and Turner, M. D., 1954, Correlation of Sedimentary Formations in Southern California: California Division of Mines Bulletin 170, Chapter III, Contribution 1, p. 5–8 and Correlation Chart of Sedimentary Formations in Southern California, prepared by Oakeshott, G. B., 1954, California Division of Mines Bulletin 170, Chapter III, Contribution 1, Plate 1.

Pagnillo, J. P., 2003, Haystack Mountain Detachment Fault, Santa Rosa Mountains, Riverside County, California [M.S. Thesis]: San Diego State University, Department of Geological Sciences, 71p.

Parcel, R. F., Jr., 1981, Structure and Petrology of the Santa Rosa Shear Zone in the Pinyon Flat Area, Riverside County, California, in Brown, A. R., and Ruff, R. W., eds., Geology of the San Jacinto Mountains: South Coast Geological Society, Annual Field Trip Guidebook No. 9, p. 139–150.

Pridmore, C. L., and Frost, E. G., 1992, Detachment Faults – California's Extended Past: California Geology, p. 3–17.

Robertson, H. S., 1984, Geology and Slope Stability of the Coastal Portion of the Malibu Beach Quadrangle, Los Angeles County, in Hollingsworth, R. A., Field Trip Chairman, and Robertson, H. S., Field Trip Co-Chairman, eds., Field Guide to Selected Engineering Geologic Features – Malibu, Los Angeles, County, California: Southern California Section of the Association of Engineering Geologists Annual Field Trip June 2, 1984, p. 68–98.

Rogers, T. H., 1966, Geologic Map of California – Santa Ana Sheet: California Division of Mines and Geology – Santa Ana Map Sheet, map scale 1:250,000.

Ruff, R. W., Bogseth, P. A., and MacGregor, B. M., 1981, Geology of the San Jacinto Mountains Field Trip Road Log, in Brown, A. R., and Ruff, R. W., eds., Geology of the San Jacinto Mountains: South Coast Geological Society, Annual Field Trip Guidebook No. 9, p. 175–219.

Sharp, R. P., 1982, Landscape Evolution (A Review): Proceedings of the National Academy of Sciences, USA, v. 79, p. 4477–4486.

Sharp, R. V., 1966, Ancient Mylonite Zone and Fault displacements in the Peninsular Ranges of Southern California [abs.]: Cordilleran Section of the Geological Society of America, 1966 annual meeting, abstracts with programs, p. 65.

Shreve, R. L., 1987, Blackhawk Landslide, Southwestern San Bernardino County, California, in Hill, M. L., ed., Centennial Field Guide Volume 1, Cordilleran Section of the Geological Society of America: Geological Society of America, Boulder, Colorado, p. 109–114.

Smith, M. B., 1964, Map Showing Distribution and Configuration of Basement Rocks in California (North Half) (South Half): United States Geological Survey, Map OM-215, map scale 1:500,000.

Stout, M. L., 1991, Landscape Evolution During the Last 35,000 Years – Perspective of a Southern California Engineering Geologist, in Abbott, P.L., and Elliott, W. J., eds., Environmental Perils – San Diego Region: Prepared for the Geological Society of America Annual Meeting at San Diego, California by the San Diego Association of Geologists, p. 161–164 (reprinted in Stout, 1993).

Stout, M. L., 1992, Mega-Landslides in Southern California, in Proceedings from the Association of Engineering Geologist Meeting in Long Beach, California, October, 1992: Star Publishing (reprinted in Stout, 1993).

Stout, M. L., 1993, Landslides – Recognition, Investigation, and Mitigation: Short Course, sponsored by the San Diego Association of Geologists and the Southern California Section of the Association of Engineering Geologists, August 28, 1993, 19p.

Sydnor, R., 1979, Chapter 8, Geology of the Northeast Border of the San Jacinto Pluton, Palm Springs, California, in Crowell, J. C., and Sylvester, A. G., eds., Tectonics of the Juncture Between the San Andreas Fault System and the Salton Trough, Southeastern California: published by the Department of Geological Sciences, University of California, Santa Barbara, for the annual Geological Society of America meeting in San Diego, California, p. 77–85.

Sylverster, A. G., 1999, Rifting, Transpression, and Neotectonics in the Central Mecca Hills, Salton Trough, California, in Pacific Section of the Society of Economic Paleontologists and Mineralogists Fall Field Trip Guidebook, p. 1–11.

Terzaghi, K., 1950, Mechanism of Landslides, in Paige, S., Chairman (ed.), Application of Geology to Engineering Practice, Berkey Volume: The Geological Society of America, Berkey Volume, p. 83–123.

Terzaghi, K., and Peck, R. B., 1948, Soil Mechanics in Engineering Practice: John Wiley & Sons, New York, 566p.

United States Department of Agriculture, 1984, black and white stereographic aerial photographs, numbers: 331615-HAP-84-F-225, 134-142, 156-165, 211-221, and 233-239, flown 8-24-1984, approximate scale 1:60,000.

United States Geological Survey, 1956, Coachella, California, 15′Topographic Quadrangle Map, scale 1″= 1 mile, contour interval 80′.

United States Geological Survey, 1958, Thousand Palms, California, 15′Topographic Quadrangle Map, scale 1″= 1 mile, contour interval 80′.

United States Geological Survey, 1959a, Idyllwild, California, 15′ Topographic Quadrangle Map, scale 1″ = 1 mile, contour interval 80′.

United States Geological Survey, 1959b, Palm Desert, California, 15′Topographic Quadrangle Map, scale 1″= 1 mile, contour interval 80′.

United States Geological Survey, 1959-69, Santa Ana, California, 1° x 2 ° Topographic Quadrangle Map, scale 1″= 4 miles, contour interval 100′.

United States Geological Survey, 1996a, Butterfly Peak, California, 7½′ Topographic Quadrangle Map, scale 1″ = 2,000′, contour interval 40′.

United States Geological Survey, 1996b, Toro Peak, California, 7½′Topographic Quadrangle Map, scale 1″= 2,000′, contour interval 40′.

Varnes, D. J., 1978, Chapter 2, Slope Movement – Types and Processes, in Schuster, R. L., and Krizek, R. j., eds., Landslides – Analysis and Control, Transportation Research Board, Commission on Sociotechnical Systems, National Research Council Special Report 176: National Academy of Sciences, Washington, D. C., p. 11–33.

Walawender, M. J., 2000, The Peninsular Rangers – A Geological guide to San Diego's Back Country: Kendall/Hunt Publishing Company, Dubuque, Iowa, 114p.

Walawender, M. J., Girty, G. H., Lombardi, M. R., Kimbrough, D., Girty, M. S., and Anderson, C., 2003, A Synthesis of Recent Work in the Peninsular Ranges Batholith, in Murbach, M. L., and Hart, M. W, eds., Geology of the Elsinore Fault Zone, San Diego Region/Hot Springs and Tourmalines of Eastern San Diego County: a joint publication of the San Diego Association of Geologists and the South Coast Geological Society, distributed by Sunbelt Publications, San Diego, California, v. 31, p.177–192.

Wallace, R. E., ed., 1990, The San Andreas Fault System: United States Geological Survey Professional Paper 1515, 283p.

West, T. R., 2007, Landslides and Society – Recognition, Impacts and Mitigation, in Turner, A. K., and Schuster, R. L., eds., Landslides and Society: Association of Engineering Geologists Special Publication No. 22, p. 117–147.

Wieczorek, G. F., 1984, Preparing a Detailed Landslide-Inventory Map for Hazard Evaluation and Reduction: Bulletin of the Association of Engineering Geologists, v. 21, n. 3, p. 337–342.

Wieczorek, G. F., 1996, Chapter 4, Landslide Triggering Mechanisms, in Turner, A. K., Schuster, R. L., eds., Landslides – Investigation and Mitigation: Transportation Research Board, National Research Council Special Report 247, National Academy Press, Washington, D. C., p. 76–90.

Wieczorek, G. F., 2002, Catastrophic Rockfalls and Rockslides in the Sierra Nevada, USA, in Evans, S. G., and DeGraff, J. V., eds., Catastrophic Landslides – Effects, Occurrences, and Mechanisms: Geological Society of America, Boulder, Colorado, Reviews in Engineering Geology, v. XV, p. 165–190.

Figure 1. Bighorn sheep standing on upper-plate granodiorite above the Asbestos Mountain-La Quinta low-angle fault.

Mylonitization and Associated Low-Angle Faulting in the Santa Rosa and San Jacinto Mountains National Monument, Southern California

Bradley Erskine, Ph.D., CEG
Kleinfelder
401 Marina Place
Benicia, CA 94510
707-738-4917

INTRODUCTION

In October 2000, the United States established the Santa Rosa and San Jacinto Mountains National Monument, incorporating 252,000 acres of undeveloped land and wilderness along the western margin of the northern Peninsular Ranges. A great part of the San Jacinto Mountains and Santa Rosa Mountains are within the monument's boundaries. The monument is spectacular. It is the home of many state and federally listed threatened or endangered species, including the Peninsular bighorn sheep (Figure 1); numerous springs and creeks supporting large stands of Washingtonia palms; outstanding exposures of the ancient shoreline of Lake Cahuilla; and desert fauna and flora of all kinds.

Equally remarkable is the geology of the monument. Exposed throughout the two ranges are undoubtedly the best and most accessible exposures in California of mylonites, deep-seated highly deformed anatexites (rocks that have undergone in situ partial melting), and low angle faults. The mylonite belt, called the eastern Peninsular Ranges mylonite zone (EPRMZ), and the associated anatexites and faults, is unique and on par with the greatest of mylonitic belts and "core complexes" present in the Western Cordillera.

Outstanding exposures of this belt are extremely accessible in the Agua Caliente Indian Canyons by Palm Springs, along Highway 74 west off Palm Desert, and west of La Quinta. One would think that rocks as spectacular as these and of great regional geological significance would have been well studied and the data available, particularly when the area is so accessible by automobile from Palm Springs. This, unfortunately, is not the case. A large portion of the belt has been mapped in detail, but most of the data are not readily accessible. The northernmost part, from Highway 74 northward to Mount San Jacinto, was mapped by the author in the early 1980s as part of a Ph.D. dissertation (Erskine, 1986), and mapping continues intermittently to the present time. The southern part, from Deep Canyon on Highway 74 to south of Martinez Mountain, has been mapped by the United States Geological Survey, and their work within the Santa Rosa Mountains continues as well. Most of this mapping is currently unavailable to the public. Other parts have been mapped by several students as part of unpublished masters theses. Still other areas, such as the western side of Martinez Mountain and the Martinez Mountain sturtzstrom were mapped intermittently by the author over the last few decades, and the geology resides on field maps. There are several excellent specialized studies within the belt, but these works are found only in scientific journals, and their relevance to the regional geological framework is lessened by the absence of geologic maps and any thorough discussion of the regional geology.

Fortunately for geologists and aficionados of the eastern Peninsular Ranges, a unified source for information and maps regarding the mylonite zone will be established in the near future. The United

States Geological Survey plans to compile, unify, and release archival 1:24,000-scale geologic maps that they and others have made on the Salton-Trough-facing flanks of the Santa Rosa and San Jacinto Mountains. There are also plans to prepare and publish several papers on the geology of the eastern Peninsular Ranges mylonite zone and associated rocks, the first of which is under review for publication in the Bulletin of the Geological Society of America (Matti et al., in preparation). In addition, there are plans for a Geological Society of America Memoir on the geology of the Peninsular Ranges that will include a more detailed paper on the eastern Peninsular Ranges mylonite zone, and may include maps of the area north of Highway 74 (Morton, D., editor, in preparation).

PURPOSE

The purpose of this paper is to provide the readers of this volume, and particularly visitors to the stops that were made on the San Diego Association of Geologists' Annual Field Trip in 2009 or Trip 1, a means to easily understand the fundamental nature of the mylonite belt and the geologic processes that created it. This will be accomplished by generalizing many of the complexities and the nuanced and sometimes conflicting studies and data, and simplifying the results of the many specialized studies. Detailed descriptions and citations of studies related to, for example, microstructural analysis, preferred orientation textures, geobarometry and geochronology, electron microscopy, and many of the complex relationships between faults and rock units that provide the details but sometimes obscure the fundamentals are avoided. Instead, this paper focuses on the key rock units, their geometrical relationship to the low angle faults, and the basic fabrics that can be observed at the stops described in the field trip log. Particular reference will be made to observations at Trip 1, Stop 4 on Highway 74 looking northward down Palm Canyon and the rocks and fault located in the Indian Canyons at Trip 1, Stops 15 and 16. By focusing on the basic elements, the fundamental structure and geometry of the belt is seen to be quite straight forward, and a basic appreciation of the geologic history is attained.

REGIONAL DISTRIBUTION OF THE EASTERN PENINSULAR RANGES MYLONITE ZONE AND RELATED ROCKS

The eastern Peninsular Ranges mylonite zone is a 100 kilometer long belt of deformation along the eastern margin of the Northern Peninsular Ranges batholith, southern California (Figure 6, Trip 1 road log). Active right-lateral strike-slip faulting along the San Jacinto fault system (Sharp, 1967) offset the eastern Peninsular Ranges mylonite zone into three geographically distinct parts. To the north is the Santa Rosa mylonite zone exposed along the eastern flanks of the San Jacinto Mountains at Palm Springs that bend eastward near Asbestos Mountain. The belt then continues to the southeast along the eastern flanks of the Santa Rosa Mountains. A relatively small and isolated fragment of the zone is exposed at Coyote Mountain east of Borrego Springs and is called the Coyote Mountain mylonite zone. The southern part of the eastern Peninsular Ranges mylonite zone continues south along the base of the mountains just west of Borrego Springs, and is called the Borrego Springs mylonite zone. The eastern Peninsular Ranges mylonite zone is presumably a part of a much larger belt of deformed rocks buried beneath alluvium of the Imperial Valley and waters of the Gulf of California.

The eastern Peninsular Ranges mylonite zone commands attention because mylonites by their very nature are beautiful and fascinating rocks, and because they are so well exposed in the area. The mylonite zone appears, at first glance, as a belt between the undeformed basement to the west and other rocks previously presumed to be unassociated with it to the east (Figures 6 and 7, Trip 1 road log). However, to the east of the eastern Peninsular

Ranges mylonite zone, at least in the north, is a structurally related group of metasedimentary rocks and anatexites called the Palm Canyon Complex. As will be documented below, this group is closely associated with the mylonites, and should be considered part of the eastern Peninsular Ranges mylonite zone. Therefore, the structural belt defined by the eastern Peninsular Ranges mylonite zone is actually much wider, more complex, and probably extends many kilometers eastward beneath the low-angle faults described below. This implies that the eastern Peninsular Ranges mylonite zone is not only a relatively thin "shear zone," but, rather, part of a major tectonic belt of regional significance.

Overlying the eastern Peninsular Ranges mylonite zone and Palm Canyon Complex are a series of imbricate low-angle faults bounding structural sheets of mostly undeformed granodiorite. The faults are spatially related to the belt throughout its length, and continue southward past the southernmost mylonite exposures.

DESCRIPTION OF THE FUNDAMENTAL ROCK UNITS OF THE EASTERN PENINSULAR RANGES MYLONITE ZONE AND ASSOCIATED ROCKS

Each unit of the eastern Peninsular Ranges mylonite zone and associated rocks is described in order as viewed from Highway 74 looking northward down Palm Canyon at Stop 4 (see Figures 6 and 7, Trip 1 road log), followed by a description of the faults that bound the units. This package is representative of the units that trend southward to at least Martinez Canyon, and probably throughout the entire zone to Borrego Springs.

The rock units and faults are described from west to east, as follows:

1. basement rocks comprised of batholitic and pre-batholithic rocks, represented here by the Desert Divide Group,
2. mylonites of the eastern Peninsular Ranges mylonite zone, represented here by the Santa Rosa mylonite zone,
3. Palm Canyon Complex,
4. upper-plate Granodiorite, and
5. Palm Canyon, Asbestos Mountain and Haystack Mountain faults.

Basement Rocks. Looking to the northwest along the crest of the San Jacinto Mountains, one will observe a group of metasedimentary rocks that are different from surrounding rocks due to a distinct banding and foliation (Figure 2). This is the Desert Divide Group, a large prebatholithic screen of quartz-dominated metasedimentary rocks exposed along the crest, or "desert divide," of the southern San Jacinto Mountains. The Desert Divide Group is an arcuate, northwest-striking screen located west and structurally below the mylonite zone. Its contact with the surrounding plutonic rocks is primarily intrusive.

Brown (1968) subdivided the Desert Divide Group into two units: a large quartzite body called the Ken Quartzite to the east, and a diverse sequence of carbonates, schists and gneisses called the Bull Canyon Formation to the west. The Ken Quartzite is a medium- to coarse-grained quartzite containing minor lenses of marble, schist and gneiss. The Bull Canyon Formation is composed of several distinct and locally mappable lithological units which include a quartz gneiss facies, a schist

Figure 2. Steeply dipping foliation in metasedimentary rocks of the Desert Divide Group.

facies consisting of quartz dominant biotite schist with interlayering of quartz schist and quartzite, and a white to grey calcite-rich marble facies.

Although no direct age determinations have been reported for the Desert Divide Group, isotopic evidence from the San Jacinto Mountains and fossils from similar rocks to the south provide some constraints on the age of sedimentation. Rubidium-strontium isotopic analysis reveals the highly radiogenic nature of these rocks and suggest a contribution of old cratonic material, or the presence of old sediments, interpreted to be late Precambrian to early Paleozoic. Uranium-lead age determinations from plutons in the northern San Jacinto Mountains provide an upper limit of 97 million years for the deposition of these rocks (Hill, 1981, 1984).

The structure of the Desert Divide Group is typical of the many prebatholithic units that outcrop in the region, but is distinct from the mylonites and associated rocks that are the subject of this paper. The Desert Divide Group has steeply dipping contacts with the surrounding plutonic rocks, and represents a screen of rocks that were enveloped by the rising plutons. The most conspicuous structural feature of the Desert Divide Group is a penetrative foliation defined by the planar alignment of platy minerals and co-parallel compositional variations. On the surface of the foliation is a strong lineation, formed by a preferred orientation of elongate minerals or mineral trains that plunge steeply to the southeast. These fabrics likely represent the minor axis of strain and direction of elongation that existed while these rocks were deformed ductilely during pluton emplacement. As will be shown below, they are distinct from the fabrics observed in the mylonites that are found structurally above this unit.

Surrounding the Desert Divide Group, and also structurally beneath the mylonite zone, are plutonic rocks represented in the area by the granodiorite of Palm Canyon. Most of these rocks are typical of the relatively undeformed plutonic rocks of the eastern Peninsular Ranges. In outcrop the unit may be fine to coarse grained, porphyritic, and non- to moderately foliated. Mafic inclusions are common to absent. Foliation tends to be concentrated along the margin of the pluton. Away from the margins, the rocks tend to be homogeneous with little evidence of deformation or flow structures (Figure 3).

Santa Rosa Mylonite Zone. As one proceeds eastward down the flank of the San Jacinto Mountains toward Palm Canyon, the plutonic rocks begin to exhibit an eastward dipping foliation with a lineation defined by oriented porphyritic hornblende crystals. At first glance the rocks appear to be undeformed; however, inclusions that were generally equant or randomly oriented become flattened parallel to the foliation (Figure 4). The orientation of the foliation and lineation is unlike those in the Desert Divide Group. The steeply dipping foliation and southeasterly plunging lineation found in the metasediments are replaced by a regional eastward dipping fabric, and a lineation that trends to the northeast. This fabric becomes more prominent to the east, until the rocks transition over a relatively short distance to rocks that are recognized as the Santa Rosa mylonite zone (Figure 5).

The rocks of the mylonite zone are autochthonous to the batholith and have not been displaced

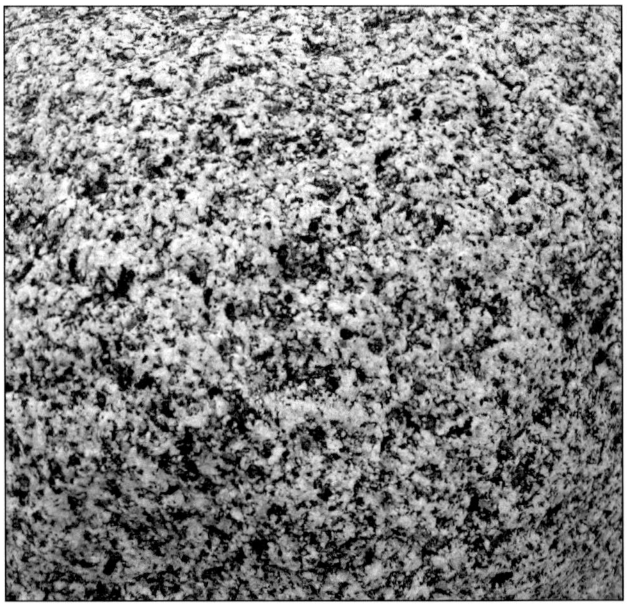

Figure 3. Homogeneous non-deformed plutonic rocks far beneath the mylonite zone.

Figure 4. Deformed xenolith in plutonic rocks closely beneath the mylonite zone.

Figure 5. Strongly foliated rocks of the Santa Rosa mylonite zone.

Figure 6. Strong stretching lineation on foliation surface of mylonite.

Figure 7. Flattened and folded leucocratic veins in mylonite.

by movements along low-angle faults. As the mylonite zone is approached from the west, rocks of the northern Peninsular Ranges batholith grade rather abruptly into mylonite gneiss, mylonite, and ultramylonite characterized by the presence of an intense foliation and a well-developed stretching lineation (Figure 6). Leucocratic dikes representing late stage syndeformational intrusion are flattened parallel to the foliation, and extended with high strains until fold axes are parallel with the regional lineation (Figure 7). The foliation is regionally consistent, striking north-northwest and dipping 35° to 55° eastward. The attitude of the lineation is extremely uniform in a given region, but its azimuth changes systematically from eastward in the southern part of the mylonite zone to about N55°E in the San Jacinto and northern Santa Rosa Mountains. Classic shear-sense indicators such as multiple foliations and asymmetric fabrics described by Platt and Vissers (1980),

S. H. White et al. (1980), and Lister and Snoke (1984) indicate an east-over-west sense of shear (thrust) throughout the zone. This sense of shear is confirmed by asymmetric quartz textures in mylonitic plutonic rocks and in quartz-rich leucocratic dikes that were transposed or refolded into the plane of the foliation (Erskine, 1986).

The contact of the mylonite zone with the underlying plutonic rocks dips less steeply (30°–40°) than the mylonitic foliation, and the geometric relations between the two are consistent with the expected geometry of a post-crystallization–sub-solidus shear zone. Based on examination of microstructures in the southern part of the eastern Peninsular Ranges mylonite zone, Simpson (1985) concluded that the mylonite zone represents a post intrusive shear zone. However, a number of structural relations between intrusive rocks both below and within the northern part of the mylonite zone suggest that deformation was, in part, contemporaneous with intrusion. The concordance of foliation, lineation, and strained mafic inclusion fabrics in plutonic rocks below the mylonite zone with that in the zone supports this view. In addition, at least one plutonic unit within the mylonite zone (leucogranite of Cactus Spring; Matti *et* al., 1983) has the geometry and structure that would be expected of an intrusion that invaded an active deformation zone and crystallized before deformation ceased. It appears likely that rocks within the region were deformed prior to, during, and following intrusion of the eastern part of the batholith.

Although the age of mylonitization is not well constrained, it probably began at some time between 97 and 62 Ma U-Pb ages of zircons from tonalite plutons in the San Jacinto Mountains indicate emplacement ages of 97 Ma (Hill, 1981, 1984), providing a lower limit for the onset of deformation. The upper limit is constrained by concordant 62 Ma apatite-sphene-zircon fission-track ages of mylonitic rocks in the northern part of the mylonite zone (Dokka, 1984). These ages suggest rapid uplift of the region at that time.

Palm Canyon Complex. Throughout its length, the eastern Peninsular Ranges batholith consists primarily of deformed plutonic rocks. In the Santa Rosa mylonite zone, outcrops of metasedimentary rocks within the mylonite belt are rare, with the only significant exposure occurring at the mouth of Palm Canyon. However, as one traverses down the flank of the Santa Rosa Mountains and across the mylonite zone, the mylonitic plutonic rocks of the Santa Rosa mylonite zone change abruptly across Palm Canyon to a different structural package of rocks called the Palm Canyon Complex (Figure 8).

The Palm Canyon Complex is the name given to a group of mylonitic metasedimentary rocks, anatexites, and orthogneiss located above the Palm Canyon fault (Figures 6 and 7, Trip 1 road log; and described below). The unit is well exposed from the mouth of Palm Canyon southward for at least 30 km to the area between Deep Canyon and Martinez Mountain, where about 5-6 km of section is exposed in a structural culmination. The structural top and bottom of the complex are truncated by

Figure 8. Banded and foliated rocks of the Palm Canyon Complex.

Figure 9. Mylonitic anatexites of the Palm Canyon Complex.

low-angle faults. The most characteristic feature of the Palm Canyon Complex is the presence of substantial volumes of mylonitic, garnetiferous leucogranite anatexites (partially melted rocks) that are associated lit-par-lit (alternately banded) with the metasedimentary rocks and orthogneiss (Figure 9). The anatexites are strongly foliated and commonly carry a well-developed down dip lineation that has the same east-northeast trend as the lineation in the mylonitic rocks below the Palm Canyon Fault. The axes of folds in carbonates within the Palm Canyon Complex and quartz veins within the mylonite zone are also parallel to the regional lineation, and reflect the high strains recorded throughout the belt (Figure 10). Metasedimentary rocks that lack anatexite textures are generally nonmylonitic, except for the calc-silicate and marble tectonites that are found throughout the Palm Canyon Complex. A variety of minor structures are present within the nonmylonitic metasedimentary rocks, but rarely do these rocks possess the well-developed northeast, down-dip

Figure 10. Folded carbonates in the Palm Canyon Complex.

lineation characteristic of the anatexites and rocks of the mylonite zone.

Perhaps the most significant rocks within the Palm Canyon Series are a wide variety of garnetiferous leucogranite anatexites apparently resulting from the in-situ partial melting of the metasedimentary rocks. The presence of such rocks implies high metamorphic grade at the peak of metamorphism, and places constraints on the conditions of deformation as well. The foliation and lineation in the Palm Canyon Complex appear to have formed during anatexis of the pelitic schists, and these structures therefore record deformation at high grades of metamorphism. The lack of significant anatexites in similar metasedimentary rocks in the Desert Divide Group to the west suggests that the Palm Canyon Complex represents the relatively deep part of a thick section of deformed rocks that was juxtaposed against shallower rocks by westward thrusting of the upper plate-Complex along a pre-mylonitic belt of deformation, and later, along the easterly-dipping mylonite zone. It is possible that some displacement occurred even later at higher crustal levels along the Palm Canyon Fault.

Upper-Plate Granitic Rocks. The term "upper-plate" granitic rocks refers to all granitic rocks that occur as parautochthonous sheets and klippen above the Palm Canyon Complex (Figures 6 and 7, Trip 1 road log; Figure 11). The most completely studied unit of these granitic rocks is the granodiorite of Asbestos Mountain, which occupies a klippe above the Asbestos Mountain fault in the northern Santa Rosa Mountains.

In outcrop the granodiorite is medium to coarse grained with a moderate to strong foliation. Locally it becomes gneissic. The most characteristic feature of this unit is the presence of large, subhedral hornblende crystals and gold-green sphene up to 0.5 cm in length. The hornblende is often dimensionally oriented producing a down dip lineation which parallels the long axis of elongate mafic inclusions which are common throughout the unit.

In the north, foliation and lineation in the "upper-plate" granitic rocks are roughly parallel to the regional mylonitic structures. However, this foliation dips less steeply than that within the underlying mylonite zone and Palm Canyon Complex. The presence of east-dipping foliations and east- to northeast-plunging lineations in the parautochthonous granitic rocks suggests that these rocks were emplaced during the episode of regional deformation that produced the mylonite zone.

The granodiorite of Asbestos Mountain is lithologically and geochemically similar to autochthonous plutonic rocks of the northern Peninsular Ranges batholith (Anderson, 1983; Erskine, 1986; Hill, 1984). These similarities suggest that

Figure 11. Asbestos Mountain, a klippe overlying the low angle Asbestos Mountain fault.

the Asbestos Mountain unit and, by inference, the other "upper-plate" granitic rocks are not exotic to the batholith. However, the "upper-plate" rocks are anomalous in relation to the eastern part of the batholith in that they are magnetite-series granitoids, whereas the granitic rocks beneath the mylonite zone belong to the ilmenite series. This suggests that either the source terrane of the upper-plate rocks lay some 60 km to the west, within the magnetite-series sub belt, or as seems more likely, a belt of magnetite-series granitoids may once have occupied the eastern part of the batholith and been tectonically displaced.

LOW-ANGLE FAULTING

Spatially associated with the eastern Peninsular Ranges batholith is a series of regionally significant low-angle faults that are distinct from and unrelated to modern San Andreas-system fault zones currently active in the region. This spatial association is significant because being representatives of the brittle deformation regime, they represent a response to strain at entirely different crustal levels than that in the basement rocks, mylonites, and anatexites. Therefore, an interesting question arises: are these faults kinematically and temporally related to the deeper deformed units, or are they younger, unrelated features with a coincidental association? To address this question, presented below is a description of the two primary faults associated with the mylonite zone and related rocks, both of which were visited on the 2009 field trip. These are: the Palm Canyon fault, and the Asbestos Mountain-La Quinta fault system.

Palm Canyon Fault. The Palm Canyon fault is marked by a breccia zone within Palm Canyon from Palm Springs to just north of Highway 74 where it assumes an east-west strike for a length of about 7 km, then bends back to a northwest-southeast strike at the head of Deep Canyon and continues southward to the southern Santa Rosa mylonite zone.

The Palm Canyon fault is relatively planar along its exposure in Palm Canyon and southward

Figure 12. Chlorite breccia zone, Palm Canyon fault. The flat surface represents the exhumed fault plane.

toward Martinez Mountain, striking north-northwest and dipping generally between 25°–35° east. This regularity is interrupted in the southern portion of the mapped area by a corrugation which, in outcrop, follows the trace of the eastward flexure of the Santa Rosa mylonite zone. Comparison of the attitude of the axis and position of the hinge line of the fault with the folds in the upper- and lower-plate reveal excellent agreement.

Where well exposed, the Palm Canyon fault consists of a surface separating rocks of contrasting lithology, structure, or both. Locally, the surface is marked by a few centimeters of strongly comminuted structureless microcataclasite often colored deep brown to black. The microcataclasite caps a zone of intense brecciation and alteration referred to as the chloritic breccia zone after references to similar rocks found beneath detachment faults in the metamorphic core complexes to the east. Near the fault surface the chlorite breccia zone is a green-colored, penetratively brecciated, and pervasively altered cataclasite with chlorite and epidote representing the bulk of secondary mineralization (Figures 12 and 13). The breccia zone contains a network of veinlets of pink alkali feldspar which often fill sub parallel to conjugate fracture sets located near the fault surface.

Figure 13. Detail of the chlorite breccia zone, Palm Canyon fault.

Although strongly brecciated material is generally confined to within 5 meters beneath the fault surface, the effects of faulting and alteration may extend deep into the footwall. Evidence for faulting begins as far as 30–40 meters within the footwall where secondary mineralization of chlorite and epidote begin to impart a green hue to otherwise gray to tan rocks. As the fault is approached, chlorite and epidote increase in abundance, and feldspar becomes pink and progressively sericitized. Brecciation begins to disrupt the mylonitic fabric eventually obliterating it entirely.

Pseudotachylite is often found associated with faulting within the footwall, and is typically present as triangular-shaped black aphanitic structures interconnected by thin dike-like veins of the same material that crosscut the existing foliation.

Several regional characteristics of the Palm Canyon Fault are particularly noteworthy. First, the fault surface represents the lowest structure of the stacked sequence of low-angle faults. The fault is not truncated by other low-angle faults, nor does it merge (branch) with any major fault segment. Second, both its hanging wall and footwall are extremely uniform in composition and structure. Rocks equivalent to the Palm Canyon complex comprise the hanging wall as far south as Martinez Canyon or beyond. The footwall is always composed of mylonitically overprinted plutonic rocks, and the mylonite zone's contact with the basement is never truncated by the fault. Third, the key structural elements- the attitudes of the foliation and lineation- are virtually identical above and below the fault. This suggests that even though the Palm Canyon fault is easily traceable for some 90 kilometers, offset along the fault may be relatively minor.

Asbestos Mountain-La Quinta Faults. The most conspicuous and structurally complex fault in the northern Santa Rosa Mountains is the Asbestos Mountain fault which places a klippe of granodiorite above the Palm Canyon Complex and the Santa Rosa mylonite zone (Figures 6 and 7, Trip 1 road log). The fault trends from Palm Springs through Deep Canyon toward the desert floor, and is replaced by the La Quinta fault which continues southward to Martinez Mountain, beneath the desert floor at La Quinta, and probably continues southward to Martinez Canyon or further.

Near the fault surface a deep-brown microcataclasite responsible for the brown color associated with the hanging wall is often seen. This microcataclasite is strongly coherent and homogeneous. The microcataclasite ultimately grades upward into a matrix-supported breccia composed entirely of granodiorite clasts and fragments. Clasts are typically 0.5 m in diameter, but some as large as 5 meters or larger are present. Sub-angular to well-rounded clasts are supported by a matrix of pulverized granitic material of all size fractions. Foliation within neighboring clasts is generally discordant. Examples of the fault breccias are shown in Figures 14 and 15.

In contrast to the relatively planar surface of the Palm Canyon fault, the Asbestos Mountain fault has been folded and warped into a series of folds, domes and basins. The surface geometry of the faults are complex, but there appears to be a general north-south orientation of flexures except to the south where the axis of the synformal warp of the Asbestos Mountain fault in the south curves from north-south to northeast-southwest down plunge. The undulatory form of the Asbestos Mountain fault is projected in cross section in Figure 7 (Trip 1 road log).

Direction of Fault Transport. The direction of displacement along the low-angle faults is of paramount importance to the understanding of the structural history of the region. A westward sense of displacement of the upper-plates would imply thrusting related to compressional tectonics, whereas an eastward-directed sense of transport for the upper-plates would suggest an extensional model for their origin.

Examination of fault-related structures along or near the primary low-angle fault surfaces provides several lines of evidence for eastward-directed transport of the upper-plates throughout the northern Santa Rosa Mountains. These are as follows:

(1) *Secondary shear-related foliations.* In a number of well exposed localities, secondary shear-related foliations, analogous to S-C structures commonly found in ductile-shear zones, are formed at high angles to, and listrically verge into, the

Figure 14. Asbestos Mountain fault showing cataclasite in the foot wall, and brecciated granodiorite in the hanging wall.

Figure 15. La Quinta fault, showing the chlorite breccia zone and brown cataclasite in the footwall, and brecciated granodiorite in the hanging wall.

fault surface. Similar structures are found in some low-angle faults within the chlorite-breccia zone below, and consistently suggest a top to the east displacement.

(2) *Drag-folding.* Along the entire length of the Palm Canyon fault, the east-dipping Palm Canyon sequence is steepened near the Asbestos Mountain fault surface. If the steepening is attributed to "drag" along the fault, then the sense of displacement is top to the east.

(3) *However, minor faults with normal displacement.* Although the upper- and lower-plate units are not correlatable across the fault surfaces, minor normal faults offsetting distinctive small-scale markers imply a down dip, top to the west displacement at least during the latest phase of faulting (Figure 16).

On a regional scale the stack of low-angle faults possess characteristics of well-known thrust belts (Boyer and Elliott, 1982) implying a westward-sense of displacement. The faults, floored by the Palm Canyon fault and roofed by the Asbestos Mountain fault resemble a west-directed duplex system, while the Haystack Mountain fault above may represent a remnant of an upper duplex.

Figure 16. Offset leucocratic dikes in the hanging wall of the La Quinta fault, indicating top to the west displacement (view is to the south).

TECTONIC INTERPRETATION

From the studies of deformation summarized above, a series of constraints on aspects of deformation has emerged which allows for a regional interpretation. Some of the conclusions, such as the sense of shear in the mylonites, are unequivocal and must constrain any interpretation, whereas others, such as the pressures and temperatures of metamorphism, are somewhat less certain.

Summarized below are the major observations and conclusions that constrain the tectonic reconstruction of the eastern Peninsular Ranges batholith and associated rocks.

1) The eastern Peninsular Ranges mylonite zone, composed of the Santa Rosa, Coyote Mountain, and Borrego Springs mylonite zones, represents a broad belt of deformation along the eastern margin of the northern Peninsular Ranges batholith. All structural sheets within the northern Santa Rosa mylonite zone, plutonic fabrics immediately below the Santa Rosa mylonite zone contact, and granodiorite occupying the hanging walls, appear to be structurally related. They are unified by a regional foliation and lineation that is distinct from, and kinematically unrelated to, structures characteristic of the basement terrain well below the zone.

(2) The east-dipping regional attitude of the mylonite zone and low-angle faults probably represents their primary orientation. There is no evidence suggesting that the mylonite zone may have originally been sub horizontal and later rotated to its present configuration.

(3) Structures within plutonic rocks below and within the mylonite zone suggest syndeformational intrusion, implying that the thrusts were active during intrusion of the eastern Peninsular Ranges batholith. The timing of the initiation of deformation is not known, but it appears to have followed the onset of regional metamorphism. Plutons intruding and crystallizing well below the zone possess deformational structures related to intrusion, whereas those intruding adjacent to

or entering the mylonite zone possess a fabric related to flattening and a west to southwestward-directed shearing.

(4) The path of deformation in the mylonitic plutonic belt is composed of flattening and shearing components. The sense of shear is west to south-westward-directed (top to the southwest). Deformation appears to have outlasted the peak of metamorphism in the Palm Canyon series.

(5) Anatexis of the Palm Canyon series is syn-deformational and restricted to rocks comprising the hanging wall of the Palm Canyon fault, implying that these rocks have been subjected to a higher grade of metamorphism than rocks below the fault (e.g., the Desert Divide Group and the mylonitic plutonic rocks). The Palm Canyon complex may represent rocks metamorphosed and deformed at greater depths and later thrust up-section to their present structural level relative to the plutons in the undeformed basement.

(6) Strain indicators within each of the fault-bounded sheets may be related to one episode of progressive deformation resulting from regional east-west compression prior to and following mylonitic deformation and anatexis.

(7) The apparent agreement between the trends of the regional Santa Rosa mylonite zone stretching lineation, stria found on fault surfaces, and the axis of large-scale corrugations in fault surfaces implies a strong relation between the overall strains involved in the ductile deformation in the mylonites and those attributed to faulting. Faulting and warping of the fault surfaces and structural sheets or plates which they envelope are thought to represent the final phases of progressive deformation as the region was subjected to decreasing temperatures and pressures.

(8) Available potassium-argon and fission-track ages are consistent with an upper limit of 62 Ma for faulting, placing the timing of faulting near that of mylonitization.

(9) The upper-plate granodiorite of Asbestos Mountain is lithologically similar to the northern Peninsular Ranges batholith, and does not appear to be exotic.

(10) The regional geometry of the fault system is suggestive of southwestward-directed thrusting consistent with the sense of shear in the mylonites. Shear indicators along fault surfaces, however, imply the opposite sense of shear. The apparent conflict may be resolved by calling for eastward-directed displacements during the reactivation of the older east-dipping thrusts. Reactivation may have occurred: a) in the mid-Tertiary associated with crustal extension observed in the metamorphic core complexes to the east, or b) during Late Cenozoic extension associated with the opening of the Gulf of California and development of the Salton Depression.

In an attempt to reconcile observed geologic data, Erskine and Wenk (1985) proposed a crustal extension model for the eastern Peninsular Ranges mylonite zone as an alternative to Sharp's (1979) proposal for thrusting. The model is based on Bott's (1971) interpretation of structural phenomena at passive margins. However, following the collection of additional data and testing the hypothesis in the field, the depiction of the eastern Peninsular Ranges batholith as a regional thrust belt was preferred by Erskine (1986) because most major observed structural relationships such as the sense of shear in mylonites and the regional thrust-system geometry of the low-angle faults are consistent, and deformation could be described in terms of a single phase of progressive deformation rather than polyphase deformation. It is relatively simple in its basic form and lends itself well to modification with further study. It is also more consistent with the general collisional and compressive nature of the batholith over time. In general, deformation is viewed as a thrust belt or crustal suture that initiated prior to or concurrent with intrusion of the eastern part of the northern Peninsular Ranges batholith which was then faulted during uplift and cooling of the region or as a result of younger Cenozoic extension related to opening of the Gulf of California, or both.

Syn-intrusive relationships suggest that the fundamental characteristics of the thrust belt may have initially developed contemporaneously with granitic intrusion in the region. Although much of the pre-intrusive structure of the belt has been modified or destroyed, it appears to have been a relatively thick (>7 kilometers) zone of deformation involving thick sequences of regionally metamorphosed miogeoclinal sediments. This interpretation is based on the character of the Palm Canyon complex that may reflect deformation along the relatively deep segment of the belt. Sediments occupying the lower crustal levels of the miogeocline were deformed during convergence and subjected to increasing temperatures and pressures as the western rocks were underthrust relative to the continental crust to the east.

If deformation began before intrusion, migration of the western volcanic arc eventually led to granitic intrusion within the deformation zone, which became strongly flattened and sheared mylonitically upon crystallization. Immediately below the belt intrusions were flattened parallel to the zone. Farther below, intrusions were unaffected by activity within the zone resulting in the intrusive-related deformational structures characteristic of northern Peninsular Ranges batholith plutonic rocks and associated metasedimentary screens.

Late Cretaceous uplift of the region exposed the eastern Peninsular Ranges batholith to decreasing temperatures and pressures, and continued east-west convergence led to a transition from mylonitic (ductile) deformation to crustal faulting. Folding of the mylonitic foliation was accompanied by development of thrust faults whose surfaces may have propagated downward from upper structural levels. These faults may have been reactivated in the Cenozoic forming the observed apparent top-to-the-east displacement. Alternatively, evidence for the faults remains with the Palm Canyon fault, and the upper fault systems (Asbestos Mountain and La Quinta faults) are normal detachment superimposed on the older thrust system.

ACKNOWLEDGMENTS

The author wishes to thank Dr. Monte Marshall for his collaboration on the 2009 SDAG field trip and review of this manuscript. Without Monte's effort and support, neither the field trip nor this manuscript would have been possible.

REFERENCES CITED

Anderson, J. R., 1983, Petrology of a portion of the eastern Peninsular Ranges mylonite zone, Southern California: Contributions to Mineralogy and Petrology, v. 84, p. 253–271.

Bott, M.H.P., 1971, Evolution of young continental margins and formation of shelf basins: Tectonophysics, v. 11, p. 319–327.

Boyer, S.E., and Elliott, D., 1982, Thrust Systems: American Association of Petroleum Geologists Bulletin, v. 66, p. 1196–1230.

Brown, A.R., 1968, Geology of a portion of the southeastern San Jacinto Mountains, Riverside County, California: unpublished M.S. thesis, University of California, Riverside, California, 95 p.

Dokka, R.K., 1984, Fission-track geochronology evidence for late Cretaceous mylonitization and early Paleocene uplift of the northeastern Peninsular Ranges, California: Geophysical Research Letters, v.11, p. 46–49.

Erskine, B.G., and Wenk, H.-R., 1985, Evidence for crustal thinning in the Santa Rosa mylonite zone, southern California: Geology, v. 13, p. 274–277.

Erskine, B.G., 1986, Mylonitic Deformation and Associated Low-Angle Faulting in the Northern Santa Rosa mylonite zone, California. Unpublished Ph.D. Dissertation, University of California, Berkeley.

Hill, R.I., 1981, Field, petrologic and isotopic studies of the intrusive complex of San Jacinto Mountain: in Brown, A.R., and Ruff, R.W., eds., Geology of the San Jacinto Mountains: South Coast Geological Society Annual Field Trip Guidebook No. 9, p. 76–89.

Lister, G.S., and Snoke, A.W., 1984, S-C mylonites: Journal of Structural Geology, v.6, p. 6 17–638.

Platt, J.P., and Vissers, R.L.M., 1980, Extensional structures in anisotropic rocks: Journal of Structural Geology, v. 2, p. 397–410.

Sharp, R.V., 1967, San Jacinto Fault Zone in the Peninsular Ranges of southern California: Geol. Soc. Am. Bull., v. 78, p. 705–730.

Simpson, C., 1985, Deformation of granitic rocks across the brittle- ductile transition: Journal of Structural Geology, v. 7, p. 503–511.

A WORD ABOUT THE TERMINOLOGY AND INTERPRETATION ASSOCIATED WITH HIGH AND LOW ANGLE FAULTS

by Monte Marshall

In Brad Erskine's paper the reader will find several discussions on whether the upper/hanging-wall block has moved up or down relative to the lower/ footwall block. This decision is usually easier in layered sedimentary rocks than in more or less homogeneous intrusive rocks. An additional complication facing geologists working in this area is interpreting the sense of shear or sense of motion in the case of the low-angle faults.

Traditionally, a fault is called high-angle if it dips more than 45 degrees, and low-angle for dips of less than 45 degrees. This nomenclature is based on a simple observation. However, the dip of a fault depends on the state of stress at the moment of brittle failure or during the interval of ductile flow as in the mylonites. In traditional theory, most faults form when the crust is being compressed or stretched horizontally.

Geology students will have fond memories of the discussion of the state of stress required to cause faulting. One of the basic tenets is that the rupture plane contains and rotates around the intermediate principal stress axis, and makes a 30-45 degree angle with the maximum principal stress axis. Another basic principle is that the plane of maximum shear stress rotates around the intermediate principal stress and lies in the middle of the maximum and minimum principal stress directions, making a 45 degree angle between the two. Of course, shear stress is what causes most faults. In the case of brittle failure, i.e., in the upper crust, compressive stress normal to the fault plane will inhibit slip—causing the actual rupture plane to make a 30 degree angle to the maximum principal stress. A plane at this angle is basically a compromise—it has a lower shear stress, but also lower normal stress. Since the shear or flow planes in ductile deformation, as in mylonite formation in the middle crust, are not affected by the normal stresses, they make a 45 degree angle to the maximum principal stress direction. Contrary to what happens when we slide two rocks past each other at the earth's surface, the ease with which planes of atoms in a crystal of quartz or feldspar slide over and past one another is not at all affected by any forces pushing the planes of atoms together.

Normal faults form when the maximum principal stress is vertical and the minimum principal stress , that is, either only slightly compressional or actually tensile, is horizontal. Theoretically, normal faults should dip between 45 and 60 degrees. In the case of thrust faults the directions of the maximum and minimum principal stresses are reversed from that of normal faults, so they should dip between 30 and 45 degrees. Traditionally geologists considered high-angle faults to be normal, and low-angle faults were assumed to be thrust faults. And often they were correct, but nature is not so simple. Many of the low-angle faults in eastern Nevada and western Utah that were mapped for decades as thrust faults were finally realized, after much debate during the 1970s, to be normal. And, as geologists began to study the Basin and Range Province in earnest, and geophysicists were able to image crustal structure with reflection seismology, they found that the steep normal faults that bound the basins and ranges at the surface commonly shallow with depth. These *listric faults* merge with an almost horizontal *detachment fault* at depth, often at the brittle-ductile transition zone. So it seems that at least in the case of crustal extension, the dip of normal faults differs from that predicted by the simple application of traditional brittle or ductile failure criteria. And, to this author's knowledge, geophysicists have yet to agree on the mechanical conditions for the formation of detachment faults.

In summary, this area probably has been at the northern end of the locus of crustal extension at least since the opening of the present Gulf of California at about 5 Ma. Many studies to the south of this area have found low-angle faults separating a relatively undeformed lower plate with a chloritized upper surface from a deformed and faulted upper plate and have called this fault a detachment fault.

Z-Fold in Deep Canyon, 2009.

A Study of the Structural Development of the Santa Rosa Mountain Section of the Eastern Peninsular Ranges Mylonite Zone, Southern California

Brian C. White
brian.white@cbifederalservices.com

INTRODUCTION

The eastern Peninsular Ranges Mylonite Zone is a more than 100-kilometer-long belt of mylonitic plutonic and metasedimentary rock (Sharp, 1979). This ductily formed tectonic belt extends from the northeastern side of the San Jacinto Mountains near Palm Springs to just south of Borrego Springs, California (Figure 1). This major tectonic feature was initially identified as having been formed by ductile shortening (thrust), but its formation has also been attributed to ductile extension or a combination of the two. To better understand the processes that produced the ductile fabric of the eastern Peninsular Ranges Mylonite Zone, a portion of the eastern central Santa Rosa Mountains, the Santa Rosa Mylonite Zone, was investigated through geologic field mapping and analysis of rock thin sections (Figure 1). This paper presents findings related to the tectonic environment that created the mylonitic rock fabric.

REGIONAL GEOLOGIC SETTING

The Santa Rosa Mountains are the southern half of a 100-kilometer long mountain range that is formed by the San Jacinto Mountains and the Santa Rosa Mountains. This northwest-southeast trending mountain range is bounded on the east by the Salton Sea Trough, to the north and northeast by the San Andreas Fault Zone and to the southwest by the San Jacinto Fault Zone (Jennings, 1994).

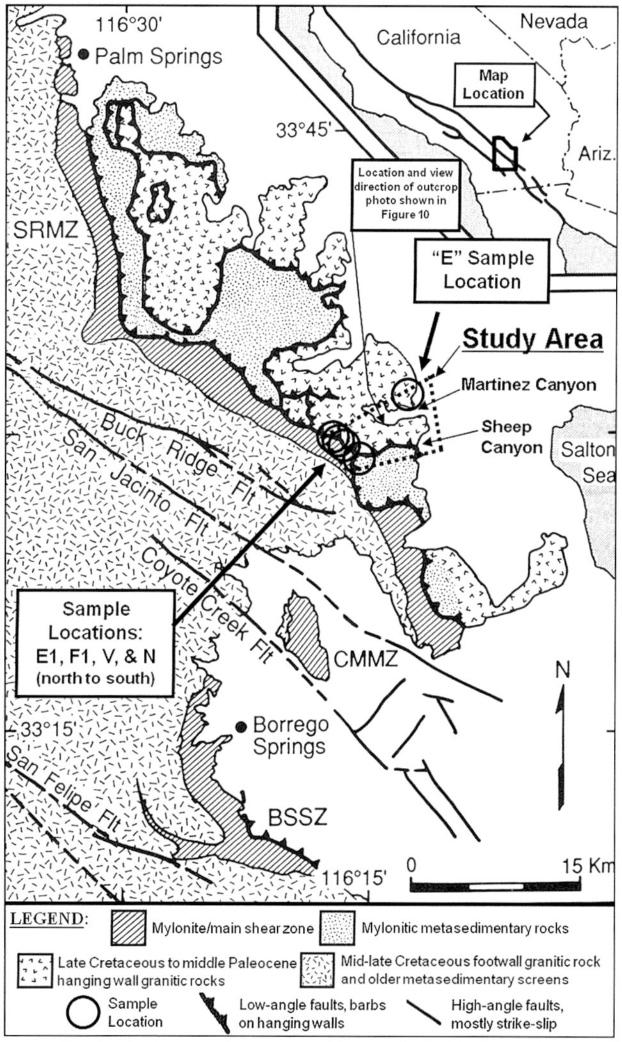

Figure 1. Generalized geologic map of the eastern Peninsular Ranges Mylonite Zone showing the locations of the Santa Rosa Mylonite Zone (SRMZ), Coyote Mountain Mylonite Zone (CMMZ)], Borrego Springs Mylonite Zone (BSSZ), study area, hand and thin section samples, and the photograph in Figure 10 (modified after Goodwin and Renne, 1991).

The San Jacinto and Santa Rosa mountains are located at the northeastern tip of the California Peninsular Ranges, adjacent and parallel to the transform margin between the Pacific and North American Plates. Transform offset between the Pacific and American Plates in this region has continued for about 5.5 million years (Larson and others, 1968; Dickinson, 1981; Curray and others, 1982; Moore and Curray, 1982; Atwater, 1989; Stock and Hodges, 1989; Lonsdale, 1989, 1991; Gans, 1997). During that period approximately 240 to 308 kilometers of right-lateral offset has accumulated, transporting the Peninsular Ranges of Southern and Baja California northwestward from western Mexico and creating the Gulf of California (Larson and others, 1968; Gastil and others, 1978, 1981, 1991; Rangin, 1978; Crowell, 1981; Moore and Curray, 1982; Lonsdale, 1989; Stock and Hodges, 1989; Bohannon and Parsons, 1995; Dickinson, 1997; Gans, 1997).

From about 26 to 17 million years ago, major extensional faulting is known to have occurred in northwest Mexico (Gans, 1997). Prior to and during the regional extension of northwestern Mexico, a zone of subduction was located along the continent to the west, where the Farallon Plate was subducting beneath the North American Plate. Subduction in that region is believed to have ceased approximately 14 to 11 million years ago (Atwater, 1970; Mammerickx and Klitgord, 1982; Lonsdale, 1989, 1991; Spencer and Normark, 1989; Stock and Hodges, 1989; Gans, 1997; Mora-Klepeis and McDowell, 2000).

PREVIOUS WORK

A great deal of geologic field research has been performed to determine the tectonics that produced and then faulted the eastern Peninsular Ranges Mylonite Zone. The mylonite zone was first recognized by Lockwood in 1961 and then studied in detail by Sharp (1966, 1967, 1968, and 1979) and Theodore (1967). Researchers investigating the eastern Peninsular Ranges Mylonite Zone have focused predominantly on the northern portion of the Santa Rosa Mylonite Zone (Dokka, 1984; English, 1985; Erskine and Wenk, 1985; Wise, 1991; George and Dokka, 1994; Goodwin and Wenk, 1995), with some work performed in the southern end of the Santa Rosa Mylonite Zone (Wallace, 1982; Wallace and English, 1982) and in the Coyote Mountain Mylonite Zone (Theodore, 1967). The primary objectives of the research projects studying the mylonite zone included determining rock-unit ages, timing the formation of the mylonite, rock cooling history, and the tectonic setting that ductily produced and then caused brittle deformation of the eastern Peninsular Ranges Mylonite Zone.

Pluton Age Dating. Since the first research detailing the Santa Rosa Mylonite Zone, plutonic rock in the vicinity of that structure has been regarded as Cretaceous in age (Lockwood, 1961). Continuing investigations and age dating have further refined the ages of the eastern Peninsular Ranges Batholith and associated igneous eastern Peninsular Ranges Mylonite Zone footwall rocks as predominantly middle to late Cretaceous (Armstrong and Suppe, 1973; Hill and Silver, 1979; Silver and others, 1979; Silver and Hill, 1986). In contrast to the age of footwall plutons, Santa Rosa Mylonite Zone hanging wall plutons are younger, ranging from latest Cretaceous to middle Paleocene (Armstrong and Suppe, 1973; Dokka, 1984). This sequence of younger, upper-plate plutonic rock above older footwall plutonic rock was noted by Axen and Fletcher (1998) and highlighted as the reverse of most detachment terrane sequences.

Age of the eastern Peninsular Ranges Mylonite Zone. Because the age of the plutonic rock is middle to late Cretaceous and formation of the eastern Peninsular Ranges Mylonite Zone followed its emplacement, mylonitization must have occurred during the late Cretaceous (Sharp, 1979; Dokka, 1984) or later. Dates suggested for formation of the eastern Peninsular Ranges Mylonite Zone range from ~80 to 62 million years ago (Goodwin and Renne, 1991), ~97 to 62 million years ago (Todd and others, 1988), ~97 to 70 million years ago (Erskine and Wenk, 1985), and ~99 to 92 million years ago (George and Dokka, 1994). Three of these four age groups begin at ~97 million years ago, which is the age of the oldest identified pluton in the vicinity (Hill and Silver, 1979; Hill, 1981). Two of the age groups end at ~62 million years ago, the minimum age of formation of the upper-plate pluton (Dokka and Frost, 1978; Dokka, 1984).

Cooling History. Age dating of upper-plate rock of the Santa Rosa Mylonite Zone identified that the easterly portion of the zone went through a brief but intense cooling event from ~63 to 61 million years ago (Dokka and Frost, 1978; Dokka, 1984). Dating of the lower plate and other eastern Peninsular Ranges Batholith rocks west of the eastern Peninsular Ranges Mylonite Zone has identified a rapid terminal cooling event from ~78 to 72 million years ago (George and Dokka, 1990; Grove and Harrison, 1992; Todd and others, 1994; George and Dokka, 1994; Grove and others, 2003) and an earlier rapid cooling event that occurred at ~93 to 92 million years ago (George and Dokka, 1994; Grove and others, 2003). Thermal equilibrium, the time when the temperatures of the upper and lower plates became similar, occurred at ~62 million years ago (Goodwin and Renne, 1991).

Tectonic Sequence. Many tectonic models have been put forth to describe the formation of the eastern Peninsular Ranges Mylonite Zone and the subsequent low angle faulting that dissected the mylonite. Most authors agree with some components of Sharp's (1979) interpretation that the mylonite zone formed due to middle-to late-Cretaceous ductile thrusting of the upper plate to the west (Simpson, 1984; O'Brien and others, 1987; Goodwin and Wenk, 1995), followed by late-Cretaceous westerly thrust faulting (Goodwin and Renne, 1991), and ended with mid-Tertiary normal faulting (Wallace and English, 1982; English, 1985; Engel and Schultejann, 1984; Goodwin and Renne, 1991; Pridmore and Frost, 1992; Goodwin and Wenk, 1995; Frost and others, 1996a and 1996b; and Axen, 1998).

In contrast to Sharp's (1979) tectonic model, other authors have suggested a multitude of differing histories. Some of the alternate models put forth to produce and then deform the mylonite zone include (1) ductile formation and brittle deformation as a result of extension (Erskine, 1984; Erskine and Wenk, 1985); (2) west-directed ductile thrusting followed by/or concomitant with ductile extensional collapse, and then brittle extension (George and Dokka, 1994); and (3) Cretaceous westerly ductile thrusting followed much later by Miocene extensional faulting (Axen, 1998).

Plate Tectonics Setting. English (1985) was the first to link the Santa Rosa Mylonite Zone to plate tectonics when he suggested that the mylonite zone formed as a result of shallowing of the Farallon plate, as it subducted beneath the American Plate. Since then other authors have fostered the idea that low-angle subduction/flattening of the subducting plate created the regional stresses that produced the eastern Peninsular Ranges Mylonite Zone and associated subsequent deformation (Todd and others, 1988; George and Dokka, 1994).

Rock Dynamics. The effect of shear stress is dependent upon lithology, temperature, pressure, and the rate of deformation. In general, rock will react to shear by breaking when the rate of deformation is high and/or temperature and confining pressure are low. With the inverse of those conditions (slow deformation and/or high temperature and pressure) rock will, in general, respond by plastic deformation. Between those two sets of conditions exists a zone where rock and the minerals within them transition from ductile to brittle deformation. In this zone, called the ductile-brittle transition zone, mineral properties dictate at what temperature and pressure each mineral will stop deforming plastically and begin to fracture when sheared. Within the ductile-brittle transition zone the foliation of the rock can be deformed plastically, while at the same time faults can form along major shear planes and individual crystals may fracture.

An important component of rock dynamics in regards to shear sense determination is the context in which the shear occurred. Mylonites are formed under intense pressure and temperatures and they produce recognizable structures that can be used to determine their sense of shear. But those structures only suggest shear sense with the sight window (i.e., scale) that is viewed. To overcome this issue, multiple measurements and samples need to be collected throughout the structure being studied to provide an overall assessment of the sense of shear.

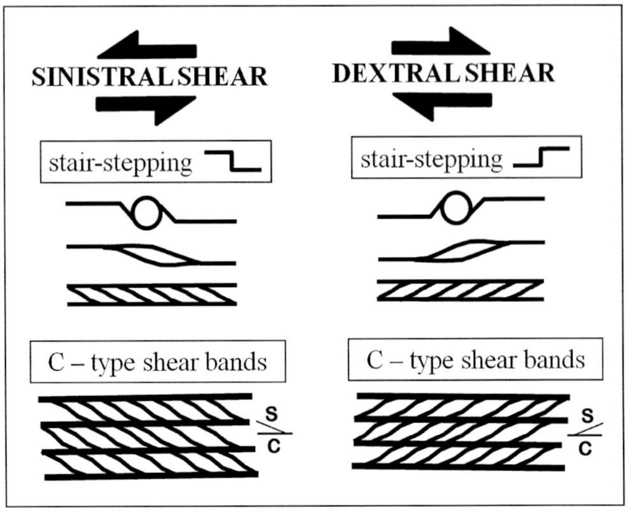

Figure 2. Common microtectonic mineral patterns/fabrics used to identify shear sense in mylonites (Passchier and Trouw, 1996). Shear patterns shown display both sinistral (top to left and bottom to right as denoted by the parallel arrows) and dextral shear sense (the opposite of sinistral shear).

SHEAR-SENSE INDICATORS

Shear stress is caused by applied forces that produce shear strain/deformation. This deformation can occur at mega, macro, and micro scales under brittle, ductile-brittle, or ductile conditions. Shear indicators are necessary to first identify that change has occurred and then to catalog the type, magnitude, and direction of that deformation.

A mylonite is a rock that has experienced ductile or crystal-plastic deformation that has resulted in grain-size reduction and conspicuous foliation and/or lineation. Shear-sense indicators used to distinguish movement in mylonitic rocks include aspects of fluid mechanics for that part of the rock more apt to flow in response to shear, and solid mechanics for the rock grains that tend to retain their shape.

Because mylonites are formed in a constrained high pressure and temperature environment, where materials behave plastically; the effect of shear is usually to compress, orient, and stretch out protolith rock fabric. One of the most obvious features seen in mylonites is a planar fabric, or foliation. One foliation is caused by the preferred orientation of mica and flattened grains of quartz, feldspar, and calcite and is called the S-foliation, from the French word for foliation, "schistosite" (Figure 2). This foliation makes a small angle to the shear zone boundary. The second kind of foliation is due to the parallel alignment of elongate grains that have been ductily stretched toward the extension axis of the finite strain ellipsoid, and is called the C-foliation, from the French word for shear, "cisaillement." C-foliation parallels the shear zone boundary. Shapes that can be used as shear-sense indicators include the presence of "tails" on rotated rock grains; the orientation and elongation of phyllosilicate minerals (e.g., mica-fish); and the "stair-stepping" of those type objects and patterns (Figure 3). Sense of shear in a stair-stepping structure is inferred to be in the direction of the upward step. Mica-fish are lensoid mica grains that have been elongated and can display stair-stepping asymmetry suggesting shear sense.

One of the most useful tools in determining the sense of shear in mylonites is the presence of 'tails' on the porphyroclasts (single crystals of a size exceeding the mean grain size in the surrounding matrix). The shapes of the tails resemble the Greek lower-case letters, delta (δ) or sigma (σ) (Figure 3). Because the porphyroclasts are relatively stronger than the crystals around them, in a simple shear flow regime, parts of the phenocrysts are brittlely- or ductily-detached and move in the direction of flow to form tails of recrystallized porphyroclast material or new minerals (Simpson, 1998). During the flow the porphyroclasts rotate and the relative rates of

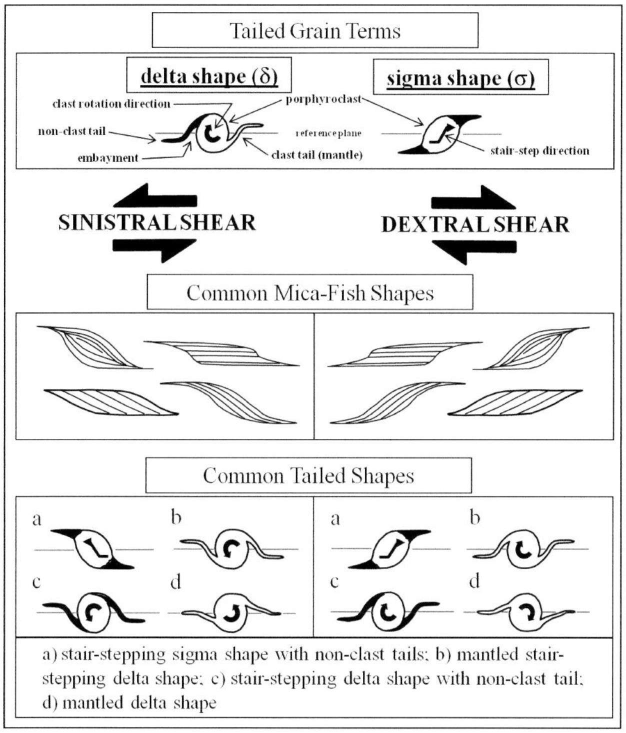

Figure 3. Common microtectonic terms and crystal shapes used to describe and identify shear sense in mylonites (Passchier and Trouw, 1996). The mica-fish and tailed shape shear forms in this figure display both sinistral and dextral shear sense as denoted by the parallel arrows.

rotation versus recrystallization and neocrystallizaion in the tails determine the shape of the tails. If the tail formation is fast compared to the rotation of the porphyrocryst, a σ-type tail will be produced. If, however, rotation of the large crystal is faster, the tail will coil around the crystal—forming a delta shape (Simpson, 1998; v.d. Pluijm and Marshak, 2004). The primary feature used to discern shear direction in δ shapes is the location of the shape's embayment(s) as shown in Figure 3. The embayment, the inlet between the porphyroclast and its tail, is formed by the dragging rotation of the porphyroclast as it spins extending/dragging the tail(s) as it spins. The spin direction is gleaned by discerning the direction the 'clast' had to spin to produce the embayment. Porphyroclast tails are either of the same mineral composition as the "clast" or, in the case of tails formed in pressure shadows, formed by another mineral or minerals (non-clast). If tails are the same mineral composition as the "clast" then the structure is referred to as a mantled porphyroclast.

INDICATORS OF A REVERSAL IN THE SHEAR SENSE

The study of mylonites is still in its infancy. There are many books and research papers describing mylonites and how they are formed, but the author has found little in the realm of what shear sense indicators might exist if the sense of shear that formed the mylonite were reversed. As a result, research attempting to assess whether shear-sense reversal has occurred is cutting-edge and must be based on interpretation of observed structures that may not be consistent with the accepted single sense shear structures normally found in the mylonites.

If shear reversal affected the original mylonite fabric, it could be described as overprinting, which is the term that is used to describe the reworking of existing rock structures. Metamorphism, re-metamorphism, or the initiation of a different shear regime is the primary mechanisms that produce overprinting. The degree of overprinting can range from not detectable to completely changing the rock such that the pre-existing structures are no longer present.

Indications the shear forces that formed a mylonite had reversed may include σ or δ type porphyroclaste shapes that have the opposite orientation than that of the main mylonite texture, the breaching of mineral flow lines, and other flow structures that are different than those that originally formed the mylonite. An important distinction to make is whether observed structures exist due to heterogeneities within the flow of the developing mylonite or whether they were produced by later stage overprinting due to shear reversal. Another important observation when assessing these structures is whether they are a localized phenomenon or if they exist throughout the mylonite.

SHEAR DIRECTION

The direction of shear in mylonites is usually called either dextral or sinistral to denote right-lateral or left-lateral shear, respectively (Figures 2 and 3). When attempting to determine whether mylonite fabric was produced by dextral or sinistral shear, a vertical rock section that is normal to the foliation and whose strike parallels the stretching lineation must be viewed. This perspective of the mylonite presents the sides of the rock that are parallel to the direction of shear, thereby, showing existing rock structures that may be used to determine shear direction. Because there are two possible sides of an outcrop, a cut rock slice, or a thin section that are parallel to stretching lineations and normal to the foliation, the mylonitic fabric must be viewed consistently from the same side when determining shear direction. A standardized view direction is required to ensure that collected shear data are reliable.

The standardized view direction selected for this study is in a southeasterly direction. The measured view directions for each sample are detailed in each figure. Because the dip of the eastern Peninsular Ranges Mylonite Zone foliation ranges from the east to northeast (Sharp, 1979), a dextral shear sense when looking to the southeast would correspond to a thrust or reverse fault at the surface, with the east side being thrust up and over the west. Whereas a sinistral shear sense would indicate normal faulting at the surface, and the east side would drop down relative to the west. Therefore, when discussing the shear sense suggested by the mylonite fabric, the term "reverse" will be substituted for "dextral," and "normal" will be used instead of "sinistral."

ROCK SAMPLE COLLECTION AND ANALYSIS

Thirty two rock samples were collected for this study. Those samples were collected along the main shear zone and from the hanging wall of the study area. Rock samples collected for laboratory analysis were selected from outcrops that had a pronounced foliation that was consistent with the surrounding area. Hydrothermal alteration and brittle deformation throughout the study area has reduced the availability of coherent rocks that are solid enough to be suitable for the production of thin sections. Sample collection was performed by measuring and labeling the sample in situ, documenting sample characteristics in field notes and photographs, and then breaking the rock sample from the outcrop.

Great care was taken in the processing of rock samples to ensure that proper orientation was maintained throughout hand-sample and thin-section production. Methods used included sequential sample names, labeling rock orientation in situ, labeling each rock slice on the northwestern-facing side of the slice, labeling thin-section slides, and notching the upper-right corner of each glass slide to distinguish the southwesterly pointing corner. All rock samples processed in the laboratory and shown in this document were cut parallel to stretching lineations and normal to the foliation and were labeled (hand sample) or mounted (thin section) to provide a southeasterly view direction.

Mylonite thin sections were examined for ductile and brittle deformation. Oriented thin sections were viewed through plane polarized light and through crossed polarizers. The photomicrographs displayed in this report were acquired through crossed polarizers with the gypsum plate inserted and the thin section inverted (i.e., microscope table/stage rotated 180 degrees) to counteract the optical inversion produced by the microscope.

FIELD AND LABORATORY RESULTS

Of the thirty two rock samples collected for laboratory examination, only the four samples from along the trace of the main shear zone and one hanging wall location (Sample E, Figure 1) displayed mylonite shear textures indicative of shear direction. Inexplicably, hanging wall rock, although well foliated did not have well enough developed C-type shear banding to permit shear direction interpretation. Prime examples of mylonite shear sense indicators are shown in hand sample Figures 4 through 6, thin section Figures 7 through 9, and outcrop photo Figure 10.

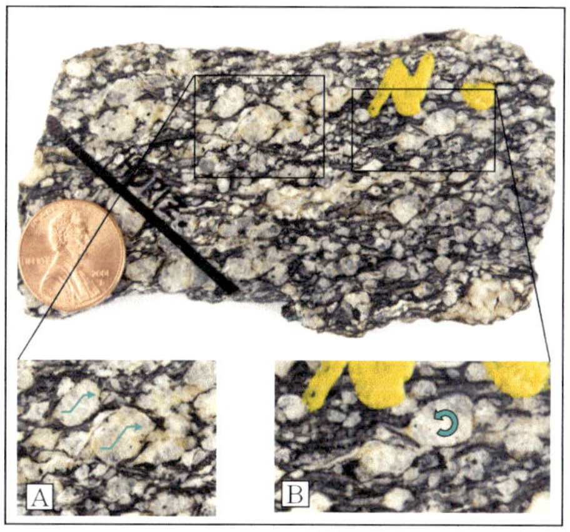

Figure 4 (left). Hand Sample N(2) displays a single, large, generally round, well-defined, mantled porphyroclast (~ 5 millimeter) in protomylonite fabric that consists of large subrounded to subangular mineral grains and biotite C-foliation. (A) Close-up of mantled porphyroclast (δ/σ-type) displaying reverse shear sense. View direction is to the south-southeast.

Figure 5 (middle). Hand Sample N(1) showing mylonite with weak C-type shear banding. (A) Two large (~ 4 and 5 millimeter) reverse stair-stepping grains with partial σ-type mantle tails displaying a reverse shear sense. (B) Porphyroclasts (~ 4 millimeter) with δ-type mantle tail indicating counterclockwise rotation/normal shear sense. View direction is to the south-southeast.

Figure 6 (right). Hand Sample V displaying mylonite to ultramylonite fabric. (A) Single δ-type porphyroclast with two opposite sense δ-type tails (~ 4 millimeter) displaying a reverse shear sense mantle tail on the right of the grain and normal shear sense biotite tail on the left side of the porphyroclast. (B) Two large (~ 3 and 4 millimeter) grains with poorly developed normal stair-stepping σ-type tails. View direction is to the southeast.

Structures interpreted as indicating reverse ductile shear in hand samples include a σ-type mantled porphyroclast (Figure 4A), two crystals with stair-stepping sigma-type tails (Figure 5A), and a δ-type porphyroclast with a single mantle tail (Figure 6A). Observed thin-section reverse motion indicators include σ-type shapes (Figures 7A, 7C, and 7D); δ-type shapes (Figures 7B and 7E); and stair-stepping mica fish (Figures 7C and 7F).

Hand samples with normal motion indicators include a δ-type single tailed mantled porphyroclast—although this crystal is located adjacent to another crystal with fairly well developed δ-type tails suggesting reverse shear (Figure 5B), a δ-type biotite tail on a single tailed mantled porphyroclast (Figure 6A), and stair-stepping σ-type tails on two mineral grains (Figure 6B). The least ambiguous indicators of normal motion seen in thin sections are two δ-type mica tails on porphyroclasts (Figures 8A and 8B). The other photomicrographs in Figure 8 are even more ambiguous motion indicators. These include a reversal of linear structures (Figure 8C), porphyroclast penetrating mica shear band (Figure 8D), crystal matrix reversed into σ-type shape (Figure 8E), and apparent mineral flow lines flowing left-laterally around a porphyroclast (Figure 8F). In summary, the indicators of a normal shear sense observed in this study were not as prevalent or as pronounced as structures/shapes that suggest a reverse sense of shear.

One thin section sample collected during this study documents simultaneous ductile and brittle deformation. A thin section produced from Sample F1 contains a plagioclase grain with a fracture, while the surrounding matrix is not broken (Figure 9). The lack of a fracture extending outward from the plagioclase grain requires that the surrounding matrix was ductile when the brittle clast fractured.

Evidence of ductile offset was observed in outcrop along the southern side of Martinez Canyon, where the canyon is perpendicular to the main ridgeline of the Santa Rosa Mountains and generally parallel to mylonite lineations (Figure 1). The observed ductile offset has displaced the rock foliation and an inclusion (Figure 10) approximately 54 centimeters and has produced small folds in the

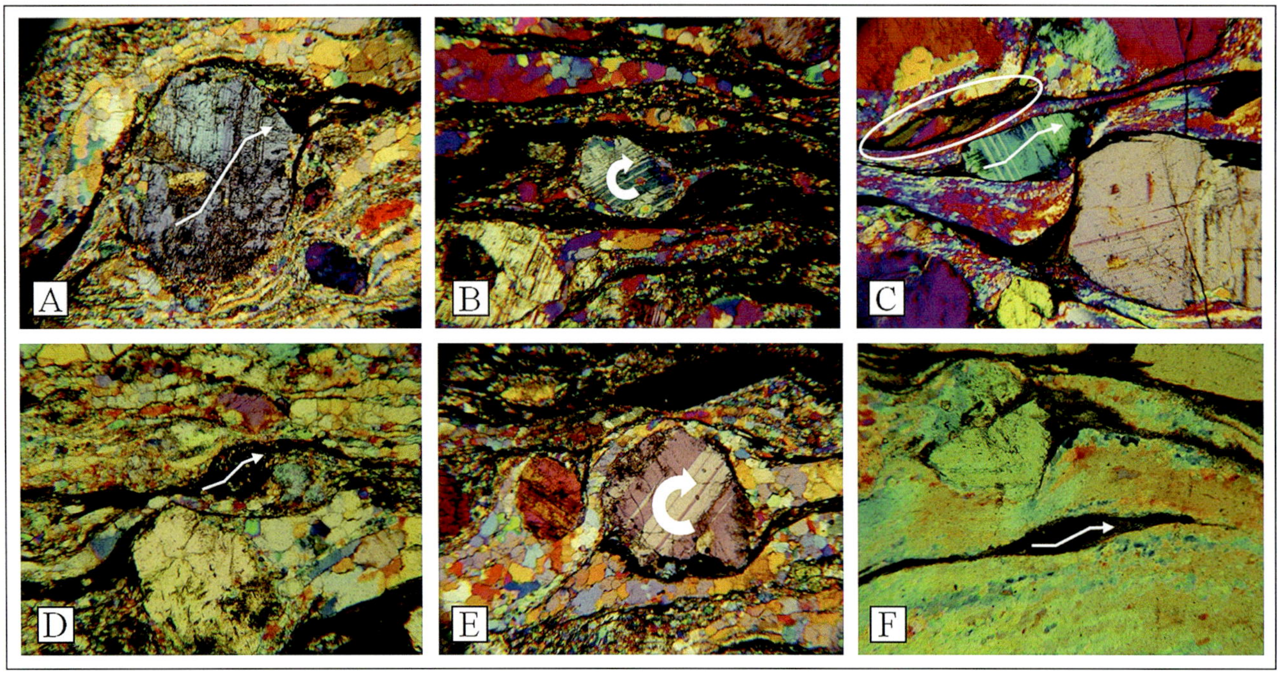

Figure 7. Thin-section micrographs displaying the predominant reverse ductile shear sense of the mylonite: (A) Stair-stepping σ-type shape with biotite tails. (B) Clockwise rotated δ-type shape with non-clast grain aggregate tail. (C) Stair-stepping mica fish (2 each in oval) and σ-type shape with biotite tails. (D) Stair-stepping σ-type shape with biotite tails. (E) Clockwise rotated δ-type shape with biotite tails. (F) Stair-stepping mica fish. View direction for micrographs (A), (C), (E), and (F) is to the southeast and to the south-southeast for (B) and (D). Micrographs (A) and (E), (B) and (D), and (C) and (F) are from rock Samples E1, F1, and E, respectively.

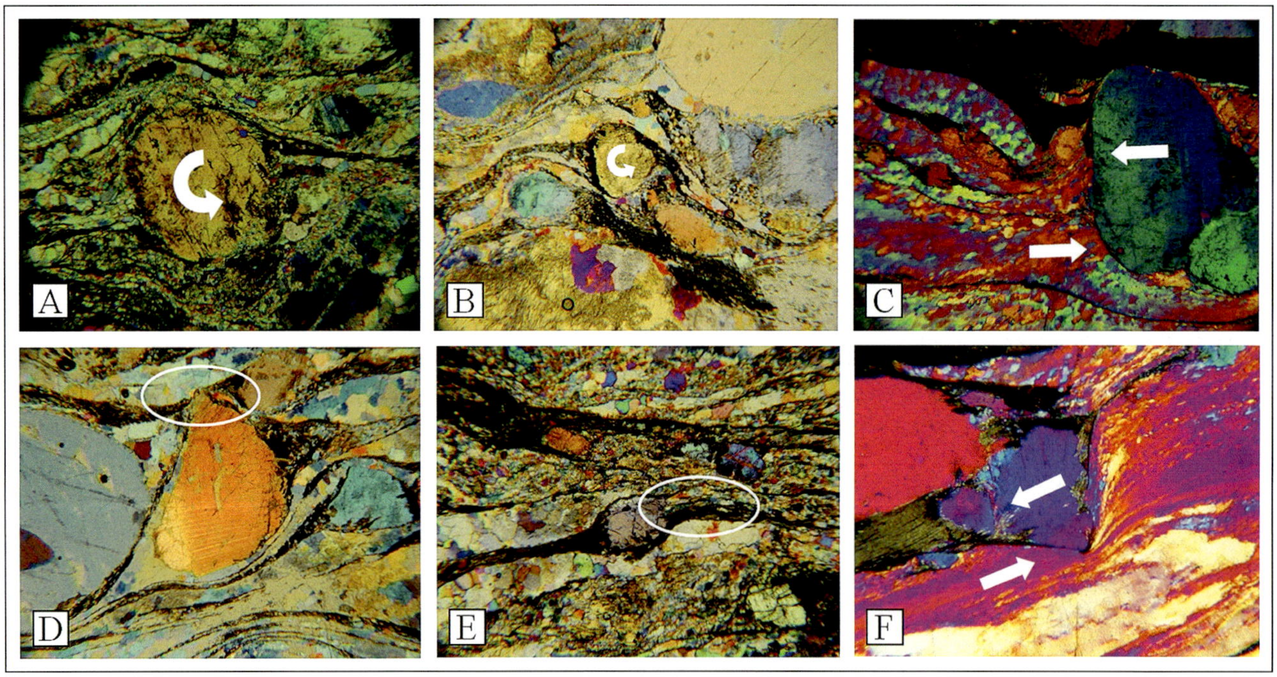

Figure 8. Thin-section micrographs displaying interpreted secondary ductile normal shear sense of the mylonite: (A) δ-type shape with biotite tails. (B) Deformed δ-type shape. (C) Biotite and other mineral flow bands which terminate and/or bend upward against the adjacent porphyroclast. (D) Porphyroclast that has penetrated/breached (oval) the mica flow line above the grain. (E) σ-type shape with biotite tails where the mineral flow band is between the clast and the upper stair-stepping biotite flow line (oval) which should be next to the grain. (F) Mineral flow lines wrapping around the lower side of porphyroclast suggesting the grain moved to the left. View direction for micrographs (A), (B), and (D) is to the south-southeast and to the southeast for (C), (E), and (F). Micrographs (A), (B) and (D), (C) and (F), and (E) are from rock Samples F1, N, E, and E1, respectively.

Figure 9. Mylonite thin-section micrograph displaying a fracture (i.e., cataclastic deformation) with sinistral offset of a plagioclase porphyroclast within ductily deformed matrix. Micrograph derived from Sample F1. View direction is to the south-southeast.

foliation. The offset is dextral at about 55 degrees to the granitic foliation and generally horizontal. The outcrop provides a three-dimensional view of the foliation revealing that the strike of the foliation is consistent above and below the offset. This is the only example of macro ductile shear found in the study area where the foliation has been crosscut. All other observed examples of macro ductile shear were parallel to the foliation.

DISCUSSION

The following provides discussion of observed mylonite structures and their suggested tectonic setting.

Ductile Shear Structures. Ductile deformation observed during this study consists of microscopic deformation of mineral grains, rock foliation, C-type shear banding of mylonitic rock, and a single example where rock foliation had been crosscut. The majority of the observed shear sense indicators point to a reverse east-over-west ductile shear as the mechanism that produced the eastern Peninsular Ranges Mylonite Zone (Figure 7). Some, possibly younger, shear sense indicators were also identified that suggest that normal ductile shear has also shaped the eastern Peninsular Ranges Mylonite Zone. Considering the limited number of study area samples that displayed mylonite shear textures indicative of shear direction, the number and variety of identified normal shear sense indicators found may be significant. However, more research is needed to further assess the nature and distribution of shear indicators to determine whether the identified normal shear is a localized phenomenon or

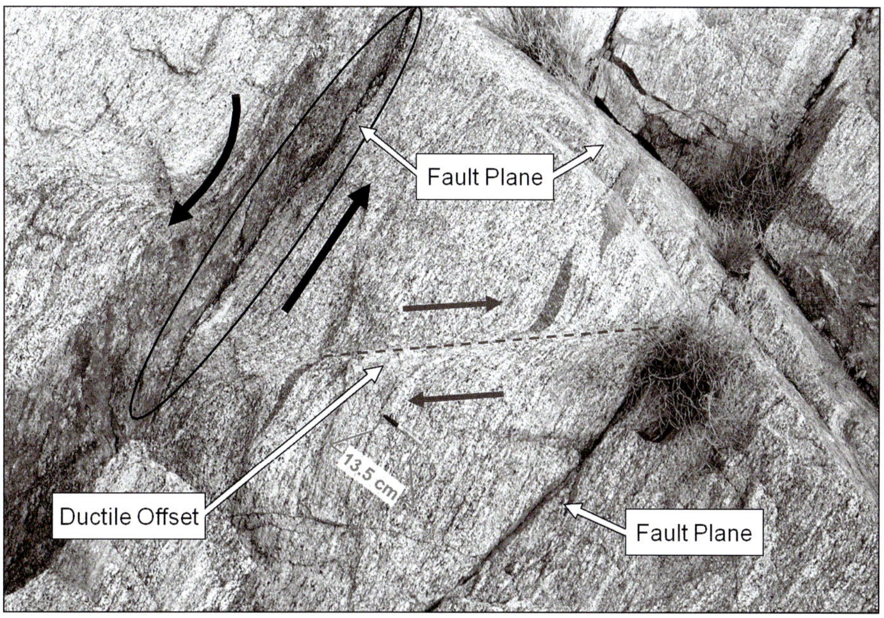

Figure 10. Outcrop photo showing east-over-west ductile offset within a fault block. Dashed line and arrows highlight the ductile shear plane and direction of offset that displaced the dark inclusion and crosscuts the original foliation. Large black arrows in the upper left bracket a fault with a possible normal sense of shear. Black-and-white pen provided for scale below the central-lower arrow tip is 13.5 centimeters long. View direction is to the southeast.

whether it represents the bulk shear sense of the eastern Peninsular Ranges Mylonite Zone.

Grain Size Enlargement Suggests a Reversal of the Shear Sense. The combined protomylonite texture and mantled porphyroclasts shown in Figures 4 and 5 suggests that grain size enlargement has altered some of the mylonite shear fabric of the eastern Peninsular Ranges Mylonite Zone. The presence of mantled porphyroclasts, which are mylonite to ultramylonite shear structures, in a protomylonite rock fabric indicates that the rock was overprinted by grain size enlargement after the mantled porphyroclasts were formed. The observed grain size enlargement within the mylonite of the eastern Peninsular Ranges Mylonite Zone is attributed to the reversal of the shear that formed the mylonite and not simply due to metamorphic recrystallization (Frost, E.G., 1998, San Diego State University, personal communication). It is postulated that a shear reversal would reduce the compressive shear stress that created the mylonite thereby permitting crystal growth, while at the same time still maintaining some of the C-foliation banding. If observed grain size enlargement was merely the result of metamorphic recrystallization under the existing temperature, pressure, and shear regime that formed the mylonite, then all of the rock throughout the mylonite zone should have been equally recrystallized/overprinted, which is not the case. If crystal growth occurred as a result of shear reversal, then grain size enlargement would been have focused in the parts of the mylonite where the greatest reduction of shear stress occurred, instead of equally throughout the mylonite zone.

Ductile-Brittle Deformation. Two structures found in the study area display the end result of ductile and brittle shear. The first, shown in Figure 9, shows a good example of simultaneous ductile and brittle deformation that occurred in the ductile-brittle part of the shear zone. The photomicrograph in Figure 9 and a similar structure found by Simpson (1984) proves that shear was still occurring when the mylonite entered the ductile-brittle zone. Although fractured porphyroclasts can be unreliable as shear sense indicators (Passchier and Trouw, 1996), given the large angle between the fracture and the foliation, many geologists would consider that the sinistral offset of the feldspar crystal suggests a local dextral or reverse shear.

The second structure found in the study area that displays both ductily sheared rock and brittle offset is the outcrop shown in Figure 10. That outcrop displays dextrally sheared foliation that is bounded by faults. Interestingly the foliation of the outcrop is ductily crosscut which is anomalous for this study area where ductile shear was only seen to exist parallel to the foliation. A good question is whether the current fault surfaces were originally ductile shear planes that transitioned to brittle behavior upon cooling, or if they formed post ductile shear. The shear plane that crosscuts the foliation was probably produced by reverse shear.

The fact that the shear surface seen in the outcrop shown in Figure 10 ductily crosscuts foliation shows that the shear surface was produced after the formation of the mylonite. To the extent that such crosscutting is unlikely under the regional shear stress that produced the large scale planar shear of the eastern Peninsular Ranges Mylonite Zone, it may have occurred when the mylonite transitioned to pressure-temperature conditions where fault block rotation could occur but still be within conditions that permitted a ductile crosscutting of the foliation.

Tectonic Setting. One of the more captivating aspects of the eastern Peninsular Ranges Mylonite Zone is the tectonic sequence that produced this regional crustal feature. As mentioned in the introduction, most authors have credited the formation of the mylonite zone to a large-scale convergent margin thrust sheet. Several authors have suggested that the production of the mylonite zone was the result of a change in the rate of subduction and a decreasing angle of subduction during the Late Cretaceous causing a regional compressional stress regime.

Within the magmatic arc portion of a subduction zone, deformation appears to be largely extensional in nature and results from slab retreat of the subducting plate (Molnar and Atwater, 1978; Dewey, 1980; Malinverno and Ryan, 1986; Royden, 1993; Collins, 2002). Regional compression does occur within a convergent margin within the fore-arc and accretionary wedge as well as in the foreland, such as in the cordilleran foreland fold and thrust belts. Compressional deformation in the accretionary wedge on one side of the magmatic arc and thrusting on the other produce the classic two-sided compression in arcs (e.g., Burchfiel and Davis, 1972, 1975; Armstrong, 1978, 1991). However, *within* the magmatic arc, *extensional* deformation appears to be the predominant structural style as shown in regions such as the Basin and Range Province (e.g., Davis, 1980; Eaton, 1980, 1982) and Colorado River Extensional Terrane (Davis, 1980; Davis and others, 1980); where mylonitic rocks that appear similar to those of the eastern Peninsular Ranges Mylonite Zone have formed during regional extensional deformation.

Determining if the eastern Peninsular Ranges Mylonite Zone is due to crustal extension or to crustal compression appears to be a significant problem, as both models have been suggested. Using the geometries and kinematics of the eastern Peninsular Ranges Mylonite Zone to infer the stress conditions at the time of formation of the mylonite has provided no consensus among geologists who have worked in the region.

But we can consider a few observations and guidelines that should be used to try to determine the stress system that produced the eastern Peninsular Ranges Mylonite Zone:

- The eastern Peninsular Ranges Mylonite Zone is basically a frozen stress field produced during Late Cretaceous time at nearly the same time as the intrusion of the major magmatic bodies (Sharp, 1979; Anderson, 1983).
- Regional metamorphism had already occurred to produce the metasedimentary country rock prior to intrusion and mylonitization of the plutons.
- Deformation within the magmatic arc portion of the convergent margin was probably produced by a regional extensional stress field; although the local stress field between the plutonic bodies and their country rocks may not have been extensional.
- Shear in a high temperature/pressure environment was necessary to produce the mylonite.
- Shear occurred during the later stages of the emplacement of the large-scale plutonic bodies.
- Rocks behaved plastically when the mylonite fabric was produced.
- The initial/primary shear that formed the mylonite appears to have been reverse--the hanging wall was thrust west over the footwall by ductile displacement.
- Subsequent reversal in the sense of shear could produce a reversal in the fabric as the hanging wall moved down-dip to the east.
- The mylonite zone transitioned upward into the ductile-brittle realm in concert with rapid uplift of the region.
- Eastern igneous hanging wall rock is younger than western igneous footwall rock.

An alternate hypothesis to explain the formation of the eastern Peninsular Ranges Mylonite Zone is that the mylonite formed as a byproduct of pluton emplacement. As such, the mylonite formed at the same time as intrusion of the plutonic bodies and was produced by the rising plutons shearing past each other and the surrounding country rocks. In this explanation of the mylonitic fabric, the asymmetry of the fabric would simply be recording the geometry and timing of which side moved upward relative to the other side of the mylonite zone. This model would not necessarily imply either compression or extension, just relative motion. Because the setting is extensional within the magmatic arc, batholithic bodies were able to be intruded into the "space" that they now occupy.

The increased shallowing of the subducting plate is believed to have produced the eastward migration of the Cretaceous magmatic arc (Dickinson, 1981; Gastil and others, 1981). This migration brought the magma plume into place to form the mylonite. The continued arc migration then shifted magma intrusion to the east, permitting extensional deformation to produce the limited shapes observed in this study that suggest a normal shear sense.

CONCLUSIONS

Thin-section analysis and associated geologic field mapping of the eastern central Santa Rosa Mountains has further documented that the eastern Peninsular Ranges Mylonite Zone formed as a result of reverse east-over-west ductile deformation. A limited number of crystals have forms that suggest that there may have been a late-stage stress field reversal while the rocks were still in the ductile state. Timing for this sequence of events includes: (1) The ductile formation of the eastern Peninsular Ranges Mylonite Zone during latest Cretaceous to early Paleocene (~74 to 62 million years ago); and (2) Initiation of a possible tectonic reversal at the end of that period. The tectonic reversal began as ductile extension and transitioned to faulting with the rapid cooling of the structure.

An alternate model for the formation of the primary mylonite is that it is due to shear between intruding plutons and shear between the magmatic bodies and the existing country rock within the magmatic arc portion of the convergent margin. Resultant plastic deformation produced the eastern Peninsular Ranges Mylonite Zone and documents late-stage intrusion into the eastern hanging wall rock.

ACKNOWLEDGEMENTS

This paper is a component of my San Diego State University Master's thesis and would not have been possible without the significant and much appreciated effort and support provided by my thesis advisor, Dr. Eric Frost, and thesis committee members, Drs. Dave Kimbrough and Bob Pozos. The review of this new manuscript by Dr. Monte Marshall is also greatly appreciated.

REFERENCES

Anderson, J.R., 1983, Petrology of a portion of the Eastern Peninsular Ranges Mylonite Zone, Southern California: Contributions to Mineralogy and Petrology, v. 84, n. 2/3, p. 253–271.

Armstrong, R.L., 1978, Cenozoic igneous history of the U.S. Cordillera from 42° to 49° N latitude, Geological Society of America Memoir, v. 152, p. 265–282.

Armstrong, R.L. and Ward, P., 1991, Evolving geographic patterns of Cenozoic magmatism in the North American Cordillera: The temporal and spatial association of magmatism and metamorphic core complexes: Journal of Geophysical Research, v. 96, p. 13201–13224.

Armstrong R.L. and Suppe, J., 1973, Potassium-argon geochronometry of Mesozoic igneous rocks in Nevada, Utah and southern California: Geological Society of America Bulletin, v. 100, p. 1375–1392.

Atwater, T., 1970, Implications of plate tectonics for the Cenozoic tectonic evolution of western North America: Geological Society of America Bulletin, v. 81, p. 3513–3536.

Atwater, T., 1989, Plate tectonic history of the northeast Pacific and western North America, in Winterer, E.L., Hussong, D.M., and Decker, R.W., eds., The Geology of North America, v. N, The eastern Pacific Ocean and Hawaii: Boulder, Colorado, Geological Society of America.

Axen, G.J. and Fletcher, J.M., 1998, Late Miocene-Pleistocene extensional faulting, northern Gulf of California, Mexico and Salton Trough, California: International Geology Review, v. 40, p. 217–244.

Bohannon, R.G. and Parsons, T., 1995, Tectonic implications of post-30 Ma Pacific and North American relative plate motions: GSA Bulletin, v. 107, n. 8, p. 937–959.

Burchfiel, B.C. and Davis, G.A., 1972, Structural framework and evolution of the southern part of the Cordillera orogen, western United States: American Journal of Science, v. 272, p. 97–118.

Burchfiel, B.C. and Davis, G.A., 1975, Nature and controls of Cordillera orogenesis, western United States, Extensions of an earlier synthesis: American Journal of Science, v. 275-A, p. 363–396.

Collins, W.J., 2002, Hot orogens, tectonic switching, and creation of continental crust: Geology, v. 30, n. 6, p. 535–538.

Crowell, J.C., 1981, An outline of the tectonic history of southeastern California, in Ernst, W.G., editor: The geotectonic development of California (Rubey Vol. I), Prentice-Hall, Inc., Englewood Cliffs, New Jersey, p. 583–613.

Curray, J.R., Moore, D.G., Kelts, K., and Einsele, G., 1982, Tectonics and geological history of the passive continental margin at the tip of the Baja California, in Initial reports of the deep sea drilling project: Washington , D.C., U.S. Government Printing Office, v. 64, p. 1089–1116.

Davis, G.A., 1980, Problems of intraplate extensional tectonics, western United States, Studies in Geophysics: The National Research Council, National Academy of Sciences, Washington, D.C., p. 84–95.

Davis, G.A., Anderson, J.L., Frost, E.G., and Shackelford, T.J., 1980, Mylonitization and detachment faulting in the Wipple-Buckskin-Rawhide Mountains terrane, southeastern California and western Arizona, in Crittenden, M.D., Coney, P.J., and Davis, G.H., editors, Cordilleran metamorphic core complexes: Geological Society of America Memoir 153, p. 79–130.

Dewey, J.F., 1980, Episodicity, sequence and style at convergent plate boundaries, in Strangway, D.W., editor, The continental crust and its mineral deposits: Geological Association of Canada Special Paper, v. 20, p. 553–573.

Dickinson, W.R., 1981, Plate tectonics and the continental margin of California, in Ernst, W.G., editor, The geotectonic development of California (Rubey Vol. I): Prentice-Hall, Inc., Englewood Cliffs, New Jersey, p. 2–28.

Dickinson, W.R., 1997, Tectonic implications of Cenozoic volcanism in coastal California: GSA Bulletin, v. 109, n. 8, p. 936–954.

Dokka, R.K. and Frost, E.G., 1978, Fission-track ages from the Santa Rosa Mylonite and its protolith and their relation to the cooling history of the Southern California Batholith: Geological Society of America Abstracts with Programs, v. 10, n. 3, p. 103.

Dokka, R.K., 1984, Fission-track geochronologic evidence for Late Cretaceous mylonitization and early Paleocene uplift of the northeastern Peninsular Ranges, California: Geophysical Research Letters, v. 11, n. 1, p. 46–49.

Eaton, G.P., 1980, Continental Tectonics, Studies in Geophysics: The National Research Council, National Academy of Sciences, Washington, D.C., p. 96–113.

Eaton, G.P., 1982, The Basin and Range Province: Origin and tectonic significance: Annual Review of Earth and Planetary Sciences, v. 10, p. 409–440.

Engel, A.E.J. and Schultejann, P.A., 1984, Late Mesozoic and Cenozoic tectonic history of south central California: Tectonics, v. 3, n. 6, p. 659–675.

English, D.J. 1985, Regional structural analysis of the Santa Rosa Mountains San Diego and Riverside Counties, California: Implications for the geologic history of southern California: Master's Thesis, San Diego State University, Spring.

Erskine, B.G., 1984, Coaxial deformation in the Santa Rosa mylonite zone; evidence for Late Cretaceous crustal thinning: Geological Society of America Abstracts with Programs, v. 16, n. 6, p. 503.

Erskine, B.G., 1985, Constraints on low-angle faulting kinematics in the eastern Peninsular Ranges Mylonite Zone, and implications on Southern California Tectonics: Geological Society of America Abstracts with Programs, v. 17, n. 7, p. 575.

Erskine, B.G. and Wenk, H.R., 1985, Evidence for Late Cretaceous crustal thinning in the Santa Rosa mylonite zone, southern California: Geology, v. 13, p. 274–277.

Frost, E.G., Fattahipour, M.J., and Robinson, K.L., 1996a, Neogene detachment and strike-slip faulting in the Salton Trough region and their geometric and genetic interrelationships, in Abbott, P.L., and Cooper, J.D., editor, Field conference guide: Pacific Section A.A.P.G, GB 73, Pacific Section S.E.P.M., Book 80, p. 263–276.

Frost, E.G., Suitt, S.C., and Fattahipour, M.J., 1996b, Emerging perspectives of the Salton Trough region with an emphasis on extensional faulting and its implications for later San Andreas deformation, in Abbot, P.L., and Seymour, D.C., editors: Sturzstroms and detachment faults, Anza-Borrego Desert State Park, California: South Geological Society, Santa Ana, California, p. 81–121.

Gans, P.B., 1997, Large-magnitude Oligo-Miocene extension in the southern Sonora: Implications for the tectonic evolution of northwest Mexico: Tectonics, v. 16, n. 3, p. 388–408.

Gastil, R.G., Morgan, G.J., and Krummenacher, D., 1978, Mesozoic history of peninsular California and related areas east of the Gulf of California, in Howell, D.G. and McDougall, K.A., editors, Mesozoic paleography of the western United States: Society of Economic Paleontologists and Mineralogists, Pacific Section, Pacific Coast Paleogeography Symposium 2, p. 107–116.

Gastil, R.G., Morgan, G.J., and Krummenacher, D., 1981, The tectonic history of peninsular California and adjacent Mexico, in Ernst, W.G., editor, The geotectonic development of California (Rubey Vol. I): Prentice-Hall, Inc., Englewood Cliffs, New Jersey, p. 284–306.

Gastil, G., Miller, R., Anderson, P., Crocker, J., Campbell, M., Buch, P., Lothringer, C., Leier-Engelhardt, P., DeLattre, M., Hobbs, J., Roldan-Quintana, J., 1991, The relation between Paleozoic strata on opposite sides of the Gulf of California, in Perez-Segura, E., and Jacques-Ayala, C., editors, Studies of Sonoran geology: Geological Society of America Special Paper 254.

George, P.G. and Dokka, R.K., 1990, Late Cretaceous uplift of the San Jacinto Mountains, California: Implications for regional thrust faulting in the Peninsular Ranges and tectonics of the Laramide Orogeny: Geological Society of

America, Abstracts with Programs, Session 135, Tectonics.

George, P.G. and Dokka, R.K., 1994, Major Late Cretaceous cooling events in the eastern Peninsular Ranges, California, and their implications for Cordilleran tectonics: Geological Society of America Bulletin, v. 106, p. 903–914.

Goodwin, L.B., and Renne, P.R., 1991, Effects of progressive mylonitization on Ar retention in biotites from the Santa Rosa mylonite zone, California, and thermochronologic implication: Contributions to Mineralogy and Petrology, v. 108, n. 3, p. 283–297.

Goodwin, L.B. and Wenk, H.-R., 1995, Development of phyllonite from granodiorite; Mechanisms of grain size reduction in the Santa Rosa mylonite zone, California: Journal of Structural Geology, v. 17, n. 5, p. 689–707.

Grove, M. and Harrison, T.M., 1992, Thermal evolution of the eastern Peninsular Ranges Batholith, California: Geological Society of America, Abstracts with Programs.

Grove, M., Lovera, O., and Harrison, M., 2003, Late Cretaceous cooling of the east-central Peninsular Ranges Batholith (33°N): Relationship to La Posta pluton emplacement, Laramide shallow subduction, and forearc sedimentation, in Johnson, S.E., Paterson, S.R., Fletcher, J.M., Girty, G.H., Kimbrough, D.L., and Martin-Barajas, A., editors, Tectonic evolution of northwestern Mexico and the southwestern USA: Geological Society of America Special Paper 374, Boulder, Colorado, p. 355–379.

Hill, R.I. and Silver, L.T., 1979, Strontium isotopic variability in the pluton of San Jacinto Peak, southern California: EOS Trans American Geophysical Union, v. 61, p. 411.

Hill, R.I., 1981, Field petrologic and isotopic studies of the intrusive complex of San Jacinto Mountains, in Brown, A.R. and Ruff, R.W., editors, Geology of the San Jacinto Mountains: South Coast Geological Society, Santa Ana, California, Annual Field Trip Guidebook, n. 9, p. 76–89.

Jennings, C.W., 1994, Fault activity map of California and adjacent areas with locations and ages of recent volcanic eruptions, scale 1:750,000, No. 6: Department of Conservation, Divisions of Mines and Geology, Sacramento, California.

Larson, R.L., Menard, H.W., and Smith, S.M., 1968, Gulf of California: A result of ocean-floor spreading and transform faulting: Science, v. 161, p. 781–784.

Lockwood, J.P., 1961, Geology of an area in the western Santa Rosa Mountains of southern California, with special emphasis on a previously unreported mylonite belt: Bachelor of Science thesis, University of California, Riverside, 75 pp.

Lonsdale, P., 1989, Geology and tectonic history of the Gulf of California, The Geology of North America, Vol. N, The eastern Pacific Ocean and Hawaii: Geological Society of America.

Lonsdale, P., 1991, Structural patterns of the Pacific floor offshore of Peninsular California, in the Gulf and Peninsular Province of the Californias, Dauphin, J.P., Simoneit, B.R.T., editors: AAPG Memoir 47.

Malinverno, A. and Ryan, W.B.F., 1986, Extension in the Tyrrhenian Sea and shortening in the Apennines as a result of arc migration driven by sinking of the lithosphere: Tectonics, v. 5, p. 227–245.

Mammerickx, J. and Klitgord, K.D., 1982, Northern East Pacific Rise; Evolution from 25 m.y. to the present: Journal of Geophysical Research, v. 87, p. 6751–6759.

Molnar, P. and Atwater, T., 1978, Interarc-spreading and Cordilleran tectonics as alternatives related to the age of the subducted oceanic lithosphere: Earth Planet, Science Letter, v. 41, p. 330–340.

Moore, D.G. and Curray, J.R., 1982, Geologic and tectonic history of the Gulf of California, in Initial reports of the deep sea drilling project: U.S. Government Printing Office, Washington, D.C., v. 64, p. 1279–1294.

Mora-Klepeis, G. and McDowell, F.W., 2000, Timing of the transition from subduction to rifting: West central Sonora, Mexico, internet: Education and planning workshop, rupturing of the continental lithosphere in the Gulf of California and the Salton Trough Puerto Vallarta, Abstract Volume, October.

O'Brien, D.K., Wenk, H.-R., Ratschbacher, L., You, Z., 1987, Preferred orientation of phyllosilicates in phyllonites and ultramylonites: Journal of Structural Geology, v. 9, n. 5/6, p. 719–730.

Passchier, C.W., and Trouw, R.A.J., 1996, Microtectonics, Springer-Verlag, Heidelberg, Germany, pp. 266.

Pridmore, C.L. and Frost, E.G., 1992, Detachment faults: California's extended past: California Geology, v. 45, p. 3–17.

Rangin, C., 1978, Speculative model of Mesozoic geodynamics, central Baja California to northeastern Sonora, Mexico, in Howell, D.G. and McDougall, K.A., editors, Mesozoic Paleogeography of the western United States: Society of Economic Paleontologists and Mineralogists, Pacific Section, Pacific Coast Paleogeography Symposium 2, p. 85–106.

Royden, L.H., 1993, Evolution of retreating subduction boundaries formed during continental collisions: Tectonics, v. 12, p. 629–638.

Sharp, R.V., 1966, Ancient mylonite zone and fault displacements in the Peninsular Ranges of southern California: Geological Society of America Special Paper 101, p. 333.

Sharp, R.V., 1967, San Jacinto fault zone in the Peninsular

Ranges of southern California: Geological Society of America Bulletin, v. 78, p. 705–730.

Sharp, R.V., 1968, The San Andreas fault system and contrasting pre-San Andreas structures in the Peninsular Ranges of southern California, in Dickinson, W.R., and Grantz, A., editors, Conference on geologic problems of the San Andreas fault system: Stanford University Geological Science Publications v. 11, p., 292–293.

Sharp, R.V., 1979, Some characteristics of the eastern Peninsular Ranges Mylonite Zone, in Conference Proceedings VIII: Analysis of actual faults in bedrock: U.S. Geological Survey Open-File Report 79-1239, p. 258–267.

Silver, L.T. and Hill, R.I., 1986, U-Th-Pb microdiscordance in young zircons; a case study in the San Jacinto Mountains, California: Eos (American Geophysical Union Transactions), v. 67, p. 339–340.

Silver, L.T., Taylor, H.P., and Chappell, B., 1979, Some petrological, geochemical and geochronological observations of the Peninsular Ranges Batholith near the international border of the USA and Mexico, in Abbott, P.L. and Todd, V.R., editors, Mesozoic crystalline rocks: San Diego State University, Department of Geological Sciences, Guidebook for Geological Society of America annual meeting, San Diego, California, p. 83–110.

Simpson, C., 1984, Borrego Springs-Santa Rosa mylonite zone: A Late Cretaceous west-directed thrust in southern California: Geology, v. 12, p. 8–11.

Simpson, C., 1998, Rotated porphyroclast systems in quartzo-feldspathic mylonites, in Fault-related Rocks, A Photographic Atlas, Snoke, A. W., J. Tullis, and V. R. Todd, Eds., Princeton University Press, Princeton, NJ.

Spencer, J.E. and Normark, W.R., 1989, Neogene plate-tectonic evolution of the Baja California Sur continental margin and the southern Gulf of California, Mexico, The Geology of North America, Vol. N, The Eastern Pacific Ocean and Hawaii: Geological Society of America.

Stock, J.M. and Hodges, K.V., 1989, Pre-Pliocene extension around the Gulf of California and the transfer of Baja California to the Pacific Plate: Tectonics, v. 8, n. 1, p. 99–115.

Theodore, T.G., 1967, Structure and petrology of the gneisses and mylonites at Coyote Mountain, Borrego Springs, California: Doctor of Philosophy thesis, University of California, Los Angeles, 268 pp.

Todd, V.R., Erskine, B.G., and Morton, D.M., 1988, Metamorphic and tectonic evolution of the northern Peninsular Ranges Batholith, Southern California, in Ernst, W.G., editors, Metamorphism and crustal evolution of the western United States (Rubey Vol. VII): Englewood Cliffs, New Jersey, Prentice-Hall, p. 894–937.

Todd, V.R., Kimbrough, D.L., and Herzig, C.T., 1994, The Peninsular Ranges Batholith from western volcanic arc to eastern mid-crustal intrusive metamorphic rocks, San Diego County, California: Geological Society of America Cordilleran Section Guidebook, Trip 9.

v. d. Pluijm, B.A. and S. Marshak, 2004, Earth Structure, Second Edition, W.W. Norton, New York, NY.

Wallace, R.D. and English, D.J., 1982, Evaluation of possible detachment faulting west of the San Andreas, southern Santa Rosa Mountains, California, in Frost, E.G. and Martin D.L., editors, Mesozoic-Cenozoic tectonic evolution of the Colorado River Region, California, Arizona, and Nevada: Cordilleran Publishers, San Diego, California, p. 502–510.

Wallace, R.D., 1982, Evaluation of possible detachment faulting west of the San Andreas Fault, southern Santa Rosa Mountains, California: Master's Thesis, San Diego State University, Fall.

The Fluid Evolution of the Eastern Peninsular Ranges Mylonite Zone, Southern California

Gregory J. Holk Department of Geological Sciences
California State University,
Long Beach, CA, 90840

ABSTRACT

Stable isotope studies (Holk et al., 2006) demonstrated infiltration of subduction-related metamorphic water at depths of at least 15-20 km during early Tertiary thrust faulting of a tonalite pluton west of Borrego Springs, CA. Major and trace element data indicate metasomatism and alteration as these deep fluids ascended through the EPRMZ. Relative to undeformed tonalite, protomylonites are enriched in SiO_2, Na_2O, Th, Nb, and Y and depleted in K_2O and most trace elements; whereas transition zone tonalites are depleted in K_2O, MgO, total Fe, and most trace elements and enriched in CaO, Sr, Zr, and Hf. Elements unaffected by deformation include Ba, Be, Sc, Sr, Ga, Hf, and Zr in the protomylonite, and Ba, Ga, and Ag in the transition zone. The increase in Na_2O and decrease in K_2O is likely the result of albitization of plagioclase as Na-rich subduction fluids leached Ca and K from feldspar, with the Ca taken up by epidote. Transition zone tonalites experienced a greater degree of trace element depletion than the protomylonites, suggesting this zone as the locus of hydrothermal fluid circulation. The transition zone is intruded by 0.5-to-10-m-thick pegmatite dikes and cut by 1-10 cm-thick mylonite-to-ultramylonite zones. Pegmatites (0.5-to-10-m-thick) may have served as an effective fluid transport medium with high-strain portions of shear zones acting as transport avenues.

INTRODUCTION

Faults and deep shear zones serve as zones of enhanced permeability that accommodate the passage of large amounts of aqueous fluids through the Earth's crust in response to cycles of microcracking and dilation (Kirschner et al., 1999; Sibson, 2001; Holk and Taylor, 2007). Deep-source metamorphic and/or magmatic fluids appear to be dominant within reverse and thrust faults (McCaig et al., 1991; Goldfarb et al., 1995), whereas meteoric-hydrothermal fluids are the dominant fluid type in extensional zones where normal and detachment faulting are prevalent (Magaritz and Taylor, 1986; Nesbitt and Muehlenbachs, 1995). We evaluate the changes in major and trace element concentrations that occurred during hydrothermal metamorphism involving seawater-sourced fluids (Holk et al., 2006) during deformation within the late Cretaceous-early Tertiary Eastern Peninsular Ranges Mylonite Zone (EPRMZ).

GEOLOGIC SETTING

The EPRMZ is a 10-km wide, east-dipping, compressional shear zone (Sharp, 1979) with a lower-plate made up of locally sheared, cataclastic, and mylonitic mid-Cretaceous plutonic rocks (Anderson, 1983) and an upper-plate of heavily sheared upper-amphibolite facies Palm Canyon Group metasediments (Todd et al., 1988). This shear zone extends from Palm Springs southward to Borrego Springs (Fig. 1) and is split into three segments by displacements along the San Jacinto and Coyote Creek faults. Displacements of at least 10 km (Sharp, 1979) along the EPRMZ are constrained to have occurred between 95 and 63 Ma (Goodwin and Renne, 1991; L.T. Silver, personal communication).

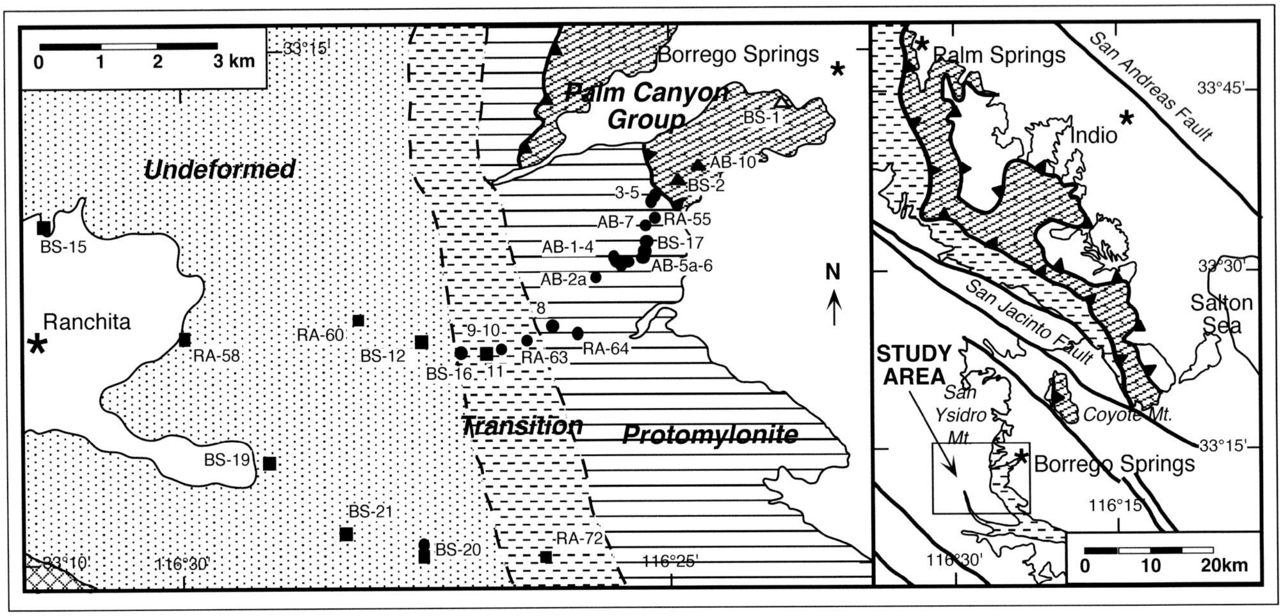

Figure 1. Simplified maps of the entire EPRMZ (modified from Todd et al, 1988, right) and the Borrego Springs segment of the EPRMZ (modified from Anderson, 1983, left), showing sample locations. The tilted brick pattern indicates the location of the metasedimentary rocks that are part of the upper plate of the EPRMZ, and the horizontal dashed line pattern shows those portions of the lower-plate tonalite plutons that are within the EPRMZ. Zones of undeformed tonalite, a transition zone, and a zone of protomylonite delineate the various degrees of deformation within the 95 Ma Ranchita pluton (Anderson, 1983). Sample types are marked as follows: undeformed tonalite (solid squares), macroscopically deformed tonalite (solid circles), metasediments (closed triangles), and leucocratic migmatite (open triangle). Heavy dark hatchured lines represent the traces of major east-dipping thrust faults associated with the EPRMZ that separate tonalites of the PRB from the upper-plate Palm Canyon Group.

Our study area is focused on the Ranchita tonalite, a 95 Ma (Silver, personal communication) pluton deformed by the EPRMZ west of Borrego Springs (Fig. 1). This pluton cooled from 450°C to less than 200°C between 75 and 70 Ma (Grove et al., 2003a). This constrains thrust deformation along the EPRMZ between 63 and 75 Ma, a time that corresponds to the lowering of subduction angle (e.g., Gastil et al., 1992). Early Tertiary thrust faulting (e.g., Engle and Schultejann, 1984) provided the final emplacement of the Palm Canyon Group onto the Ranchita Pluton (Fig. 1). Pliocene detachment faults (e.g., Stinson and Gastil, 1996) offset the EPRMZ south of the study area.

Anderson (1983) defined four deformational zones within the Ranchita pluton (Fig. 1). Zone A is undeformed tonalite (Fig. 2a) with mafic inclusions that display very slight elongation. In thin section (Fig. 3a), quartz shows only minor undulatory extinction, and dark-brown biotite and plagioclase appear unaltered. Zone B tonalite is undeformed in outcrop, but thin sections reveal strained quartz, kinked biotite, and fractured plagioclase. Collectively, Zone A and zone B tonalites comprise undeformed tonalite (Fig. 1). Transition zone tonalite (Zone C) exhibits unevenly developed mylonitic foliation (Fig. 2b), numerous fault surfaces, shredded and recrystallized greenish biotite (Fig. 3b), and undulating highly strained planar deformation zones of quartz (e.g. ribbon quartz and localized, elongate pressure-solution features, Fig. 3c), and slightly deformed feldspar. Pervasive mylonitic foliation defines the protomylonite (Zone D). Mineralogical changes include recrystallized and shredded green biotite porphyroblasts, uniform pressure-solution recrystallization of quartz, alteration epidote and sphene, and bent, fractured feldspars (Fig. 3d). Smeared E-W trending biotite porphyroblasts define lineation and foliation dips 30° NE (Fig. 2c).

Figure 2. Outcrop photos of the EPRMZ. (A) Undeformed tonalite at RA-58. Small mafic clots of hornblende and biotite are common in this felsic rock. (B) Bands of ultramylonite and pseudotachylite that cut foliation at BS-16. (C) Close-up view of biotite porphyroblasts, mylonitic foliation and lineations that trend E-W and dip 30° E at BS-17. (D) View looking north from the Montezuma Highway in the transition zone. White areas are 5–10-m-thick pegmatite dikes that cut variably deformed tonalite.

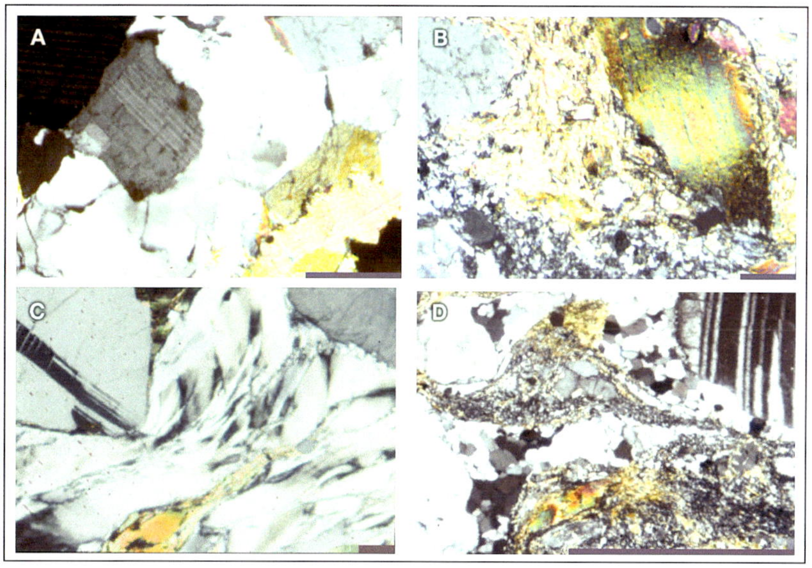

Figure 3. Photomicrographs of alteration and mineral textures formed during deformation at the EPRMZ. (A) Undeformed and unaltered tonalite from BS-15 showing unaltered biotite and plagioclase, and unstrained quartz. (B) Sheared, recrystallized, in an annealed quartz matrix and shredded biotite porphyroblast from RA-64. (C) High-strain ribbon quartz and sheared biotite that is characteristic of the transition zone. Feldspar is unaffected by deformation. (D) Protomylonite with recrystallized quartz matrix, bent and fractured feldspar, fine-grained epidote, and fine-grained shredded biotite.

Zones of extreme high strain characterized by thin layers (0.5-1.0 cm) of ultramylonite and pseudotachylite are common near the boundary with the undeformed tonalite. Numerous 1-10 meter-thick pegmatite dikes (Fig. 2d) that display variable degrees of transposition and attenuation cut the EPRMZ within a belt that extends from the eastern part of undeformed zone and extends through the entire deformed zone all the way to the fault contact with metasediments.

SUMMARY OF STABLE ISOTOPE RESULTS AND FLUID SOURCES

Stable isotope studies (Fig. 4) confirm large-scale infiltration of metamorphic water at depths of at least 15 km during the late Cretaceous or earliest Tertiary into the EPRMZ (Holk et al., 2006). Whole-rock $\delta^{18}O$ values in undeformed tonalite are +11.2±0.4 and in mylonitized tonalite are +12.3±0.3, indicating a 1‰ increase in $\delta^{18}O$ values during open-system water-rock exchange during mylonitization. Quartz, feldspar, and biotite $\delta^{18}O$ values in undeformed tonalite are +12.8±0.4, +11.0±0.7, and +7.7±0.7, compared with +13.4±0.2, +11.7±0.3, and +9.2±0.3 in mylonitized tonalite. The extremely uniform quartz-feldspar and quartz-biotite ^{18}O-fractionations in the mylonitized tonalite indicate equilibration at 500–550°C, using the calibrations of Clayton et al. (1972) and Clayton and Kieffer (1991). These temperatures are consistent with Anderson's (1983) estimate for water-rock interaction in the EPRMZ. Using the equilibrium plagioclase-water $^{18}O/^{16}O$ fractionation factor of O'Neil and Taylor (1967), a composition of An_{30} for plagioclase, and the above temperature, water in equilibrium with the mylonitic tonalites would have had $\delta^{18}O = +11.1$ to $+12.2$ (Fig. 4). This is consistent with H_2O derived from a deep crustal high-^{18}O source.

Biotite and hornblende δD values from undeformed tonalite are −73±4, compared with −52 to −68 in mylonitic tonalite. Constant biotite and hornblende Fe/Mg (≈ 1.25; Anderson, 1983) in both kinds of tonalite and the application of the Suzouki and Epstein (1968) equilibrium hydrogen isotope biotite-H_2O fractionation factor allow for the estimation of water δD values. The calculated δD of waters (−40 to −60) that would have been in equilibrium with undeformed tonalite are within the range of primary magmatic waters (e.g. Sheppard, 1986). The high-^{18}O waters that infiltrated the EPRMZ had high δD values (−10 to −30). The hydrogen isotope composition of these syndeformation waters match those from the Franciscan Formation (Magaritz and Taylor, 1976) and the Catalina Schist (Bebout, 1991) and were likely derived by dehydration of subducted oceanic crust or underplated volcanogenic sediments.

Syndeformational tonalitic pegmatites have quartz and feldspar $\delta^{18}O$ values of +13.5±0.2 and +12.2±0.3, and biotite and muscovite δD values of −64±1 and −42, suggesting a genetic relationship between pegmatites and these deep-seated fluids. Higher whole-rock $\delta^{18}O$ values (+13.0 to +16.1) and much lower δD values (−98 to −114) occur where the EPRMZ cuts upper-plate amphibolite-facies Palm Canyon Group metasediments. The shift to lower δD values in these finer-grained rocks probably occurred during interactions with heated meteoric groundwaters during Pliocene detachment faulting associated with the opening of the Gulf of California.

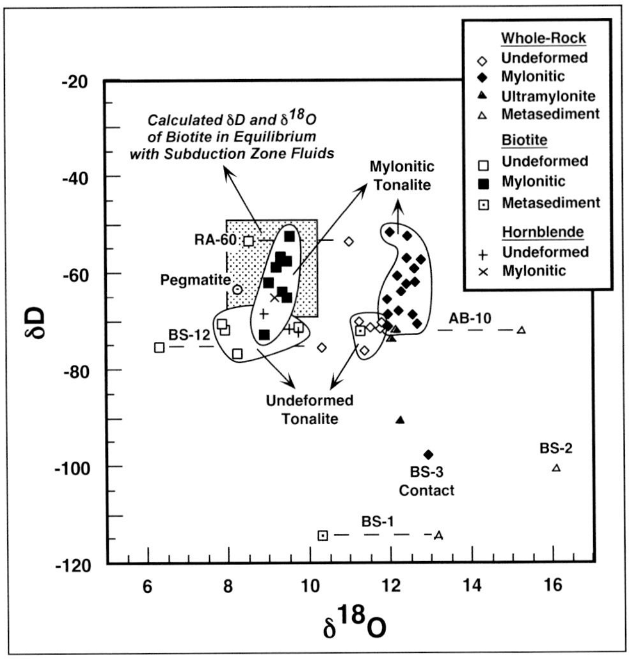

Figure 4. Plot of $\delta^{18}O$ values vs δD values for biotite, hornblende, and whole-rock from the Borrego Springs portion of the EPRMZ. Note that the plotted whole-rock δD values are those from coexisting biotite for those samples for which no actual whole-rock δD anayses were carried out. The bulk of the hydrogen in all of these rocks is provided by the biotite, which is typically much more abundant (5–10%) than the hornblende (1–4%), and which also contains twice as much hydroxyl. The isotopic composition of biotite in equilibrium at 500°C with subduction zone fluids having a +12‰ < $\delta^{18}O$ < +14‰ and −30‰ < δD < −10‰ (e.g. Bebout, 1991) is shown by the shaded rectangle. The parameters of this rectangle were calculated using the biotite-quartz (Javoy, 1977) and quartz-water (Clayton et al., 1972) $^{18}O/^{16}O$ fractionation factors, the Suzuoki and Epstein (1968) biotite-H_2O D/H fractionation factor and the Fe-Mg-Al compositions of EPRMZ biotites from Anderson (1983). Dotted tielines join biotite and whole-rock $\delta^{18}O$ values from two of the upper plate metasedimentary rocks (AB-10, BS-1), as well as from two undeformed tonalites (RA-60 and BS-12) that do not fit within the range of data defined by the other tonalites.

ANALYSIS OF ELEMENTAL DATA

Elemental concentrations from the different zones of the Ranchita Pluton affected by the EPRMZ are evaluated as follows. First, the average composition of the undeformed tonalite is compared with the average compositions of plutons from Silver and Chappell's (1988) Eastern Peninsular Ranges Batholith. These plutons are also considered by Walawander et al. (1990) to be the La Posta type. Any deviation greater than 10% from the average for this suite of plutons is considered to be either an enrichment or depletion. A similar analysis is done between the average values for undeformed tonalite, transition zone rocks, and protomylonites with the former serving as the baseline.

Elemental mobility within the EPRMZ is evaluated by comparing an element's concentration with another element considered immobile in hydrothermal systems. The immobile major element species is taken to be Al_2O_3 while Zr is the immobile trace element species (e.g., Grant, 1986). Al_2O_3 concentrations of undeformed tonalite (16.87%), transition zone tonalite (16.99%), and protomylonite (16.52%) indicate no change in the concentration of this element during deformation. The plotting of major element/Al_2O_3 ratios provides insight into major element mobility during deformation. Zirconium concentrations for undeformed tonalite (142 ppm) and protomylonite (136 ppm) are similar, allowing for the use of ratio of a trace element/Zr ratios for evaluation of trace element mobility in the deformed rocks. However, this kind of analysis breaks down in the transition zone because much higher Zr concentrations (218 ppm) are observed there, suggesting either addition of Zr, or the removal of 35% rock mass by hydrothermal fluids.

MAJOR ELEMENT SYSTEMATICS

The major element composition of the Ranchita Pluton (Table 1) is similar to the average compositions for the La Posta type (e.g., Walawender et al., 1990) Eastern Peninsular Ranges Batholith plutons (Silver and Chappell, 1988). However, the undeformed tonalite is enriched in Fe_2O_3, FeO, MnO, and P_2O_5 and slightly depleted in SiO_2 and Na_2O relative to the other La Posta type plutons. This contrasts with the deformed tonalites that are slightly more siliceous and less potassic than the La Posta type plutons.

	Eastern Batholith	Undeformed Tonalite	Transition Zone	Proto-Mylonite
SiO_2	66.05	64.63	66.96	66.81
TiO_2	0.65	0.80	0.69	0.64
Al_2O_3	16.23	16.87	16.99	16.52
Fe_2O_3	0.78	1.22	0.98	1.04
FeO	2.71	3.12	2.29	2.69
MnO	0.06	0.12	0.04	0.05
MgO	1.49	1.67	1.34	1.48
CaO	4.41	4.62	4.60	4.34
Na_2O	3.82	3.50	3.52	3.61
K_2O	2.26	1.88	1.68	1.77
P_2O_5	0.15	0.23	0.19	0.16
Ba	863	1124	1035	1009
Rb	73	78	44	69
Cs	N/A	2.50	0.73	1.27
Sr	501	530	753	518
Be	N/A	1.7	1.4	1.8
Pb	12	10	6	7
Ta	N/A	0.9	0.3	0.6
Th	8.6	3.6	1.3	6.3
U	1.7	1.4	0.4	0.7
Zr	150	142	218	136
Hf	N/A	4.1	5.6	3.8
Nb	8.6	7.3	5.8	8.7
Y	12	14	4	17
La	20	18	10	20
Ce	43	40	20	41
Nd	16	19	10	22
Sc	8.0	6.1	3.2	5.9
V	50	58	23	42
Cr	24	34	13	14
Co	N/A	31	15	19
Mo	N/A	1.9	0.6	0.9
Sn	N/A	5.5	0.8	2.0
Cu	3	34	19	25
Zn	88	97	76	93
Ga	20.5	22.2	23.5	21.9
Stable Isotopes (‰)				
$\delta^{18}O$		11.3	12.2	12.0
δD		−73	−67	−61

TABLE 1. Average Whole Rock Major Element, Trace Element, and Stable Isotope Values from the Ranchita Pluton and EPRMZ.

A comparison of major element concentrations between the undeformed tonalite, the transition zone tonalites, and the protomylonite indicate slight changes in the concentrations of these elements (Table 1). The increased SiO_2 concentrations for the transition zone rocks (66.96%) and the protomylonites (66.81%) relative to undeformed tonalite (64.63%) indicate a slight degree of silicification during deformation. Lower average concentrations of TiO_2, Fe_2O_3, FeO, MnO, MgO, K_2O and P_2O_5 observed in both the transition zone and the protomylonites relative to the undeformed tonalite indicate loss of these elements during deformation.

Differences in major element concentrations (Fig. 5) between undeformed tonalite, the transition zone tonalites, protomylonites, and ultramylonite indicate that aqueous fluids were

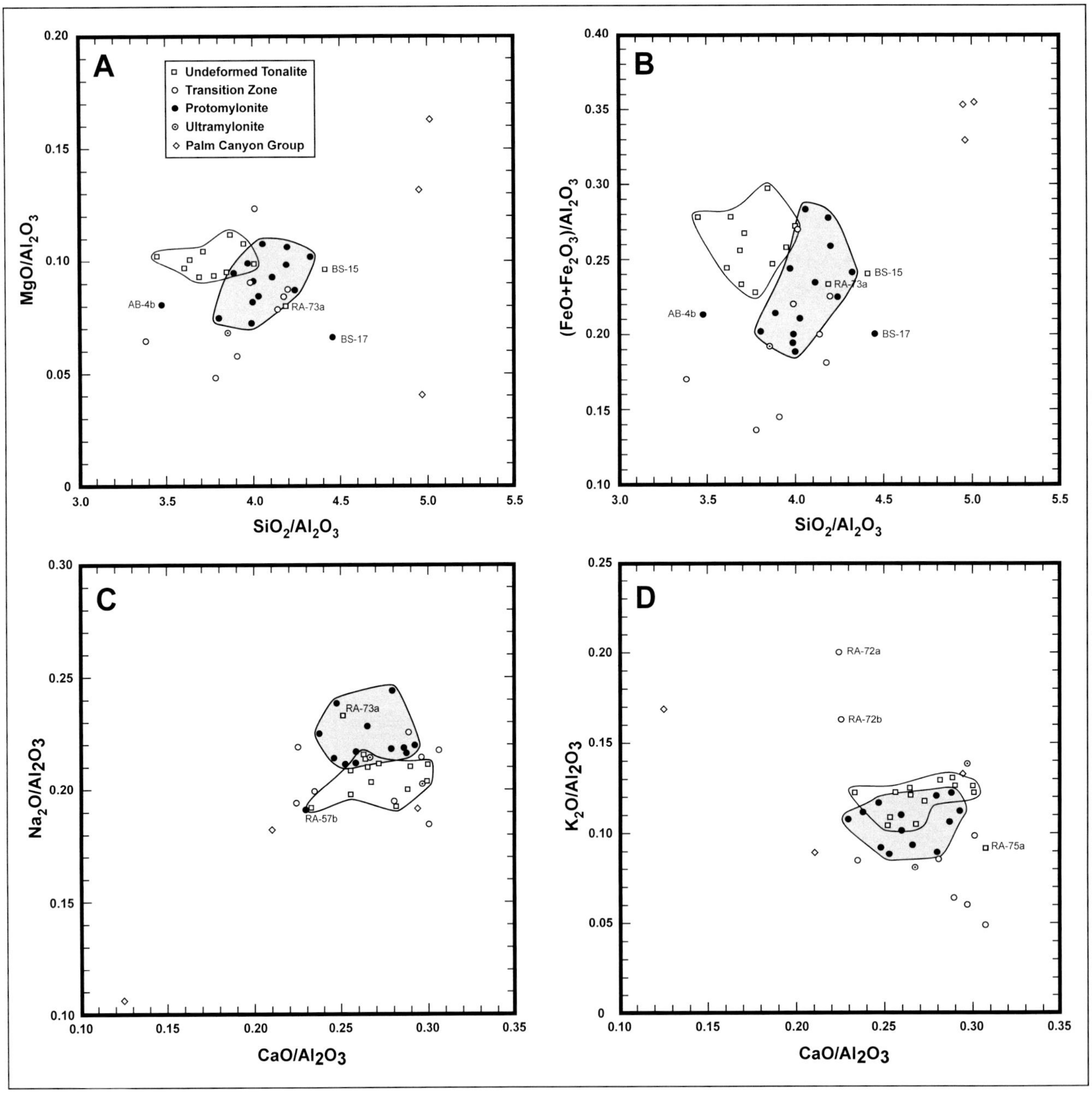

Figure 5. Plots of major element/Al_2O_3 ratios for undeformed tonalite (open squares), the transition zone (open dots), protomylonite (closed dots), ultramylonite (bulls-eyes), and the upper plate Palm Canyon Group (diamonds). Plots include: (a) MgO/Al_2O_3 versus SiO_2/Al_2O_3, (b) total Fe versus SiO_2/Al_2O_3, (c) $Na2O/Al_2O_3$ versus CaO/Al_2O_3, and (d) K_2O/Al_2O_3 versus CaO/Al_2O_3. Data are from this study and Anderson (1980; 1983).

an effective medium for mass transport and alteration. A comparison of the concentrations of SiO_2 and the mafic elements MgO, FeO and Fe_2O_3 illustrate differences between the different styles of deformation observed at the EPRMZ, especially as it relates to silica mobility and the alteration of minerals such as biotite and hornblende. Protomylonites have higher SiO_2/Al_2O_3 ratios than undeformed tonalite (Fig. 5a and 5b) whereas the transition zone rocks exhibit a range in SiO_2/Al_2O_3 that overlap both the range in values for both the undeformed and protomylonitic tonalities. Except for a few samples, MgO/Al_2O_3 ratios (Fig. 5a) are similar for both the undeformed tonalite and the protomylonite. Transition zone tonalities have depleted MgO/Al_2O_3 ratios than the undeformed tonalite. Trends for $(FeO + Fe_2O_3)/Al_2O_3$ ratios for the EPRMZ are similar to those of MgO/Al_2O_3 (Fig. 6b), indicating similar behavior for these elements.

Variations in the concentrations of CaO, Na_2O, and K_2O provide information about alteration styles involving feldspar during deformation along the EPRMZ. CaO/Al_2O_3 ratios are unchanged between all rock types within the tonalite (Fig. 5c and 5d), but the occurrence of epidote after plagioclase in many of the protomylonites suggest a decrease in the abundance of plagioclase during deformation. Higher values of Na_2O/Al_2O_3 for the protomylonites (Fig. 5c) relative to the undeformed tonalites and the transition zone tonalites indicate albitization of plagioclase as Na is introduced into the EPRMZ during deformation. This is consistent with the increased SiO_2/Al_2O_3 ratios in the protomylonites relative to the undeformed tonalite. Transition zone tonalites have similar Na_2O/Al_2O_3 values as those from the undeformed tonalite. Protomylonites have lower K_2O/Al_2O_3 ratios than most of the undeformed tonalites (Fig. 5d). Except for sample RA-72, transition zone tonalites are depleted in K_2O/Al_2O_3 relative to both protomylonite and undeformed tonalite.

TRACE ELEMENT SYSTEMATICS

Most trace element concentrations for the undeformed Ranchita Pluton are similar to those from the other Eastern Peninsular Ranges plutons (Table 1; Silver and Chappell, 1988). Exceptions include enrichments of Ba, Cr and Cu, and depletions of Th and Sc.

The transition zone displays the greatest changes in trace element concentrations relative to the undeformed tonalite (Table 1). Few trace elements (Ba and Ga) in the transition zone were unaffected ($< \pm 10\%$ concentration change) by the EPRMZ. Elements with large enrichments ($> +25\%$) relative to undeformed tonalite include Zr, Hf, and Sr. The 53.5% increase in average Zr concentrations indicates a volume reduction or loss of mass of approximately 30% for these variably strained rocks during deformation along the EPRMZ. All other trace elements are depleted in the transition zone relative to the undeformed tonalite. Slightly depleted (between −10% and −25%) elements include Be, Nb, and Zn. Highly depleted ($> -25\%$) elements include Rb, Cs, Pb, Ta, Th, U, Y, La, Ce, Nd, Sc, V, Cr, Co, Mo, Sn, and Cu.

Significant trace element metasomatism also occurred within the protomylonites, but the magnitude of elemental mobility is less than that observed at the transition zone (Table 1). Unchanged ($< \pm 10\%$) trace element concentrations are Ba, Sr, Be, Zr, Hf, Ce, Sc, Zn, and Ga. Elements with slightly elevated (+10-25%) concentrations relative to undeformed tonalite include Nb, Y, La, and Nd. Only Th (+75%) concentrations are highly elevated relative to undeformed tonalite, most likely due to the abundance of allanite in some protomylonites. Highly depleted ($> -25\%$) elements are Cs, Pb, Ta, U, V, Cr, Co, Mo, Sn, and Cu. Only Rb (−11.5%) displays slight depletions relative to the undeformed tonalite.

A comparison of trace element/Zr ratios illustrates relative geochemical changes during deformation along the EPRMZ (Figs. 6 and 7). In most cases, elements chosen for comparison are predicted to behave similarly in hydrothermal systems based on their position on the periodic

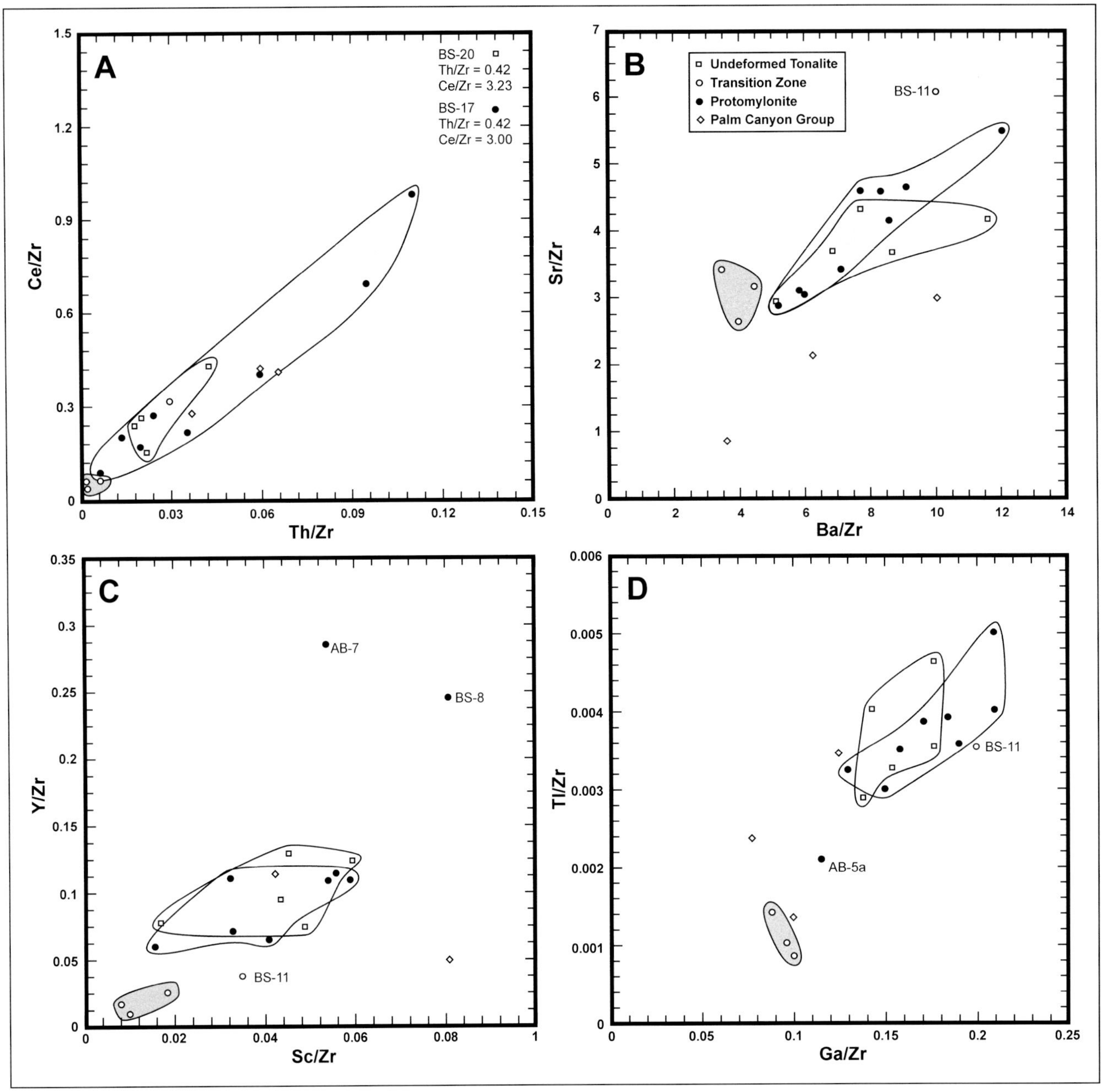

Figure 6. Trace element/Zr ratios for elements that are either unchanged, or have a variable kinds of interaction with hydrothermal fluids. Plots include: (a) light rare earth Ce/Zr versus actinide Th/Zr, (b) alkali earth elements Sr/Zr versus Ba/Zr, (c) transition metals Y/Zr versus Sc/Zr, and (d) metalloids Tl/Zr versus Ga/Zr.

table. The most striking feature of these plots is that these ratios for the transition group tonalites are consistently smaller than those of both the undeformed tonalite and the protomylonite. This is an artifact of the elevated Zr concentrations observed in this group of rocks. If the protomylonites and transition zone tonalites follow the same trend, then the mechanism for trace element mobility is congruent for each of these domains. Incongruence between the trends in element ratio for protomylonite and the transition zone indicates different styles of trace element mobility for these two domains. Furthermore, congruence in the ratio trends and value ranges between the undeformed tonalite and protomylonite indicates that deformation along the EPRMZ had no effect on those trace elements.

The overlap of trace element concentrations observed for protomylonite and undeformed tonalite (Fig. 6) indicate no transport of these elements by hydrothermal fluids. The large range of actinide

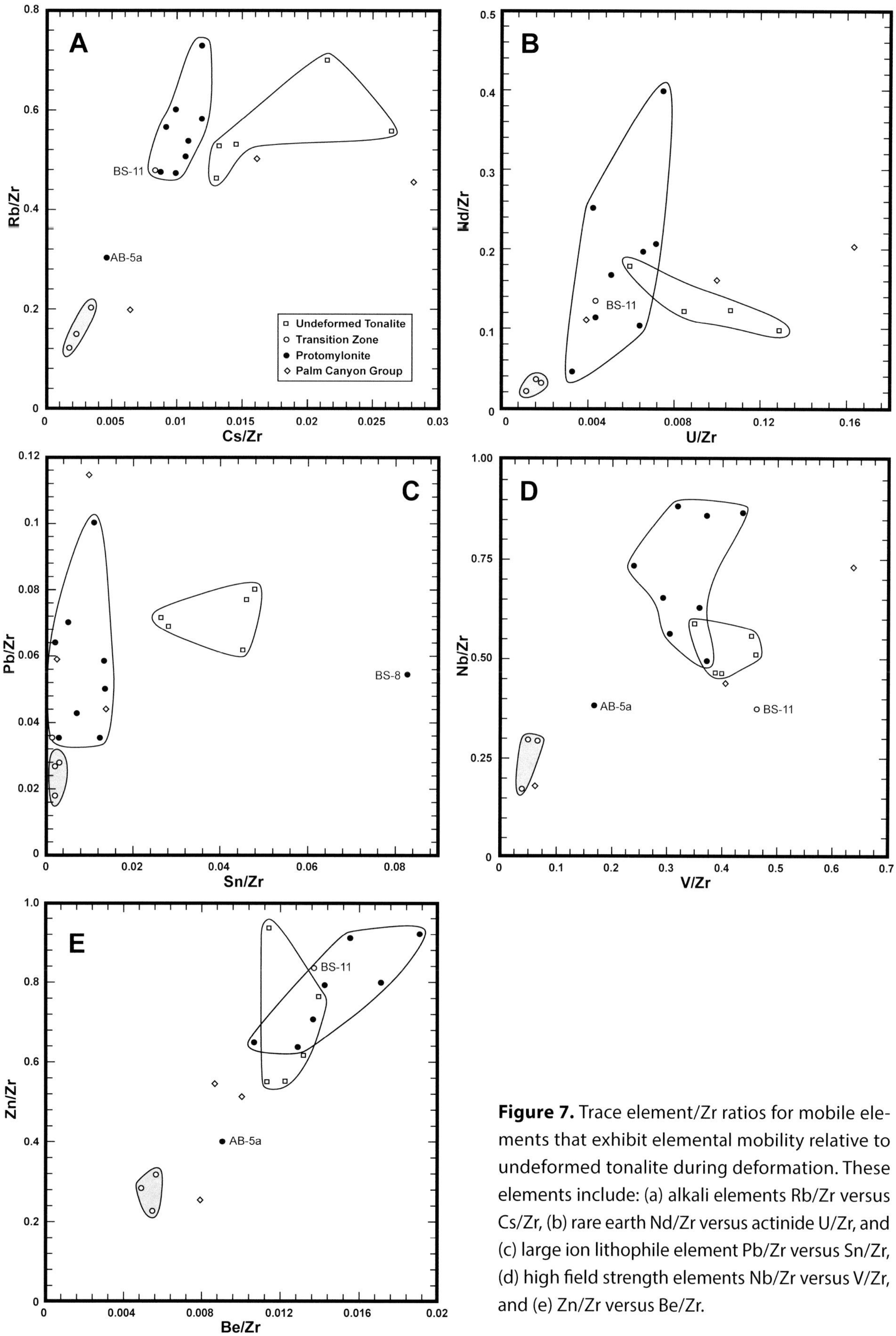

Figure 7. Trace element/Zr ratios for mobile elements that exhibit elemental mobility relative to undeformed tonalite during deformation. These elements include: (a) alkali elements Rb/Zr versus Cs/Zr, (b) rare earth Nd/Zr versus actinide U/Zr, and (c) large ion lithophile element Pb/Zr versus Sn/Zr, (d) high field strength elements Nb/Zr versus V/Zr, and (e) Zn/Zr versus Be/Zr.

Th/Zr and lanthanide Ce/Zr (Fig. 6a) observed in these rocks is due to the presence or absence of the epidote-group mineral allanite and is independent of deformation. Slightly elevated alkali earth Sr/Zr values relative to Ba/Zr values (Fig. 6b) observed in some protomylonites and all transition zone tonalites suggest slight enrichments in Sr relative to Ba during deformation. Overlap between transition group elements Y/Zr and Sc/Zr values (Fig. 6c) from the undeformed tonalite and protomylonite indicate no mobility of these elements during deformation, but a deviation toward lower Y/Zr relative to Sc/Zr in the transition zone rocks suggest loss of Y by hydrothermal fluids. A similar relationship is observed for the Tl/Zr and Ga/Zr pair of elements (Fig. 6d), with a loss of Tl in the transition zone rocks. Gallium is immediately below Al on the periodic table and is expected to be immobile in hydrothermal systems, and there is no evidence for Ga mobility in this system.

Differences in the values and ranges of trace element concentrations between the undeformed tonalites and protomylonites indicate migration of these elemental species either into or out of tonalite during deformation along the EPRMZ (Fig. 7). Protomylonite is markedly depleted in Cs relative to undeformed tonalite (Fig. 7a), but range of Rb/Zr ratios exhibited by the protomylonite is equal to that observed for the undeformed tonalite. Cs/Zr and Rb/Zr ratios from the protomylonite and transition zone tonalite define a consistent trend separate from data from the undeformed tonalite. Uranium is depleted in both protomylonite and transition zone tonalite relative to undeformed tonalite (Fig. 7b). Concentrations of the rare earth element Nd are highly variable in the protomylonite; this large range is likely due to variable amounts of LREE-rich allanite in these rocks, similar to our observations for similar elements Th and Ce. Both protomylonites and transition zone tonalites are significantly depleted in Sn relative to undeformed tonalites (Fig. 7c). Most protomylonites are depleted in Pb relative to undeformed tonalite, but one sample (AB-2a) is enriched. Niobium is one of the rare elements that are enriched in the protomylonite relative to the undeformed tonalite (Fig. 7d), in contrast with the depletions in V observed for the protomylonite. Slight enrichments in Zn/Zr and Be/Zr are observed in the protomylonite relative to the undeformed tonalite (Fig. 7e).

COMPARISON WITH STABLE ISOTOPE RESULTS

Major element metasomatism during deformation of the EPRMZ is demonstrated by comparing oxygen isotope trends to K_2O/SiO_2 (Fig. 8) with the San Jacinto Intrusive Complex (Hill et

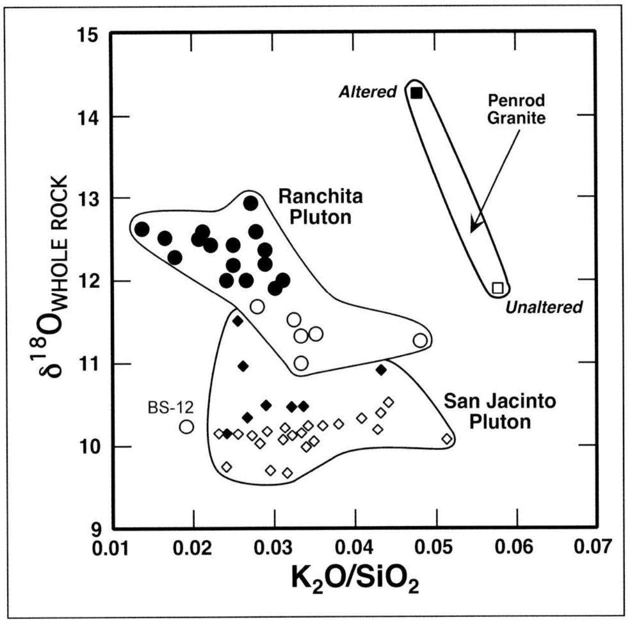

Figure 8. Plot of whole rock $\delta^{18}O$ versus K_2O/SiO_2, comparing our data with data from the San Jacinto intrusive complex and the Penrod granite (Hill et al., 1986; 1988), a small intrusion affected by the EPRMZ. The $\delta^{18}O$ of undeformed Ranchita tonalite is 1.0‰ higher than primary San Jacinto tonalite of similar K_2O/SiO_2. This is expected because the Ranchita pluton is in a higher-^{18}O part of the PRB relative to the San Jacinto block (Taylor and Silver, 1978). Mylonitic tonalite has $\delta^{18}O$ values 1.5‰ higher than for undeformed equivalents having similar K_2O/SiO_2. A similar increase in $\delta^{18}O$ is observed for altered Penrod granite. These data contrast with the much smaller core-to-rim elevation of $\delta^{18}O$ values (<1‰) for a system unaffected by the EPRMZ (San Jacinto). This strongly suggests that $\delta^{18}O$ values of deformed tonalite were elevated in response to the passage of high-^{18}O fluids during deformation. The high-^{18}O metamorphic fluids that infiltrated the EPRMZ apparently produced a significant loss of K_2O during deformation.

al., 1986; 1988), a La Posta type plutonic system distal from the EPRMZ. The San Jacinto data display a slight increase in $\delta^{18}O$ with increasing K_2O/SiO_2 at the rim of this pluton, a trend suggesting a small degree of magmatic assimilation with the high-^{18}O metapelitic country rocks (Hill et al., 1988). These data contrast sharply with the opposite trend observed for the Ranchita Pluton where $\delta^{18}O$ values rise with decreasing K_2O/SiO_2. Instead, the trend observed in Figure 5 indicates that the heated, high-^{18}O aqueous fluids that infiltrated the EPRMZ resulted in a loss of K_2O and a gain in SiO_2 from the deformed tonalites. This interpretation is consistent with that of Simpson and Wintsch (1989), who showed that loss of K_2O accompanied myrmekitization of alkali feldspar during the development of mylonitic fabrics in tonalites within the EPRMZ.

A comparison involving the most depleted and most enriched elements (Cs and Nb, respectively)

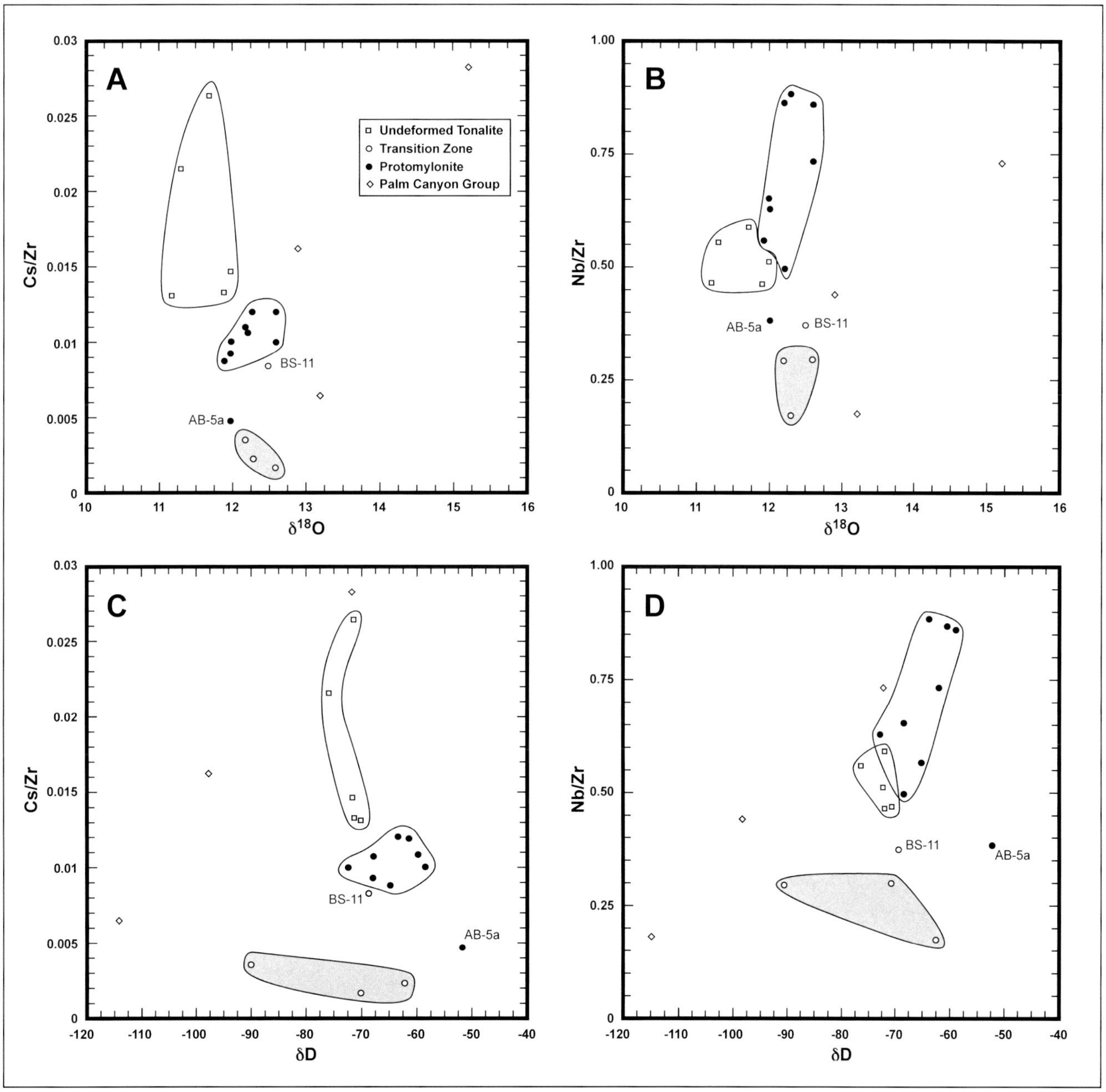

Figure 9. Plots that compare Cs/Zr and Nb/Zr variations with $\delta^{18}O$ and δD values from undeformed tonalities (open squares), protomylonites (closed dots), the transition zone (open dots), and the upper plate Palm Canyon Group (diamonds). These plots include: (a) Cs/Zr versus $\delta^{18}O$, (b) Nb/Zr versus δD, (c) Cs/Zr versus δD, and (d) Nd/Zr versus $\delta^{18}O$.

with $\delta^{18}O$ and δD values (Fig. 9) indicate that trace element metasomatism occurred due to the infiltration of subduction zone-related metamorphic fluids into the EPRMZ. Values of both δD and $\delta^{18}O$ are enriched in both protomylonites and transition zone tonalite relative to undeformed tonalite. These isotopic relationships are accompanied by depletions in Cs/Zr and enrichments in Nb/Zr in the protomylonites. The transitional tonalites exhibit depletions in both Cs/Zr and Nb/Zr relative to the undeformed tonalite.

FLUID SOURCES AND TECTONIC MECHANISMS

The EPRMZ formed during the shallowing of the Farallon Plate subduction zone during the late Cretaceous and early Tertiary (Gastil et al., 1992). This was accompanied by an eastward migration of magmatism along the western margin of North America (Coney and Reynolds, 1977). The stable isotope composition of waters in equilibrium with the EPRMZ provides clues about fluid source. Accepting this tectonic model, there are at least two possible sources for the high-$\delta^{18}O$, high-δD metamorphic water that infiltrated the EPRMZ during its formation at 65–75 Ma: (1) Dehydration of underplated volcaniclastic sediment (e.g., Pelona or Orocopia Schist; Grove et al., 2003b) or (2) Dehydration of the subducted slab of oceanic crust itself. In either of these scenarios the ascendant fluids would never encounter rocks held at temperatures greater than the wet solidus of either basalt or tonalite, they would have a $\delta^{18}O \sim +13\pm2‰$ and a $\delta D \sim -20\pm15‰$ (Bebout, 1991).

Assuming a subduction angle of 20° and a distance of 120 km from the trench, the depth of dehydration would have been about 30–40 km. Hornblende geobarometry of the eastern PRB suggests that the current exposures of the Ranchita Pluton originally crystallized at depths of about 25 km (George and Dokka, 1994). However, thermochronologic data from the nearby San Jacinto Plutonic Complex indicate cooling below 100°C and uplift by 80 Ma (Wolf et al., 1997), indicating reburial and reheating of an already cold Ranchita Pluton during at least 11 km of displacement along the EPRMZ thrust (Sharp, 1979). The EPRMZ fluids would have ascended more than 15 km through the crust, implying a flow path of perhaps 30 km or more, judging by the 30° dip of lineations observed in the mylonitic tonalite. Water-rock ratios within the EPRMZ were variable (from <0.1 to >1.0), an amount more than sufficient to allow for degree of elemental transport observed along the shear zone.

Reheating of those portions of the Ranchita Pluton that were affected by the EPRMZ would have been the result of: (1) frictional heating during deformation (e.g., Graham and England, 1976), (2) conduction and/or advection of heat from the hot upper plate downward into the lower plate (e.g., England and Thompson, 1984), or (3) introduction of the heat carried upward by the aqueous fluids themselves (e.g. Barnett and Chamberlain, 1991). The presence of pseudotachylite in the study area supports the idea that, at least locally, heat was supplied to the EPRMZ by shear heating. Note that these shear heating episodes would be transient and would likely occur only during earthquakes; the frictional melting associated with the pseudotachylites would serve to lubricate and further weaken the fault zone (Scholz, 1990).

Considering the thermal models for the evolution of overthrust belts by England and Thompson (1984) and the high temperatures of the upper plate (>600°C; Theodore, 1970), temperatures within the lower plate of the EPRMZ probably would have risen to at least 500°C within 10 My after the onset of thrust faulting, forming an inverted metamorphic gradient in the footwall Ranchita pluton. Inverted metamorphism has been documented at other subduction-related thrust systems coeval with the EPRMZ in southern California (e.g., in the Sierra Pelona, Ehlig, 1968; and in Santa Catalina Island, Bebout, 1991).

CONCLUSIONS

These data demonstrate metasomatism of both major and trace elements during deformation along the EPRMZ. Elemental and alteration trends are consistent with an oxidizing environment. Zones of most intense metasomatism correspond to those parts of the EPRMZ that experienced the most strain (i.e., transition zone). The EPRMZ may have served as a major conduit of fluid release for the subducting Farallon plate and associated accretionary wedge sediments during the Late Cretaceous and Early Tertiary.

ACKNOWLEDGEMENTS

We thank James R. Anderson, L. Peter Gromet, and Leon T. Silver for generous access to sample materials, numerous discussions, and invaluable guidance in the field, and Conrad Gregoire for assistance with the acquisition of major and trace element data. We also thank Laurel Goodwin and Dan Francis for discussions, and the staff of Anza-Borrego State Park for permitting sample collection and field study.

REFERENCES

Anderson, J.R., 1980, A study of mylonitic rocks from major fault zones. US Geological Survey Open-File Report 80-1154.

Anderson, J.R., 1983, Petrology of a portion of the Eastern Peninsular Ranges mylonite zone, Southern California. Contributions to Mineralogy and Petrology, v. 84, p. 253–271.

Bebout, G.E., 1991, Field-based evidence for devolatilization in subduction zones: Implications for arc magmatism. Science, v. 251, p. 413–416.

Barnett, D.E., and Chamberlain, C.P., 1991, Relative scales of thermal- and fluid migration-driven metamorphism in fold nappes, New England, U.S.A. American Mineralogist, v. 76, p. 713–727.

Clayton, R.N., O'Neil, J.R., and Mayeda, T.K., 1972, Oxygen isotope exchange between quartz and water. Journal of Geophysical Research, v. 77, p. 3057–3067.

Clayton, R.N., and Kieffer, S.W., 1991, Oxygen isotopic thermometer calculations. In: Stable Isotope Geochemistry: A Tribute to Samuel Epstein. Geochemical Society Special Publication, v. 3, p. 3–10.

Coney, P.R., and Reynolds, S.J., 1977, Cordilleran Benioff zones. Nature, v. 170, p. 403–406.

Ehlig, P.L., 1968, Causes and distribution of Pelona, Rand, and Orocopia Schist along the San Andreas and Garlock faults. Stanford University Publications in Geological Sciences, v. 11, p. 294–305.

England, P.C., and Thompson, A.B., 1984, Pressure-temperature-time paths of regional metamorphism I. Heat transfer during the evolution of regions of thickened continental crust. Journal of Petrology, v. 25, p. 894–928.

Engle, A.E.J., and Schultejann, P.A., 1984, Late Mesozoic and Cenozoic tectonic history of south central California. Tectonics, v. 3, p. 659–675.

Gastil, G, Wratcher, M, Strand, G., Kear, L.L., Eley, D., Chapman, D., and Anderson, C., 1992, The tectonic history of the Southwestern United States and Sonora, Mexico, during the past 100 m.y. Tectonics, v. 11, p. 990–997.

George, P.G., and Dokka, R.K., 1994, Major late Cretaceous cooling events in the eastern Peninsular Ranges, California and their implications for Cordilleran tectonics. Geological Society of America Bulletin, v. 106, p. 903–914.

Goldfarb, R.J., Newberry, R.J., Pickthorn, W.J., and Gent, C.A., 1995, Oxygen, hydrogen, and sulfur isotope studies in the Juneau gold belt, Southeastern Alaska: Constraints on the origin of hydrothermal fluids. Economic Geology, v. 86, p. 66–80.

Goodwin, L.B., and Renne, P.R., 1991, Effects of progressive mylonitization on Ar retention in biotites from the Santa Rosa mylonite zone, California, and thermochronologic implications. Contributions to Mineralogy and Petrology, v. 108, p. 283–297.

Grant, J.A., 1986, The isocon diagram: a simple solution to Gresen's equation for metasomatic alteration. Economic Geology, v. 81, p. 1976–1982.

Graham, C.M., and England, P.C., 1976, Thermal regimes and regional metamorphism in the vicinity of overthrust faults: An example of shear heating and inverted metamorphic zonation from southern California. Earth and Planetary Science Letters, v. 31, p. 142–152.

Grove, M., Lovera, O., and Harrison, M., 2003a, Late Cretaceous cooling of the east-central Peninsular Ranges batholith (33°N): Relationship to La Posta pluton emplacement, Laramide shallow subduction, and forearc sedimentation. In: Tectonic Evolution of Northwestern Mexico and the Southwestern USA (Johnson, S.E. et al., eds), Geological Society of America Special Paper 374, p. 355–380.

Grove, M., Jacobson, C.E., Barth, A.P., and Vicic, A., 2003b, Temporal and spatial trends of late Cretaceous-early Tertiary underplating of Pelona and related schist beneath southern California and southwestern Arizona. In: Tectonic Evolution of Northwestern Mexico and the Southwestern USA (Johnson, S.E. et al., eds), Geological Society of America Special Paper 374, p. 381–406.

Hill, R.I., Silver, L.T., and Taylor, H.P., Jr., 1986, Coupled Sr-O isotope variations as an indicator of source heterogeneity in the northern Peninsular Ranges Batholith. Contributions to Mineralogy and Petrology, v. 92, p. 351–361.

Hill, R.I., Chappell, B.W., and Silver, L.T., 1988, San Jacinto Intrusive Complex, 2. Geochemistry. Journal of Geophysical Research, v. 93, p. 10349–10372.

Holk, G.J., Taylor, H.P., and Gromet, L.P., 2006, Stable isotope evidence for large-scale infiltration of metamorphic fluids generated during shallow subduction into the Eastern Peninsular Ranges Mylonite Zone (EPRMZ), Southern California. International Geology Review, v. 48, p. 209–222.

Holk, G.J., Taylor, H.P., Jr., 2007, $^{18}O/^{16}O$ studies of regional metamorphism, anatexis, extensional magmatism, meteoric-hydrothermal activity and ore deposition in the Valhalla metamorphic core complex, British Columbia. Economic Geology, v. 102, p. 1063–1078.

Javoy, M., 1977, Stable isotopes and geothermometry. Geological Society of London Journal, v. 133, p. 609–636.

Kirschner, D.L., Masson, H., and Sharp, Z.D., 1999, Fluid migration through thrust faults in the Helvetic nappes (Western Swiss Alps). Contributions to Mineralogy and Petrology, v. 136, p. 169–183.

Magaritz, M., and Taylor, H.P. Jr., 1976, Oxygen, hydrogen, and carbon isotope studies of the Franciscan Formation, Coast Ranges, California. Geochimica et Cosmochimica Acta, v. 40, p. 215–234.

Magaritz, M., and Taylor, H.P. Jr., 1986, Oxygen18/Oxygen16 and D/H studies of plutonic granitic and metamorphic rocks across the Cordilleran batholiths of Southern British Columbia. Journal of Geophysical Research, v. 91, p. 2193–2217.

McCaig, A.M., Wickham, S.M., and Taylor, H.P. Jr., 1991, Deep fluid circulation in alpine shear zones, Pyrenees, France: Field and oxygen isotope studies. Contributions to Mineralogy and Petrology, v. 106, p. 41–60.

Nesbitt, B.E., and Muehlenbachs, K., 1995, Geochemical studies of the origins and effects of synorogenic crustal fluids in the Southern Omineca Belt of British Columbia. Geological Society of America Bulletin, v. 107, p. 1033–1050.

O'Neil, J.R., and Taylor, H.P. Jr., 1967, The oxygen isotope and cation exchange chemistry of feldspars. American Mineralogist, v. 52, p. 1414–1437.

Scholz, C.H., 1990, The mechanics of earthquakes and faulting. Cambridge University Press, Cambridge, 439 p.

Sharp, R.V., 1979, Some characteristics of the eastern Peninsular Ranges mylonite zone. US Geological Survey Open-File Report 79-1239, p. 258–267.

Sheppard, S.M.F., 1986, Characterization of isotopic variations in natural waters. In: Stable Isotopes in High Temperature Geological Processes, Reviews in Mineralogy, Valley, J.W., Taylor, H.P. Jr., and O'Neil, J.R., eds., v. 16, p. 165–181.

Sibson, R.H., 2001, Seismogenic framework for hydrothermal transport and ore deposition. Reviews in Economic Geology, v. 14, p. 25–50.

Silver, L.T. and Chappell, B.W., 1988, The Peninsular Ranges batholith: An insight into the evolution of the Cordilleran batholiths of southwestern North America. Royal Society of Edinburgh Transactions, v. 79, p. 105–121.

Simpson, C., and Wintsch, R.P., 1989, Evidence of deformation-induced K-feldspar replacement by myremekite. Journal of Metamorphic Geology, v. 7, p. 261–275.

Stinson, A.L., and Gastil, R.G., 1996, Mid-to-Late Tertiary detachment faulting in the Pinyon Mountains, San Diego County, California: A setting for long-runout landslides in the Split Mountain area. South Coast Geological Society Annual Field Trip Guide, v. 24, p. 221–245.

Suzouki, T., and Epstein, S., 1976, Hydrogen isotope fractionation between OH-bearing minerals and water. Geochimica Cosmochimica Acta, v. 40, p. 1229–1240.

Taylor, H.P., Jr. and Silver, L.T., 1978, Oxygen isotope relationships in plutonic igneous rocks of the Peninsular Ranges Batholith, southern and Baja California. United States Geological Survey Open-File Report 78-701, p. 423–426.

Theodore, T.J., 1970, Petrogenesis of mylonites of high metamorphic grade in the Peninsular Ranges of southern California. Geological Society of America Bulletin, v. 81, p. 435–450.

Todd, V.R., Erskine, B.G., and Morton, D.M., 1988, Metamorphic and tectonic evolution of the northern Peninsular Ranges batholith, southern California. In: W.G. Ernst, ed, Metamorphism and Crustal Evolution of the Western United States, Rubey volume VII, p. 894–936.

Walawender, M.J., Gastil, R.G., Clinkenbeard, J.P., McCormick, W.V., Eastman, B.G., Wernicke, R.S., Wardlaw, M.S., Gunn, S.H., and Smith, B.M., 1990, Origin and evolution of the zoned La Posta-type plutons, eastern Peninsular Ranges Batholith, southern and Baja California. In: Anderson, J.L., ed., The nature and origin of Cordilleran magmatism: Geological Society of America Memoir 174, p. 1–18.

Wolf, R.A., Farley, K.A., and Silver, L.T., 1997, Assessment of (U-Th)/He thermochronometry: The low temperature history of the San Jacinto mountains, California. Geology, v. 25, p. 65–68.

Possible Sources of Natural Materials Used by the Cahuilla Indians Living in the Pinyon Flats Area of Riverside County, California

Harry M. Quinn

Many natural materials were utilized by the local Cahuilla Indians. Some of these materials are not found locally and had to be imported into the Pinyon Flats area. The following is a listing of some of these materials and their potential source areas. This provenance study provides a better understanding of the local culture and its relationship to other peoples in the region.

OBSIDIAN

Debitage, projectile points, and point fragments made from obsidian are known from the Pinyon Flats area. Most of the obsidian is of a hydrated type containing flecks of crystobilite. A few specimens appear to be of a non-hydrated variety. There is no known source of obsidian in the Pinyon Flats area. Peters (1975) included fused shale in her listing of California obsidian sources, but no fused shale had been identified from the Pinyon Flats area. Fused shale is a glassy material formed when deposits of hydrocarbon-rich shale catch fire and burn in situ. This material closely resembles the hydrated obsidian found at Obsidian Butte near the southeast end of the Salton Sea. The three nearest listed obsidian sources to the Pinyon Flats area shown by Peters (1976) are as follows:

1. Obsidian Butte Area, Imperial County. This is the source of the hydrated type obsidian with white flecks and being located near the bottom of the Salton Sink, which was completely underwater during periods when an ancient lake was present.

2. Santa Clara River and its tributaries near the town of Fillmore in Ventura County is the source of deposits of fused shale.

3. Coso Mountains Area in Inyo County is where a number of obsidian quarries have been found. With the new typing of obsidian sources, this area has been found to have supplied obsidian to much of southern California.

Other potential sources for obsidian include the obsidian nodules and Apache Tear, localities near the towns of Bagdad and Ludlow, in Section 16 of Township 6 North, Range 11 East; and Sections 3 and 4 of Township 7 North, Range 7 East, San Bernardino Base and Meridian (SBB&M), in San Bernardino County.

SOAPSTONE (STEATITE)

A few artifacts and fragmented artifacts of soapstone, or soapstone-like material, are known from the Pinyon Flats area. While there are no known soapstone deposits in the Pinyon Flats area, there are some local outcrops of serpentine near the old asbestos mines. Peters (1976) does not include talc, serpentine, or soapstone in her listings of steatite sources, the three nearest to the Pinyon Flats area shown by Peters (1976) are:

1. Santa Catalina Island near Little Springs and Pots Valley in Los Angeles County.

2. Along the northeastern side of Jacumba Valley in San Diego County.

3. Santiago Canyon about 15 miles south of Buena Vista Lake in Kern County.

Other potential sources of steatite-like materials include serpentine deposits in Section 30 of Township 5 South, Range 1 West (near Murrieta), Sections 2 and 32 of Township 6 South, Range 5 East (at Pinyon Flats), and Sections 27, 28, and

35 of Township 3 South, Range 14 East (at the Eagle Mountain Iron Mine) in Riverside County. Additional sources include some soapstone-talc schist deposits in Section 8 of Township 5 North, Range 13 West (near Sleepy Valley) in the Pelona Schist Belt of the Sierra Pelona.

Additional deposits are known from other portions of the Sierra Pelona Mountains. There may be others in the Pelona Schist Belt along the north flank of the San Gabriel Mountains in Los Angeles and San Bernardino Counties. Some of the talc deposits in the Death Valley Area and the soapstone deposits along the western foothills of the Sierra Nevada may have also supplied soft, carveable stone to southern California, including the Pinyon Flats area.

BASALT

Artifacts made from vesicular (scoria) and a dense variety of basalt are known from the Pinyon Flats area, but there are no known local sources of basalt in the area. The nearest sources of basalt to the Pinyon Flats area are:

1. Basalt outcrops and basalt clasts in alluvial deposits present in the Painted Hills Area of Riverside County. These deposits are north of Interstate 10 and west of Highway 62 and are the closest known sources.

2. Basalt outcrops and basalt clasts in alluvial deposits near Yucca Valley and Twentynine Palms and within the Joshua Tree National Park, San Bernardino County.

3. Basalt outcrops in the Amboy and Pisgah Crater areas in San Bernardino County.

4. Basalt outcrops and basalt clast in alluvial deposits near Blythe in Riverside County.

There are many additional basalt sources scattered throughout Mojave Desert, mainly in San Bernardino County. These source areas are farther from the Pinyon Flats area than those listed above, and since basalt is dense and heavy were less likely carried there.

PUMICE

Artifacts made from pumice are known from the Pinyon Flats area. These pumice artifacts commonly exhibit wear patterns that indicate they were used as abrading tools. There are no known local sources of pumice in the Pinyon Flats area, bur some potential sources include:

1. Obsidian Butte and Rock Hill (Pumice Hill) near the southern end of the Salton Sea in Imperial County. This area is near the bottom of the Salton Sink, so would not have been available when lake levels exceeded the top of the hills (-130 feet for Obsidian Butte, -138 for Rock Hill).

2. The eastern shoreline of past lakes filling the Salton Sink, such as the Holocene Lake Cahuilla series that filled the basin up to a height of 42 feet (12 meters) above mean sea level and lasted until around AD 1680-1690.

3. The Coso Mountains Area and Bishop Area in Inyo County.

FINE-GRAINED FELSITIC MATERIALS

Debitage and artifacts made from a fine-grained felsitic material, much of it now called wonderstone, are known from the Pinyon Flats area. While most of the artifacts made from this material are in the form of chipped stone tools, a few have been shaped by grinding. The term "fine grained felsitic material" as used in this paper includes rhyolite, dacite, andesite, wonderstone (also known as Truckhaven Rhyolite), and aplite.

1. The closest known deposit of felsitic material appears to be in the Wonder Canyon Area, at the southeast end of the Santa Rosa Mountains just northwest of the old town of Truckhaven, Imperial County. The material was originally mapped as the Truckhaven Rhyolite by Dibblee (1954) and was described as "a lens of varicolored felsitic rock that was extruded along an adjacent east-west fault." Subsequent mapping by Weismeyer (1968) found that what had been called the Truckhaven Rhyolite was actually silicified sediments along the fault zone and not an igneous rhyolite intrusion. Because it is

a zone of altered sediments, it is quite variable in appearance and composition. It includes silicified shales, siltstones, sandstones, and even some conglomerates. Where the silica-rich waters reached the surface they precipitated out deposits of agate and chalcedony. Some of the silicified shales and siltstones have the appearance of jasper and are collected today by rockhounds to make "jasper" cabochons (Mitchell 1996: 18). Detrital grains and relic depositional textures can be found in thin sections of this material and some of the coarser materials can be distinguished in hand specimen.

2. Another close potential source is the Dos Palmas Rhyolite mapped by Dibblee in the Orocopia Mountains northeast of the Salton Sea in Riverside County.

3. A third potential source is the Alverson Andesite found in the western foothills of Coyote Wells Valley in San Diego County. This material is described as a dark brown basic andesite, with the outcrops consisting of andesite lavas, breccia, and tuffs.

4. Some andesite and rhyolite dikes have been reported to be present in the Orocopia Mountains, other than the Dos Palmas Rhyolite.

5. Aplite dike material suitable for flaking tools can be found in the Joshua Tree National Park and at Alligator Ridge south of Desert Center. Aplite dikes are present in the Pinyon Flats area, but all of them examined so far have been too coarse to be suitable for most tool making. With the large number of aplite dikes present in the Santa Rosa and San Jacinto Mountains, some of it is likely suitable for tool making.

QUARTZ AND QUARTZ-RELATED MATERIALS

Both debitage and artifacts made from quartz and quartz-related material are known from the Pinyon Flats area. White, or milky quartz, appears to be the most common form of quartz used and it is probably of local origin.

A. Quartz. The most common quartz artifacts found in the Pinyon Flats area are made from milky quartz. These are followed by clear quartz (quartz crystal), rose quartz, and way down the line by smoky quartz, rutilated quartz, and tourmalinated quartz. Some potential sources are as follows:

1. Milky quartz can be found locally in quartz veins and in quartz-feldspar pegmatites. It is commonly found in scattered veins and pegmatites throughout the granitic rocks that make up the Peninsular Ranges and Transverse Ranges.

2. Clear (crystal) quartz can be found locally, usually in quartz-feldspar pegmatites, but sources are rare. Good crystals of clear quartz have been found in the Thomas Mountain, Cahuilla Mountain, and San Jacinto Mountain areas, which are relatively close. Additional potential source areas include the San Bernardino Mountains, Lakeview Mountains, Tule Mountain Area, Ash Hill Area near the town of Ludlow, Crystal Hill Area near Quartzsite, Arizona, and at pegmatite localities throughout San Diego County.

3. Rose (pink) quartz has been found locally as float, but is quite rare. No nearby described localities for rose quartz were found.

4. Smoky quartz has been found in pegmatites in the Idyllwild Area of the San Jacinto Mountains and in the Thomas Mountain Area. However, it is uncommon in both areas.

5. No local source for rutilated and tourmalinated quartz has been found. This material probably comes from the San Diego County region.

B. Chalcedony. While debitage and artifacts made from chalcedony have been found in the Pinyon Flats area, no local sources for this material have been found. Potential sources close to the area include:

1. The nearest chalcedony source is in the Wonder Canyon Area where it occurs with the wonderstone deposits. Here it can be found as vein material, alluvial nodules, and as chalcedony roses.

2. Another potential source for chalcedony is in the Chocolate Mountain Area, southerly of the Hayfield Pumping Station. Here it is found as chalcedony roses.

3. Additional sources include the Little Chuckawalla Mountains, the Mule Mountains, and the Palo Verde Mountains. Chalcedony is also reported to be present in the Turtle Mountains, Barron Mountains, Cargo Muchacho Mountains, Bristol Mountains, Cady Mountains, and in the Ash Hills near Ludlow.

C. Agate. While debitage and artifacts made from agate have been found in the Pinyon Flats area, no local sources for this material have been found. Potential sources close to the area include:

1. The nearest agate source is the Wonder Canyon Area where it occurs with the Wonderstone deposits. Here it can be found as vein material and alluvial nodules.

2. Another potential source for agate is in the Chocolate Mountain Area, southerly of the Hayfield Pumping Station. Here it is found as float in alluvial deposits.

3. Additional sources include the Little Chuckawalla Mountains, the Mule Mountains, and the Palo Verde Mountains. Chalcedony is also reported to be present in the Turtle Mountains, Barron Mountains, Cargo Muchacho Mountains, Cady Mountains, and the Hackberry Wash Area east of the Providence Mountains.

D. Agatized wood (petrified wood). Debitage of what appears to be agatized wood has been found in the Pinyon Flats area and there is no known local source. This material is rare, consisting of only a couple of pieces total. The nearest potential sources found for agatized wood are the Sacramento Mountains, Providence Mountains, and the Hackberry Wash Area east of the Providence Mountains.

E. Jasper. Debitage and artifacts made from red or yellow jasper or both have been found in the Pinyon Flats area. Most of this material is probably from the Wonder Canyon area and is actually a type of Wonderstone. A single outcrop consisting of a 3 to 6 inch thick seam of a yellow to yellowish brown jasper-like material was found in the metasedimentary rock sequence on the northwest flank of Santa Rosa mountain over 30 years ago. It was found while cutting cross country returning to our car one late afternoon. No definite quarry operations or lithic scatters was observed at the site and it has not been visited since. While no other local sources of jasper are known, the presence of this site suggests that there might be others in the area as well.

Other locations found were the Twentynine Palms Marine Base area, Castle Dome area in Arizona, the Earp Badlands over near Vidal Junction, the Bristol Mountain area, and the Turtle Mountain area. A yellow jasper can be found near the mining town of Tumco in Imperial County (Bohan, personal communiqué). There is a large amount of both yellow and red jasper near Barstow in San Bernardino County.

F. Slate. Debitage and artifacts made from slate have been found in the Pinyon Flats area. At least two types of slate are present, the most common is a dark gray to black material and the other is a siliceous gray green to green slaty material that can have gray to gray brown weathered rind. Both of these slate types have been recovered from sites in the Coachella Valley (Quinn, personal observations). The dark slate has been used to make projectile points, knives, and scrapers and the greenish type only appears to have been used for projectile points.

The dark slate is of local origin, as this type of slate can be found within the metasedimentary rock belt that runs up Palm Canyon and then makes a nearly right angle turn and runs across the north face of Santa Rosa Mountain. This band then turns southward near the east end of Santa Rosa Mountain and extends down to the Borrego area. There are probably a number of local slate sources within this metasedimentary rock belt, and while the source of the greenish slate has not been found, it probably comes from some place within this belt as well.

G. Quartzite. Debitage and artifacts made from quartzite have been found in the Pinyon Flats area. Most of the quartzite has been derived locally, as quartzite can be found in the same metasedimentary rock sequence as the slate. Most of this locally derived quartzite is a gray-green color with a reddish brown weathered surface. This type of quartzite has been recovered from several sites in the La Quinta-Indian Wells Area (personal observations). A few pieces of quartzite that do not appear to have come from the local sources have also been found. Other possible sources include the San Bernardino Mountains, the Marble Mountains, and the McCoy Mountains. The most common artifacts made from quartzite are knives and scrapers.

H. Pine pitch. One olla is known from the Pinyon Flats area that had a crack mended with pine pitch. The crack was in the rim and pitch was applied to the exterior to keep the crack from spreading. There have also been some ceramic shards found in the area that have pine pitch on them. The pine pitch was no doubt obtained locally, as the Pinyon Flats area was used primarily as a pine nut gathering area.

I. Asphaltum. One olla that had been repaired with asphaltum is known to have come from the Pinyon Flats area. There are no known local sources for asphaltum. The potential sources found include the following:

1. A reported seep down in the Split Mountain area. The actual location of this seep, or if it actually even exists, has not been found.
2. From natural seeps along the Whittier Fault in the Puente Hills area on Los Angeles County.
3. From natural seeps in the Newport Beach Area in Orange County
4. From the La Brea Tar Pit area in Los Angeles County
5. From seeps along the Santa Barbara Coast in Ventura and Santa Barbara Counties.
6. From tar globules that collected along the rocky beaches in the Palos Verdes Hills area from offshore seeps in Los Angeles County.

J. Clay. The pottery found in the Pinyon Flats area appears to be made from two distinct types of clay, a residual clay and a sedimentary clay. The most common type is that made from residual clay, as this clay can be derived locally. The residual clays produce a ceramic that commonly referred to as "brownware." The brownware ceramics tend to be thick, to have coarse temper, and fire to a brown to reddish brown color. These residual clays can be found in low areas within alluvial fans, around springs, and in low areas in the stream beds.

According to Mr. Art Guanche, the deposits in Bull Canyon were extensively used as a local source of residual type clay (personal communiqué). The Bull Canyon source area is located on the Santa Rosa Indian Reservation and was being used in 1995 by Mr. Tony Largoas a source for some of his clay.

This residual clay is a terra cotta type clay. It is interesting to note that the Luiseño Indians, who occupied the Elsinore-Temescal Canyon area, were not known for their pottery making. Yet the Elsinore-Temescal Canyon area contained one of the largest deposits of residual clay to be found in southern California. The area has produced industrial clay for brick, tile, and pipe for over one hundred years. In comparison, the Cahuilla Indians, who are known for their pottery, had only small and localized deposits of residual clay that could be found mainly in the Santa Rosa and San Jacinto Mountains.

The sedimentary clays are commonly found associated with old lake bed deposits and along old river courses. No sources of sedimentary clay are known to be present in the Pinyon Flats area. The sedimentary clays tend to produce pottery that is thin, fine grained, has rounded sand temper, and fires a light gray to buff color. This type of ceramics is commonly referred to as bufiware ceramics. The best sources for this type of clay appear to be within the Salton Sink portions of the Coachella Valley and Imperial Valley areas.

Good deposits of sedimentary clay are reported to be present in the Salton City and Fish Creek Wash areas, both of which are within the Salton Sink. Any clay deposits within the Salton Sink that were below the high water level of any of the old lakes that filled the Salton Basin would not have been available when a lake was present. This means that many of the known clay sources were only usable during times when there was no lake present. However, additional sources of sedimentary clay are known to be present in the Mecca Hills, Indio Hills, and along the Whitewater River course, none of which would have been affected by lakes in the Salton Basin.

Painted pottery is rarely found in the area, but when it is found it is almost exclusively bufiware ceramics that are painted. Incised marking can be found on both bufiware and brownware and these impressions are commonly in the rims and consist mainly of small holes, fingernails impressions, and small cut grooves.

K. Marine shell material. Beads and pendants made from marine shell material have been found in the Pinyon Flats area. Most of this material is too small to be identified as to genera, let alone species. A couple of larger shell fragments were found, but these were never made available to me for analysis. The identified marine shell material is as follows:

1. Abalone shell (*Haliotis* sp.) from the Pacific Ocean. Alien (1969:55) shows abalone to be found along the coast of Australia, Japan, and North America and to be commonly found in intertidal and subtidal habitats along the southern California coast.

2. Olive shell *(Olivella biplicata)* from the Pacific Ocean. Alien (1969:65) shows *Olivella* sp. to be found on sandy-mud substrate, along the southern California coast south of Point Conception. While all of the identifiable forms were *Olivella biplicata*, one broken spirelopped bead may have been made from *Olivella baetica*.

3. Clam shell (unknown) probably from the Pacific Ocean. A couple of clam shell disc beads are known to have come from the Pinyon Flats area. However, because of the ground surfaces none of them could be identified to genera, so no specific location of origin can be assigned to them. It is possible that they came from the Gulf of California, but it is more likely that they are from the Pacific Ocean.

REFERENCES

Abbott, A. L., 1978, Gem Trails in California. Abbott & Abbott, Anaheim, California.

Alien, Richard K., 1969, Common Intertidal Invertebrates of Southern California, Peek Publications.

Anonymous, 1966, Mineral Resources of California. California Division of Mines and Geology, Bulletin 191, Sacramento, California.

Bailey, Thomas L. and Richard H, Jahns, 1954, Geology of the Transverse Range Province, Southern California. In: Jahns, J. H., ed., Geology of Southern California, California Division of Mines, Bulletin 170, San Francisco, California.

Bohan, Daniel M., 1991, Personal communiqué, Notes in Files of the Author.

Bortugno, E. J. and T. E. Spittler, 1986, Geologic Map of California – San Bernardino Sheet. California Division of Mines and Geology Geologic Map Series, Sacramento, California.

Brown, Arthur R., 1981, Structural History of the Metamorphic, Granitic and Cataclastic Rocks in the Southeastern San Jacinto Mountains. In:, Brown, A. R. and R. W. Ruff, eds., Geology of the San Jacinto Mountains, South Coast Geological Society, Annual Field Trip Guide Number 9, pp. 100–138, Costa Mesa, California.

Brown, Vinson, David Allan, and James Stark, 1972, Rocks and Minerals of California. Naturegraph Publishers, Healdsburg, California.

Calzia, J. F., D. J. Madden-McGuire, H. W. Oliver, and R. A. Schreiner, 1988, Mineral Resources of the Santa Rosa Mountains Wilderness Study Area, Riverside County, California. US Geological Survey Bulletin 1710-D, Washington, D.C.

Crowell, John C., 1975, Geologic Sketch of the Orocopia Mountains, Southeastern California. In: Crowell, J. C., ed., San Andreas Fault in Southern California, California Division of Mines and Geology, Report 117, pp. 99–110, Sacramento, California.

Dibblee, Jr., Thomas W., 1954, Geology of the Imperial Valley Region, California. In: Jahns, R. H., ed., Geology of Southern California, California Division of Mines, Bulletin 170, San Francisco, California.

Freeman, T. A., 1992, Prehistoric Rockhounds in California - A Look at Southern California's Earliest Rock Collectors. Indian Artifact Magazine, v. 11, n. 2, pp. 10–12, and 57.

Gunther, G. and J. Gunther, 1971, Favorite Field Trips For Los Angeles Gem Hunters. Tepee Rock Shop, Saugus, California.

Henry, Darold, J., 1948, California Gem Trails: A Field Guide for the Gem Hunter, The Mineral Collector and the Tourist. The Mineralogist Publishing Company, Portland, Oregon.

Henry, Darold, J., 1957, California Gem Trails: A Field Guide for the Collector of California Gems and Minerals, Lowell

R Gorden, Long Beach, California.

Hodgson, Susan F., 1987, Onshore Oil and Gas Seeps in California. California Division of Oil and Gas, Publication TR26, Sacramento, California.

Jahns, Richard H., 1954, Geology of the Peninsular Range Province, Southern California. In:, Jahns, R. H., ed., Geology of Southern California, California Division of Mines, Bulletin 170, San Francisco, California.

Jennings, Charles W., 1967, Geologic Map of California, Salton Sea Sheet. California Division of Mines and Geology Geologic Map Series, Sacramento, California.

Johnson, H. Cyril, 1966, Western Gem Hunters Atlas. H. Cyril Johnson, Susanville, California.

Keen, A. Myra, 1963, Marine Molluscan Genera of Western North America – An Illustrated Key. Stanford University Press, Stanford, California.

Mitchell, James R., 1996, Gem Trails of Southern California. Gem Guides Book Co., Baldwin Park, California.

Morton, Paul K., 1977, Geology and Mineral Resources of Imperial County, California. California Division of Mines and Geology, County Report 7, Sacramento, California.

Murdoch, J. and R W. Webb, 1956, Minerals of California. California Division of Mines Bulletin 173, San Francisco, California.

Noller, Jay S., 1986, Mojave Desert Xenolith Suites – Malapai Hill and Dish Hill Field Trip Roadlog. In: Ehlig, Perry L., ed., Southern California Fieldtrips Guidebook, The Geological Society of America, Cordilleran Section.

Parcel, Jr., Rodney F., 1981, Structure and Petrology of the Santa Rosa Shear Zone in the Pinyon Flats Area, Riverside County, California. In:, Brown, A. R. and R. W. Ruff, eds., Geology of the San Jacinto Mountains, South Coast Geological Society, Annual Field Trip Guide Number 9, pp. 139–150, Costa Mesa, California.

Parker, Ronald B., 1963, Recent Volcanism at Amboy Crater, San Bernardino County, California. California Division of Mines, Special Report 76, San Francisco, California.

Peters, Martha L., 1976, California Gemstones: Gems and Indians. Lapidary Journal, April Issue, pp. 238–249.

Proctor, Richard J., 1968, Geology of the Desert Hot Springs-Upper Coachella Valley Area, California. California Division of Mines, Special Report 94, San Francisco, California.

Quinn, Harry M., 1992, A Collection of Papers on the Santa Rosa Mountain Area, Riverside County, Paper Number One. Self Published Pamphlet.

Quinn, Harry M., 1995, An Introduction to the Geology, Hydrogeology, Economic Potential and Geologic Features of the Santa Rosa Wilderness Area and Santa Rosa Mountain National Scenic Area, Riverside County, California. Self Published Pamphlet.

Quinn, Harry M., 1997, Possible Sources of Natural Materials Used by the Cahuilla Indians Living in the Pinyon Flats Area, Riverside County, California. In: Denning, Jean, ed., Observations on the Cahuilla Indians: Past and Present, Coachella Valley Archaeological Society Occasional Paper No. 1, pp. 3–12, Palm Springs, California.

Quinn, Harry M., 2005, Coachella Valley Obsidian Sources. In: Mouriquand, Leslie, ed., Coachella Valley Archaeological Society, Occasional Paper No. 3, pp. 31–23, Palm Springs, California.

Quinn, Harry M., 2005, Wonderstone: AKA Truckhaven Rhyolite. In: Mouriquand, Leslie, ed., Coachella Valley Archaeological Society, Occasional Paper No. 3, pp. 35–38, Palm Springs, California.

Quinn, Harry M., 2005, Palm Canyon Slates and Santa Rosa Mountain Slates. In: Mouriquand, Leslie, ed., Coachella Valley Archaeological Society, Occasional Paper No. 3, pp. 39–43, Palm Springs, California.

Quinn, Harry M., 2005, Species of the Gastropod Genus *Olivella* Swainson, 1831 from the Pacific Coast of North America. In: Mouriquand, Leslie, ed., Coachella Valley Archaeological Society, Occasional Paper No. 3, pp. 44–49, Palm Springs, California.

Rogers, John J. W., 1961, Igneous and Metamorphic Rocks of the Western Portion of Joshua Tree National Monument, Riverside and San Bernardino Counties, California. California Division of Mines, Special Report 68, San Francisco, California.

Rogers, Thomas H., 1965, Geologic Map of California - Santa Ana Sheet. California Division of Mines and Geology, Geologic Map Series, Sacramento, California.

Steese, Collis H., 1952, Illustrated Guide – Imperial and Coachella Valleys. Stanford University Press, Stanford, California.

Strand, Rudolph G., 1962, Geologic Map of California – San Diego-El Centro Sheet. California Division of Mines and Geology Geologic Map Series, Sacramento, California.

Strong, Mary J., 1966, Desert Gem Trails – A Guide to the Gems and Minerals of the Mojave & Colorado Desert. Gembooks, Mentone, California.

Strong, Mary J., 1971, Desert Gem Trails - A Field Guide to the Gems and Minerals of the Mojave & Colorado Deserts and Adjacent Areas of Nevada and Arizona. Gembooks, Mentone, California.

Trent, D. D. and Richard W. Hazlett, 2002, Joshua Tree National Park Geology. Joshua Tree National Park Association, Twentynine Palms, California.

***California Geology* cover photo:** View from the Indio Hills, Riverside County. Photo by Robert M. Norris. See: The Indio Hills: "Desert Badlands," pages 155–159.